Structural Analysis and Synthesis

A Laboratory Course in Structural Geology

Stephen M. Rowland
University of Nevada, Las Vegas

Blackwell Scientific Publications, Inc., Palo Alto

For my parents

Editorial Offices

667 Lytton Avenue, Palo Alto, California, 94301
Osney Mead, Oxford, OX2 0EL, UK
8 John Street, London WC1N 2ES, UK
23 Ainslie Place, Edinburgh, EH3 6AJ, UK
52 Beacon Street, Boston, Massachusetts 02108
107 Barry Street, Carlton, Victoria 3053, Australia

Distributors

USA and Canada
Blackwell Scientific Publications
P.O. Box 50009
Palo Alto, California 94303

Australia
Blackwell Scientific Publications (Australia) Pty Ltd
107 Barry Street, Carlton
Victoria 3053

UK
Blackwell Scientific Publications
Osney Mead
Oxford OX2 0EL

Sponsoring Editor: John Staples

Manuscript Editor: Andrew Alden

Production Coordinator: Robin Mitchell

Interior and Cover Design: Gary Head

Composition: Graphic Typesetting Service

Library of Congress Cataloging-in-Publication Data

Rowland, Stephen Mark.
Structural analysis and synthesis.

Includes index.
1. Geology, Structural—Laboratory manuals.
I. Title.
QE501.R73 1986 551.8'07'8 86-13631
ISBN 0-86542-308-3

On first examining a new district, nothing can appear more hopeless than the chaos of rocks; but by recording the stratification and nature of the rocks and fossils at many points, always reasoning and predicting what will be found elsewhere, light soon begins to dawn on the district, and the structure of the whole becomes more or less intelligible.

CHARLES DARWIN

Contents

Preface

This book is intended for use in the laboratory portion of a first course in structural geology. There is throughout it a strong emphasis on geologic maps. Because students begin the study of structural geology with a wide variety of skills and experiences, I have assumed a minimum of previous experience with geologic maps. Instructors whose students receive extensive training with maps and structure sections in earlier courses may be able to skip some of the early chapters or exercises.

Structural geology, like all courses, is taught differently by different people. I have tried to strike a balance between an orderly sequence of topics and a collection of independent chapters that can be flexibly shuffled about to suit the instructor. Chapter 5 on stereographic projection, for example, may be moved up by those instructors who like to expose their students to stereonets as early as possible, and Chapter 12 on rheological models may be moved up by those who start with an introduction to stress and strain.

There is, however, an underlying strategy in the organization of the material. As is explicit in the title, this book is concerned with both the analysis and the synthesis of structural features. Most of the first ten chapters involve some interaction with a contrived geologic map of the mythical Bree Creek Quadrangle, which is in six separate sheets at the back of the book. Before beginning work on Chapter 3 the student is asked to assemble and color the Bree Creek Quadrangle map. More than mere busywork, this map coloring requires the student to look carefully at the distribution of each rock unit. The Bree Creek Quadrangle becomes the student's "map area" for the remainder of the course. Various aspects of the map are analyzed in Chapters 2 through 10 (except for Chapter 6), and in Chapter 11 these are synthesized into a written summary of the structural history of the quadrangle. Some instructors will choose to skip this synthesis, but I hope that most do not—students need all the writing practice they can get.

I have placed the synthesis report in Chapter 11 so that it would not be at the very end of the semester, to allow some writing time. Chapters 12 through 14, in any case, contain material that is not conducive to this teaching approach.

Most courses will not cover all of the material contained in this book. The first nine chapters contain the material most commonly covered in structural geology labs. The remaining chapters contain material that is often covered in the lecture portion of the course but less commonly in the lab. If the lab time is not available, some of these chapters may be profitably perused by the student anyway.

I have written each chapter with a three-hour laboratory period in mind. In probably every case, however, all but the rarest of students will require additional time to complete all of the problems. The instructor must, of course, exercise judgment in deciding which problems to assign. Many instructors have their own favorite lab and field exercises to intersperse with those in the lab book. An appendix on the use of the Brunton compass has been included to facilitate the use of supplemental field work.

This laboratory manual evolved over a period of several years, first at the University of California, Santa Cruz (where I was a graduate student and teaching assistant), and then at the University of Nevada, Las Vegas. Many students suffered through earlier versions, and I thank them for their patience, enthusiasm, and suggestions. I have been unusually fortunate to have learned structural geology from some very talented geologists, without whom I could never have written this lab manual. These include Edward A. Hay, Othmar Tobisch, Edward C. Beutner, and James Dieterich.

I am grateful to Sam Longiaru, Rod Newman, Mike Ryan and R. A. Hoppin for many useful suggestions, to C. G. Scroggins, Angela Sewall, and Chris Barton for reviewing portions of the final manuscript, to Andrew L. Alden who improved the clarity throughout, and to Gail Cummins for her cheerful word processing wizardry. The Blackwell staff, especially Production Coordinator, Robin Mitchell, were wonderfully tolerant and supportive. My wife DeeAnn, while secretly convinced that I would never get this finished, provided constant encouragement to the contrary as well as the recipe for play dough.

In spite of all this help, there are doubtless some omissions, misleading statements, and explanations that some students will find abstruse. I will be delighted to have these pointed out to me so that they may be rectified in future editions.

STEPHEN M. ROWLAND

Read This First

You are about to begin a detailed investigation of the basic techniques of analyzing the structural history of the earth's crust. Structural geology, in my view, is the single most important course in the undergraduate curriculum (with the possible exception of field geology). There is no such thing as a good geologist who is not comfortable with the basics of structural geology. This book is designed to help you become comfortable with the basics—to help you make the transition from naive curiosity to perceptive self-confidence.

Because self-confidence is built upon experience, in an ideal world you should learn structural geology with real rocks and structures, in the field. The field area in this lab manual is the Bree Creek Quadrangle. The geologic map of this quadrangle is located at the back of the book on six pages (which will be assembled into one large map). This map will provide continuity from one chapter to the next, so that the course will be more than a series of disconnected exercises.

Most of the things that you will do in this laboratory course are of the type that, once done, the details are soon forgotten. A year or two from now, therefore, you will remember what kinds of questions can be asked, but you probably won't remember exactly how to get the answers. A quick review of your own solved problems, however, will allow you to recall the procedure. If your solutions are neat, well labeled, and not crowded together on the paper they will be a valuable archive throughout your geologic career.

In most of the chapters I have inserted the problems immediately after the relevant text, rather than putting them all at the end of the chapter. The idea is to get you to engage yourself with certain concepts, and master them, before moving on to the next concepts. We all learn best that way. One unfortunate consequence, however, is that pages containing problems may have to be removed from the book in the middle of the chapter. I recommend that you buy a three-ring binder, rip out all of the pages in this book, and put them in the binder. That way, when an exercise is graded by your lab instructor and returned to you, you can insert it back into the proper position for future reference. Tape a large envelope to the inside cover of the binder to keep your Bree Creek map in.

You will need the following equipment in this course:

- colored pencils (at least 15)
- ruler (centimeters and inches)
- straightedge
- graph paper (10 squares per inch)
- tracing paper
- protractor
- drawing compass
- masking tape
- transparent tape
- 4H or 5H pencils with cap eraser
- thumbtack (store it in one of the erasers)
- drawing pen (e.g. Rapidograph or Mars)
- black drawing ink
- pocket calculator with trigonometric functions

A zippered plastic binder bag is a convenient way to keep all of this in one place.

If structural geology is the most important course in the curriculum, it should also be the most exciting, challenging, and meaningful. My sincere hope is that this book will help to make it so.

1

Attitudes of Lines and Planes

OBJECTIVE

Solve apparent dip problems using orthographic projection, trigonometry, polar tangent diagrams, and alignment diagrams

This chapter is concerned with the orientations of lines and planes. Lines and planes are the elements of the earth's structures that permit us to describe and analyze the structures themselves. In this chapter we will examine several graphical and mathematical techniques for solving apparent dip problems. Each technique is appropriate in certain circumstances. The examination of various approaches to solving such problems serves as a good introduction to the techniques of solving structural problems in general.

The following terms are used to describe the orientations of lines and planes:

Attitude The orientation in space. The attitude of a plane is defined by its *strike* and *dip;* the attitude of a line is defined by its *trend* and *plunge.*

Bearing The horizontal angle between a line and a specified coordinate direction, usually true north or south; the compass direction or azimuth.

Strike The bearing of a horizontal line in an inclined plane (Fig. 1-1).

Dip The vertical angle between an inclined plane and a horizontal line perpendicular to its strike. The direction of dip can be thought of as the direction water would run down the plane (Fig. 1-1).

Trend The bearing of a line (Fig. 1-2). Non-horizontal lines trend in the down-line direction.

Plunge The vertical angle between a line and the horizontal (Fig. 1-2).

Pitch The angle *within* an inclined plane between a horizontal line and the line in question (Fig. 1-3). Also called **rake.**

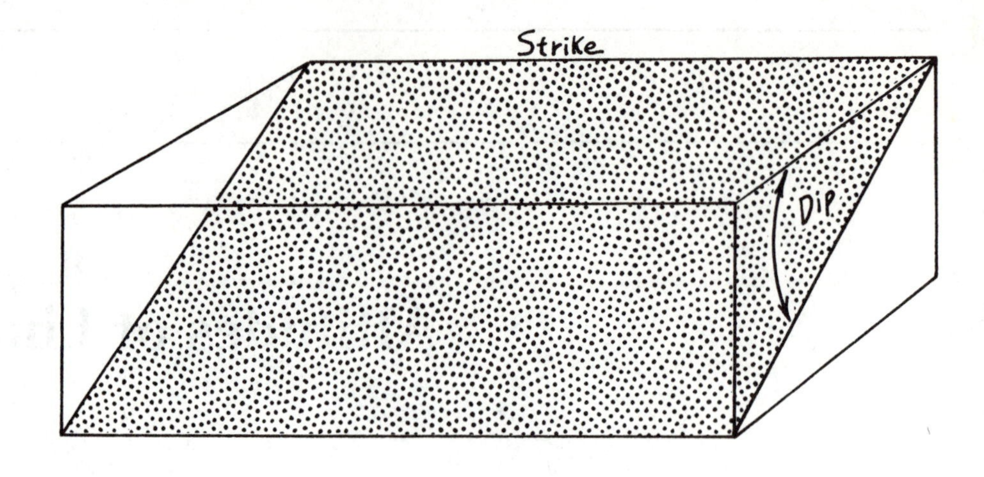

Figure 1-1
Strike and dip of a plane.

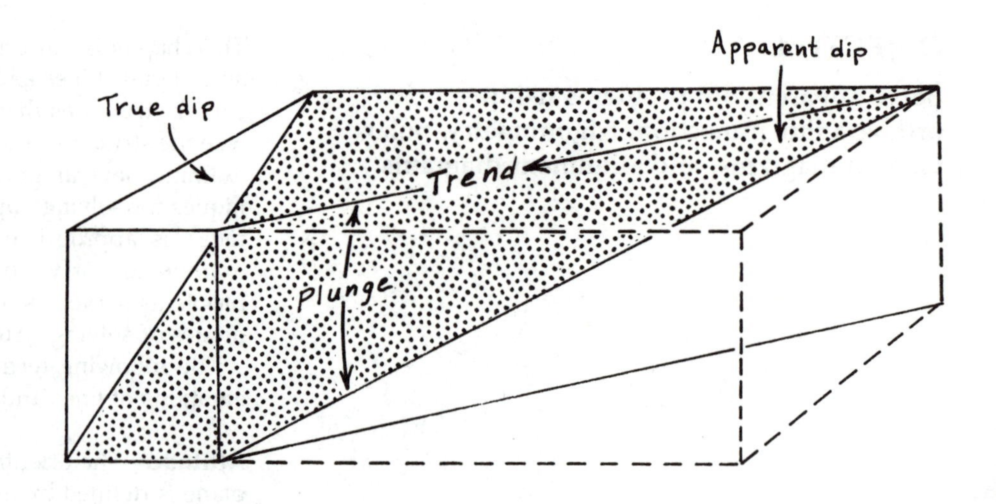

Figure 1-2
Trend and plunge of apparent dip.

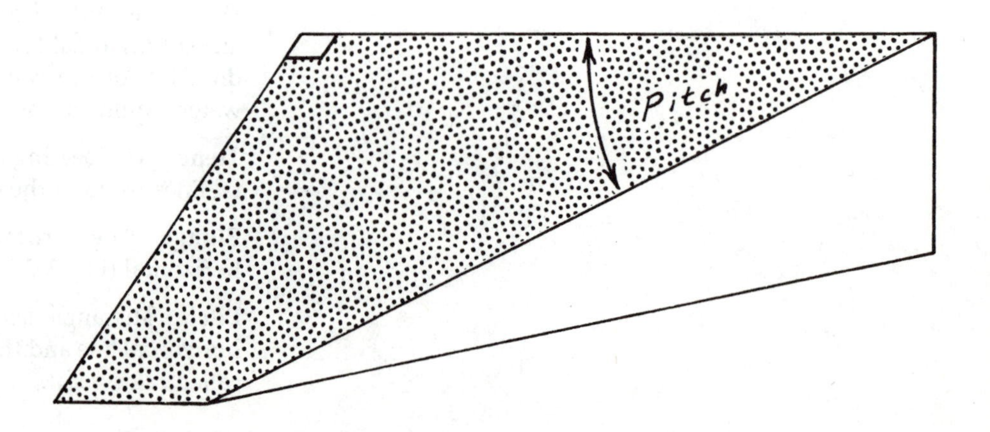

Figure 1-3
Pitch (or rake) of a line in an inclined plane.

Apparent dip The vertical angle between an inclined plane and a horizontal line which is not perpendicular to the strike of the plane (Fig. 1-2). For any inclined plane (except a vertical one), the true dip is always greater than any apparent dip. Note that the apparent dip is defined by its trend and plunge or by its pitch within a plane.

There are two ways of expressing the strikes of planes and the trends of lines (Fig. 1-4). The **azimuth** method is based on a 360° clockwise circle; the **quadrant** method is based on four 90° quadrants. A plane that strikes northwest-southeast and dips 50° southwest could be described as 315, 50SW (azimuth) or N45W, 50SW (quadrant). Similarly, a line that trends due west and plunges 30° may be described as 30, 270 or 30, N90W. By convention, the strike is given before the dip, and the plunge is given before the trend.

Notice that because the strike is a horizontal line, either direction may be used to describe it. Thus a strike of N45W (315) is exactly the same as S45E (135). Usually, however, the strike is given in reference to north. The dip, on the other hand, is usually not a horizontal line, and the down-line direction must be given. Because the direction of dip is always perpendicular to the strike, the exact bearing is not needed, and the dip direction is approximated by giving the quadrant in which it lies or the cardinal point to which it most nearly points. The attitude of a plane could be specified by recording the plunge and trend of the dip without recording the strike at all.

Apparent Dip Problems

There are many situations in which the true dip of a plane cannot be accurately measured in the field or cannot be drawn on a cross-section view. Any cross-section that is not drawn perpendicular to strike, for example, displays an apparent dip rather than the true dip of a plane (except for horizontal and vertical planes).

Apparent dip problems involve determining the attitude of a plane from the attitude of one or more apparent dips, or vice versa. The strike and dip of a plane may be determined from either (1) the strike of the plane and the attitude of one apparent dip, or (2) the attitudes of two apparent dips.

There are five major techniques for solving apparent-dip problems. These are: (1) orthographic projection, (2) trigonometry, (3) polar tangent diagrams, (4) nomograms (alignment diagrams), and (5) stereographic projection. Stereographic projection is described in Chapter 5. The other four techniques are discussed in this chapter.

Throughout this and subsequent chapters the following symbols will be used:

α (alpha) = plunge of apparent dip
β (beta) = angle between the strike of a plane and the trend of an apparent dip
δ (delta) = plunge of true dip
θ (theta) = direction (trend) of apparent dip

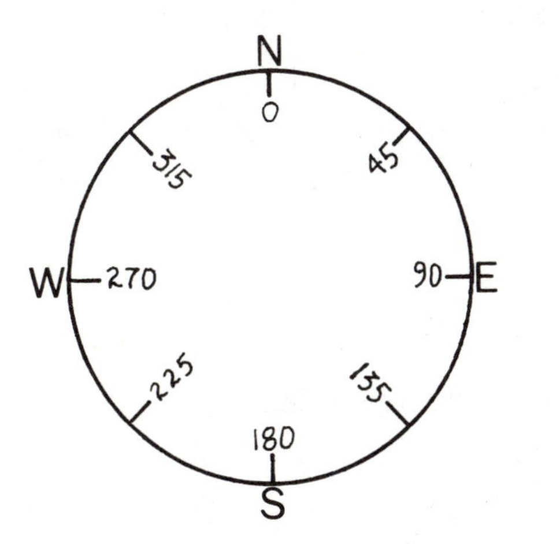

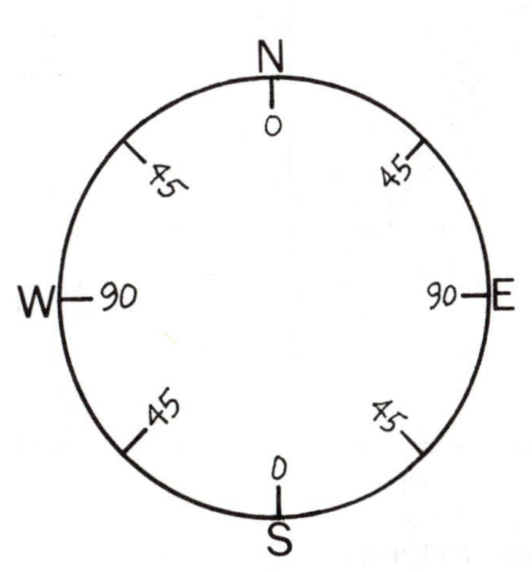

Figure 1-4
Azimuth and quadrant methods of expressing compass directions.

Orthographic Projection

One way to solve apparent dip problems is to carefully draw a layout diagram of the situation. This technique, called orthographic projection, is more time-consuming than the other approaches, but it helps you to develop the ability to visualize in three dimensions and to draw precisely.

Example 1: Determine true dip from strike plus attitude of one apparent dip

Suppose that a quarry wall faces due north and exposes a quartzite bed with an apparent dip of 40, N90W. Near the quarry the quartzite can be seen to strike N25E. What is the true dip?

Before attempting a solution, it is crucial that you visualize the problem. If you can't draw it, then you probably don't understand it. Figure 1-5a shows the elements of this problem.

Solution:

1. Carefully draw the strike line and the direction of the apparent dip in plan (map) view (Fig. 1-5b).
2. Add a line for the direction of true dip. This can be drawn anywhere, perpendicular to the strike line (Figs. 1-5c and d).
3. Now we have a right triangle, the hypotenuse of which is the apparent dip direction. Imagine that you are looking down from space and that this hypotenuse is the top edge of the quarry wall. Now imagine folding the quarry wall up into the horizontal plane. This is done graphically by drawing another right triangle adjacent to the first (Figs. 1-5e and f). The apparent dip angle, known to be 40° in this problem, is measured and drawn directly adjacent to the direction of apparent dip line. Since the apparent dip is to the west, the angle opens to

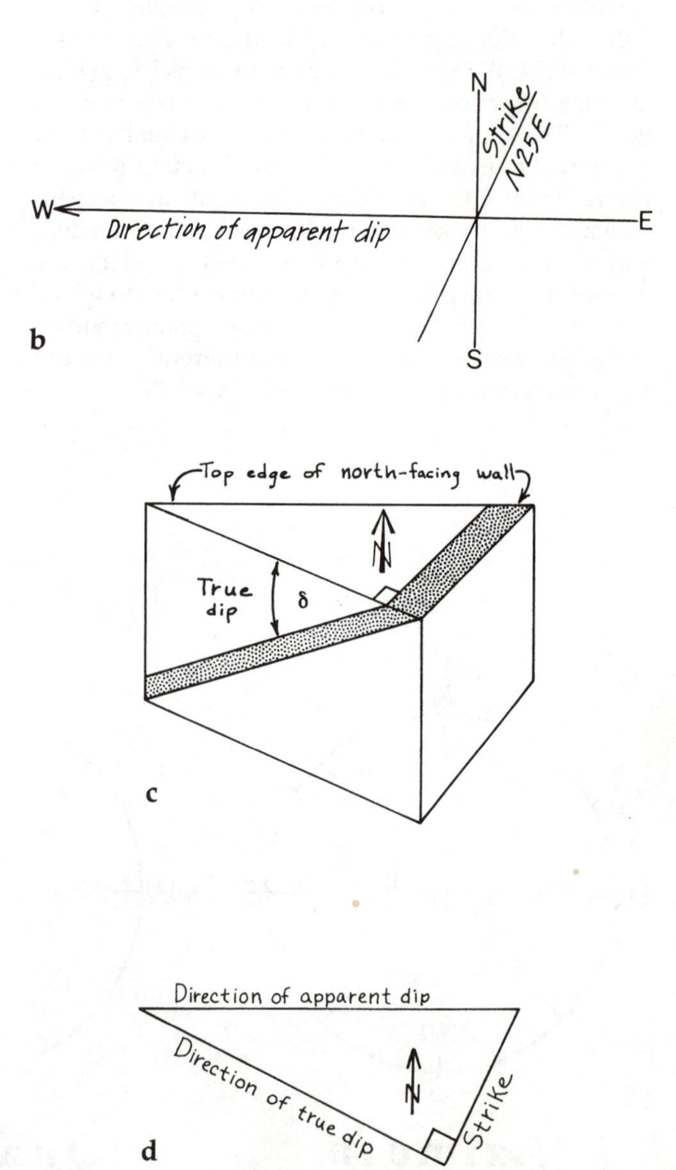

Figure 1-5
Solution of example Problem 1. (a) Block diagram. (b) Step 1 of orthographic solution. (c) Block diagram looking north. (d) Orthographic projection of step 2. (e) Block diagram of step 3. (f) Orthographic projection of step 3. (g) Block diagram of step 4. (h) Orthographic projection of step 4.

the west on the drawing. The line opposite angle α is of length *d*, or the height of the wedge shown in Figure 1-5a.

4. Finally, the direction of true dip is used as a fold line, and another line of length *d* is drawn perpendicular to it (Figs. 1-5g and h). The true dip angle δ is then formed by connecting the end of this new line to the strike line. Since the true dip is to the northwest, angle δ opens toward the northwest. Angle δ is measured directly off the drawing to be 43°.

If you have trouble visualizing this process, try making a photocopy of Figure 1-5h, fold the paper along the fold lines, and reread the solution to this problem.

Problem 1-1

Along a railroad cut, a bed has an apparent dip of 20° in a direction of N62W. The bed strikes N67E. Using orthographic projection, find the true dip.

Problem 1-2

A fault has the following attitude: N80E, 48S. Using orthographic projection, determine the apparent dip of this fault in a vertical cross-section striking N65W.

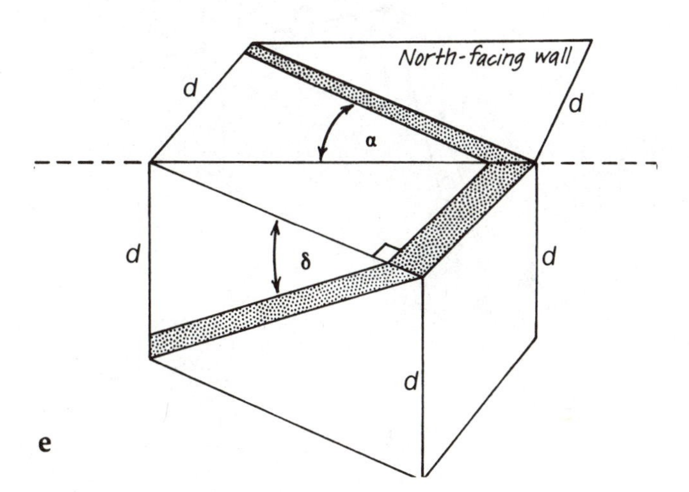

e

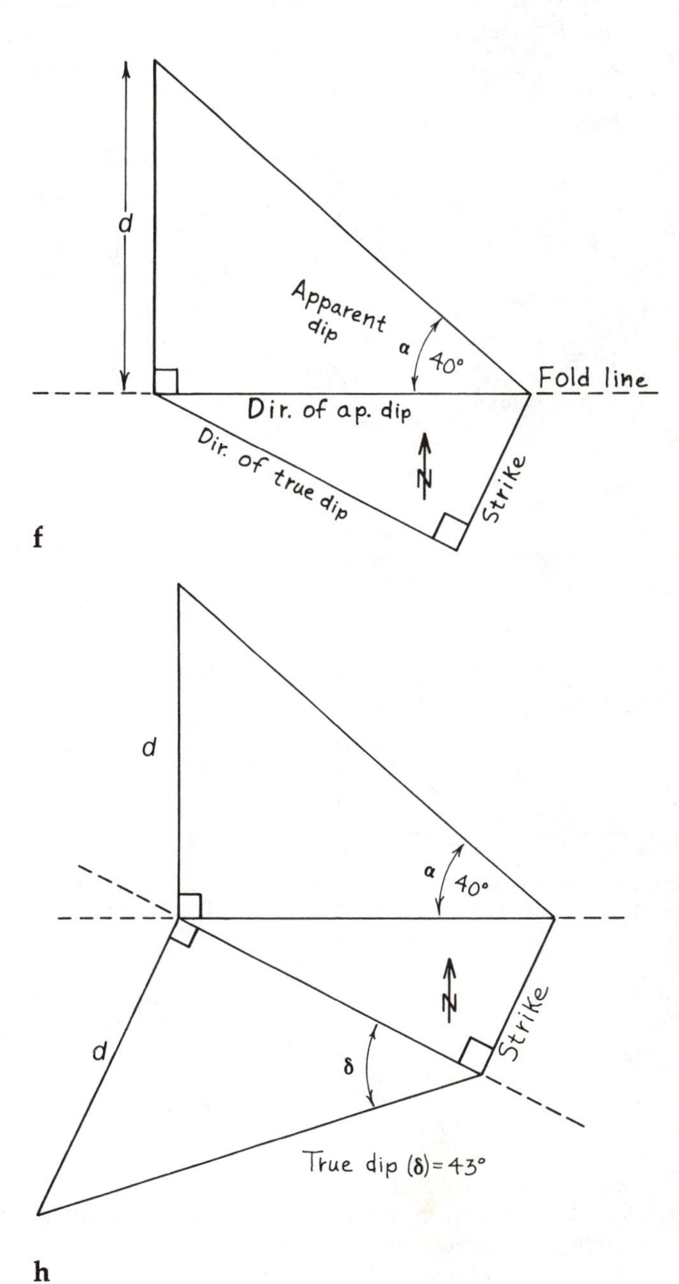

f

h

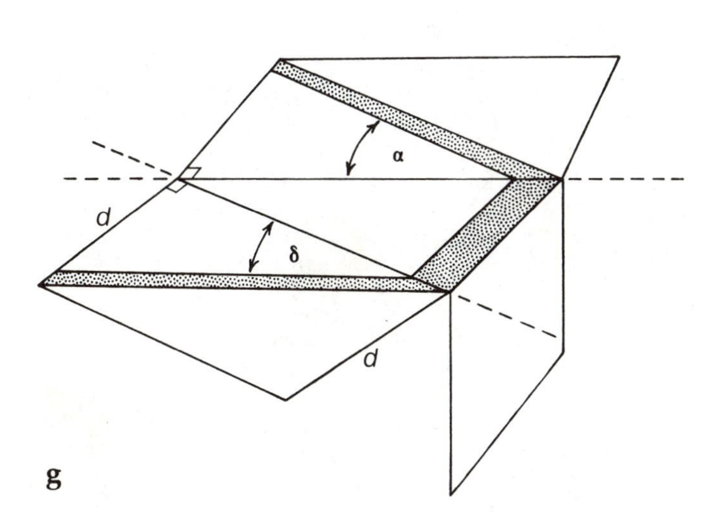

g

Example 2: Determine strike and dip from two apparent dips

Suppose that a fault trace is exposed in two adjacent cliff faces. In one wall the apparent dip is 15, S50E, and in the other it is 28, N45E (Fig. 1-6a). What is the strike and dip of the fault plane?

Solution:

1. Visualize the problem as shown in Figure 1-6b. We will use the two trend lines, OA and OC, as fold lines, and as in Example 1 we will use a vertical line of arbitrary length *d*. Draw the two trend lines in plan view (Fig. 1-6c).

2. From the junction of these two lines (point O) draw angles α_1 and α_2 (Fig. 1-6d). It does not really matter on which side of the trend lines you draw your angles, but drawing them outside the angle between the trend lines results in a minimum of clutter on your final diagram.

3. Draw a line of length *d* perpendicular to each of the trend lines to form the triangles COZ and AOX (Fig. 1-6e). Find these points on Figure 1-6b. The size of *d* is not important, but it must always be drawn exactly the same length.

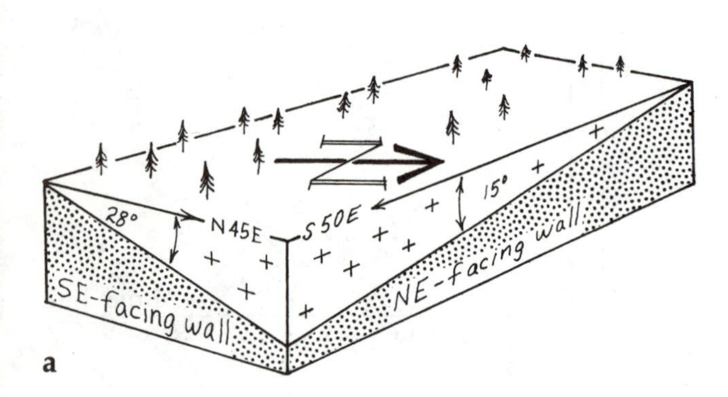

a

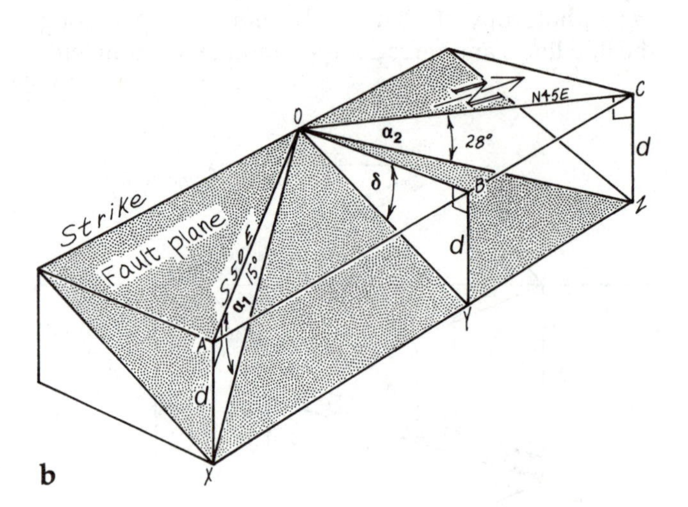

b

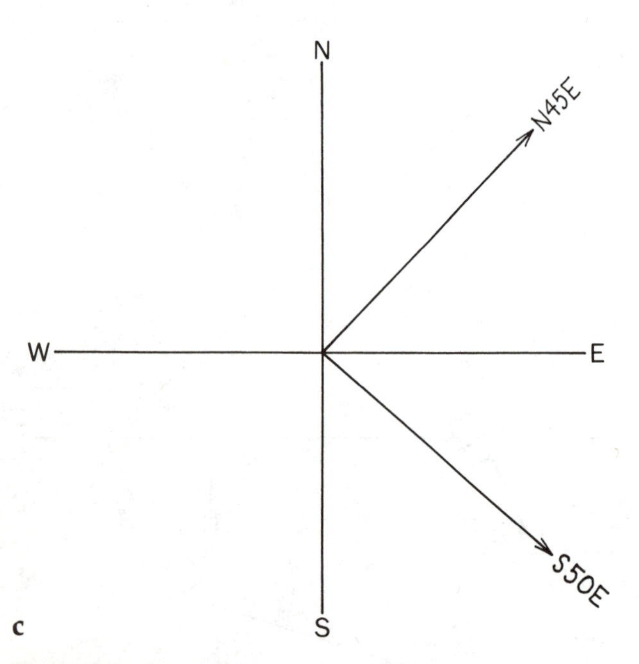

c

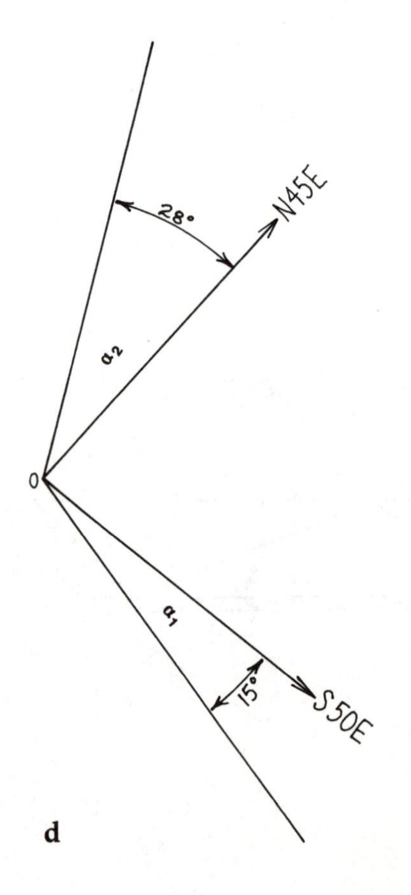

d

Figure 1-6
Solution of example Problem 2. (a) Block diagram. (b) Block diagram showing triangles involved in orthographic projection and trigonometric solutions. (c) Step 1 of orthographic solution. (d) Step 2. (e) Step 3. (f) Step 4. (g) Steps 5 and 6.

4. Figure 1-6e shows triangles COZ and AOX folded up into plan view with the two apparent dip trend lines used as fold lines. As shown in Figure 1-6b, line AC is horizontal and parallel to the fault plane; therefore it defines the fault's strike. We may therefore draw line AC on the diagram and measure its trend to determine the strike (Fig. 1-6f); it turns out to be N22W.
5. Line OB is then added perpendicular to line AC (Fig. 1-6g).
6. Using line OB as a fold line, triangle BOY (as shown on Fig. 1-6b) can be projected into the horizontal plane, again using length *d* to set the position of point Y (Fig. 1-6g). The true dip δ can now be measured directly off the diagram to be 30°.

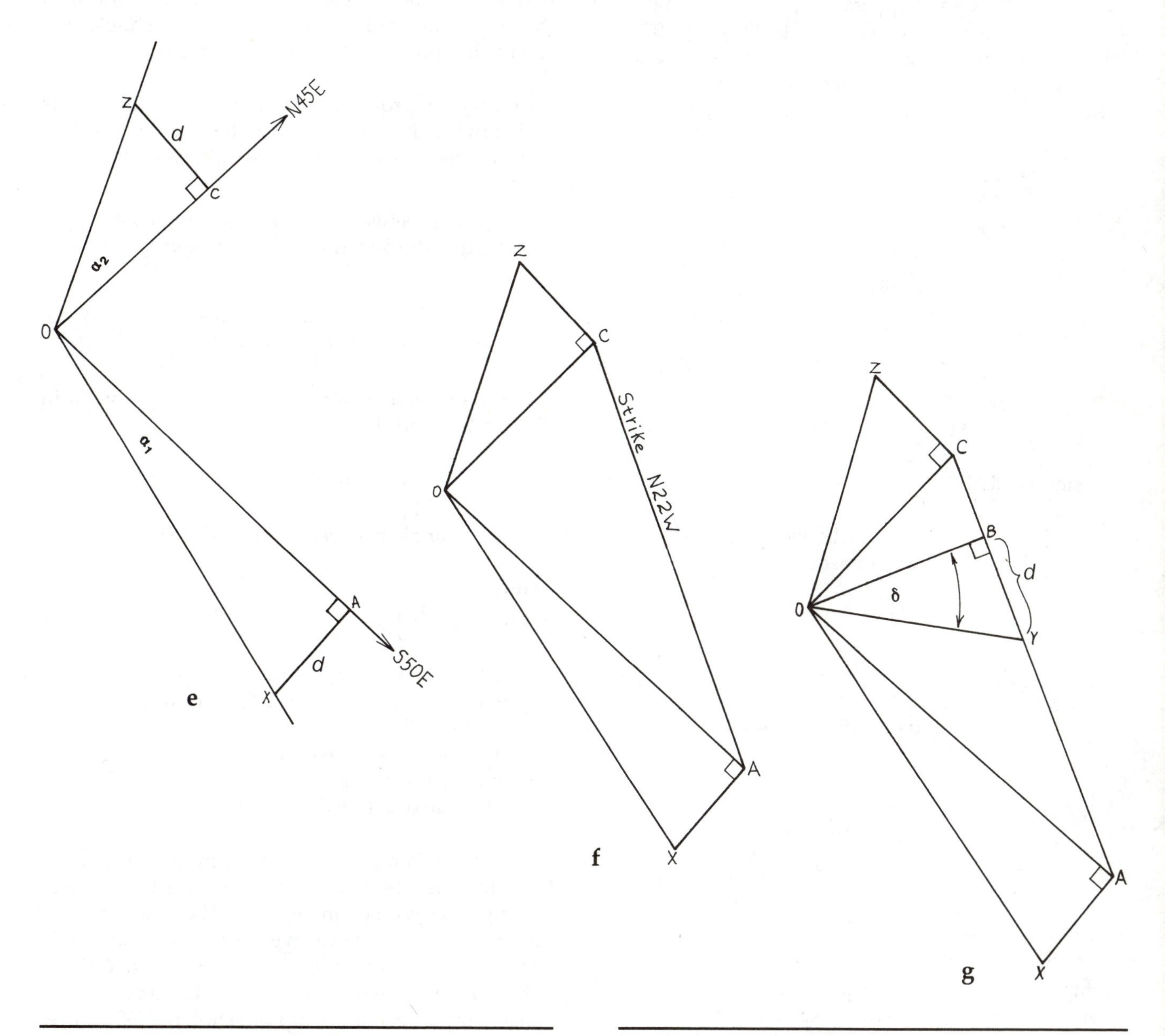

Problem 1-3

A fault plane is intersected by two mine drifts. In one drift the plunge and trend of the apparent dip is 20, N10W, and in the other it is 32, N85W. Use orthographic projection to determine the attitude of the fault plane.

Problem 1-4

A bed strikes N75E and dips 40° to the south. Two vertical cross-sections need to be drawn through this bed, one oriented north-south and the other oriented east-west. By orthographic projection determine the apparent dip on each cross-section.

Trigonometric Solutions

Apparent dip problems can be done much faster and more precisely trigonometrically, especially with a calculator. This method is particularly suitable when very small dip angles are involved. Even when the angles are not drawn orthographically, however, you should sketch a block diagram to clearly visualize the problem. Programs to solve apparent dip problems on programmable calculators are discussed by De Jong (1975). Trigonometric functions are listed in Apppendix B for those whose calculators don't contain them.

Refer to Figure 1-6b for the following derivation:

$$\begin{aligned} AX &= BY \\ \tan AOX &= \frac{AX}{OA} = \frac{BY}{OA} \\ \tan AOX &= \frac{BY}{OB\ (\sec AOB)} \\ \tan AOX &= \frac{OB\ (\tan BOY)}{OB\ (\sec AOB)} \\ &= \frac{\tan BOY}{\sec AOB} \\ &= \tan BOY \cos AOB \end{aligned}$$

or, using symbols,

$$\tan\alpha = (\tan\delta)\begin{pmatrix}\text{cos angle between true and}\\ \text{apparent dip directions}\end{pmatrix} \quad (1)$$

or

$$\tan\delta = \frac{\tan\alpha}{\text{cos angle between true and apparent dip directions}} \quad (2)$$

or

$$\tan\delta = \frac{\tan\alpha}{\sin\beta} \quad (3)$$

Example 3: Determine true dip from strike plus attitude of one apparent dip

Example 1 is a convenient problem of this type to solve trigonometrically. The strike of a bed is known to be N25E but we don't know the dip. An apparent dip is 40, N90W.

$$\alpha = 40^\circ \quad \tan 40^\circ = .839$$
$$\beta = 65^\circ \quad \sin 65^\circ = .906$$

Solution:

From equation 3,

$$\tan\delta = \frac{\tan\alpha}{\sin\beta} = \frac{.839}{.906} = .926$$
$$\delta = 42.8^\circ$$

Example 4: Determine strike and dip from two apparent dips

Because two apparent dips with trend θ are involved, they will be labeled θ_1 and θ_2, which correspond with the two apparent dip angles α_1 and α_2. θ_1 should represent the more gently dipping of the two apparent dips.

This type of problem has two steps. The first step is to determine the angle between the true dip direction and θ_1. The relevant trigonometric relationships are

$$\begin{matrix}\text{tan angle between } \theta_1\\ \text{and true dip direction}\end{matrix} = \begin{pmatrix}\text{csc angle between}\\ \theta_1 \text{ and } \theta_2\end{pmatrix} \begin{bmatrix}(\cot\alpha_1)(\tan\alpha_2) - (\cos\\ \text{angle between } \theta_1 \text{ and } \theta_2)\end{bmatrix} \quad (4)$$

Using Example 2, we have the situation shown in Figures 1-6c and d.

$$\theta_1 = 130\ (\text{S50E}) \quad \alpha_1 = 15^\circ$$
$$\theta_2 = 45\ (\text{N45E}) \quad \alpha_2 = 28^\circ$$
$$\text{angle between } \theta_1 \text{ and } \theta_2 = 85^\circ$$

Solution:

From equation 4,

$$\begin{aligned} &\text{tan angle between } \theta_1 \text{ and true dip direction} \\ &= (\csc 85^\circ)[(\cot 15^\circ)(\tan 28^\circ) - (\cos 85^\circ)] \\ &= 1.004\ [(3.732)(.532) - (.087)] \\ &= 1.004\ [1.985 - .087] = 1.91 \\ &\text{angle between } \theta_1 \\ &\text{and true dip direction} = 62.4^\circ \end{aligned}$$

This angle is measured from θ_1 in the direction of θ_2. In this case the computed angle (62.4°) is less than the angle between θ_1 and θ_2 (85°). The true dip direction, therefore, lies between θ_1 and θ_2. θ_1 is 130° (S50E) so the direction of true dip is 130° − 62° = 68° (N68E). Examination of Figure 1-6b shows that this is a reasonable dip direction. A dip direction of N68E corresponds to a strike of N22W, which agrees with our orthographic projection solution.

If the angle between θ_1 and the true dip direction is determined to be greater than the angle between θ_1 and θ_2, then the angle is measured from θ_1 toward and beyond θ_2.

Once the true dip direction (and therefore the strike direction) has been determined, equation 3 is used to determine δ:

$$\tan \delta = \frac{\tan \alpha}{\sin \beta}$$

$\alpha = 15°$ $\tan \alpha = .268$

β = angle between 130° (S50E) and 158° (S22E) = 28°

$\sin \beta = .469$

$$\tan \delta = \frac{.268}{.469} = .571$$

$\delta = 30°$

Problem 1-5

Solve Problem 1-1 using trigonometry.

Problem 1-6

Solve Problem 1-2 using trigonometry.

Problem 1-7

A coal seam dips 2° due east. A mining company wants its mining adits to slope at least 1° so that water will drain out. In what directions can adits be driven without sloping less than 1°?

Problem 1-8

The apparent dip of a fault plane is measured in two trenches. Toward 220° the apparent dip is 4°. Toward 100° the apparent dip is 7°. Trigonometrically determine the direction and amount of true dip.

Polar Tangent Diagrams

Another method for solving apparent dip problems is on a polar tangent diagram. This is a polar coordinate graph on which the tangent of dip is plotted as a vector radiating from the center of the graph (Fig. 1-7).

The tangent of dip is a true vector (a quantity with direction and magnitude) that obeys the cosine law of vector addition and resolution. Tangent diagrams utilize this law to permit a two-dimensional analysis of dip angles and directions. In addition to being much faster than orthographic projection, this method allows the geometrical relationships between true and apparent dips to be clearly visualized. Unlike solutions by trigonometry, however, this technique is not suitable for problems with small dip angles. A large polar tangent diagram may be found at the back of this book. Photocopy the diagram or use tracing paper over it when using this technique.

Example 5: Determine true dip from strike plus attitude of one apparent dip

We will use the same example problem that we solved by orthographic projection and trigonometry: $\alpha = 40$, N90W; strike = N25E (Fig. 1-5). What is the true dip?

Solution:

1. Plot $\mathbf{V}_a$, the apparent dip, as a vector from the center of the diagram (Fig. 1-7). The direction is read at the circumference of the circle, and the length (proportional to the tangent of dip) is read on the concentric circles.
2. Draw a line in the direction of true dip.
3. From the terminus of $\mathbf{V}_a$ draw a line perpendicular to $\mathbf{V}_a$. The point where this line intersects the true dip direction line defines the terminus of $\mathbf{V}_t$, the true dip vector. True dip is read off the diagram to be 43°.

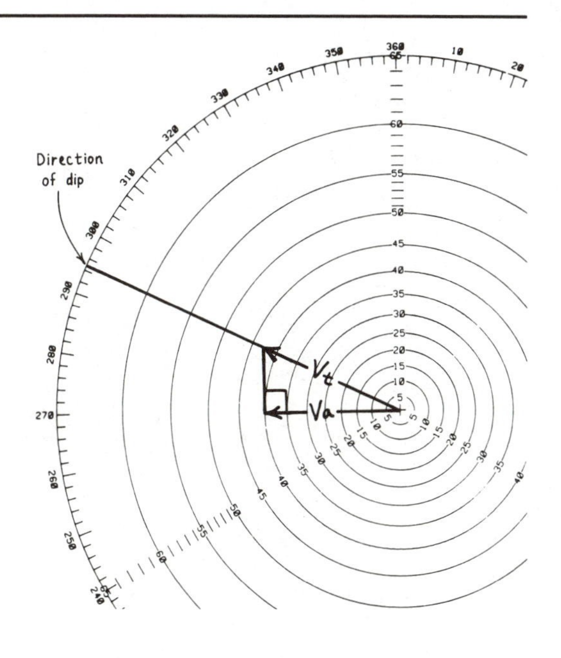

Figure 1-7
Polar tangent diagram solution to example Problem 5. $\mathbf{V}_a$ is apparent dip, $\mathbf{V}_t$ is true dip.

Example 6: Determine strike and dip from two apparent dips

Suppose that a plane has apparent dips of 55, 330 and 59, 37. What is the true dip?

Solution:

1. Plot $\mathbf{V}_{a1}$ and $\mathbf{V}_{a2}$, the two apparent dip vectors, on the tangent diagram (Fig. 1-8).
2. Draw perpendicular lines through their end points.
3. Vector $\mathbf{V}_t$, representing the true dip, is drawn from the origin to the intersection of the two perpendiculars. The true dip is determined to be 62°. The direction of dip is read off the circumference of the graph to be 10°, corresponding to a strike of 100°.

Problem 1-9

A bed strikes N60W and has an apparent dip of 52, N80E. Use the polar tangent diagram to find the true dip.

Problem 1-10

A mineralized fault plane has the following apparent dips: 48, N80W and 65, S20W. A mining engineer wants to excavate the steepest possible shaft within the mineralized fault zone. Use the polar tangent diagram to determine the attitude of this shaft.

Problem 1-11

In the mine described in Problem 1-10, a ventilation tunnel with a trend of S10E is to be drilled within the mineralized zone. Use the polar tangent diagram to determine the plunge of this second shaft.

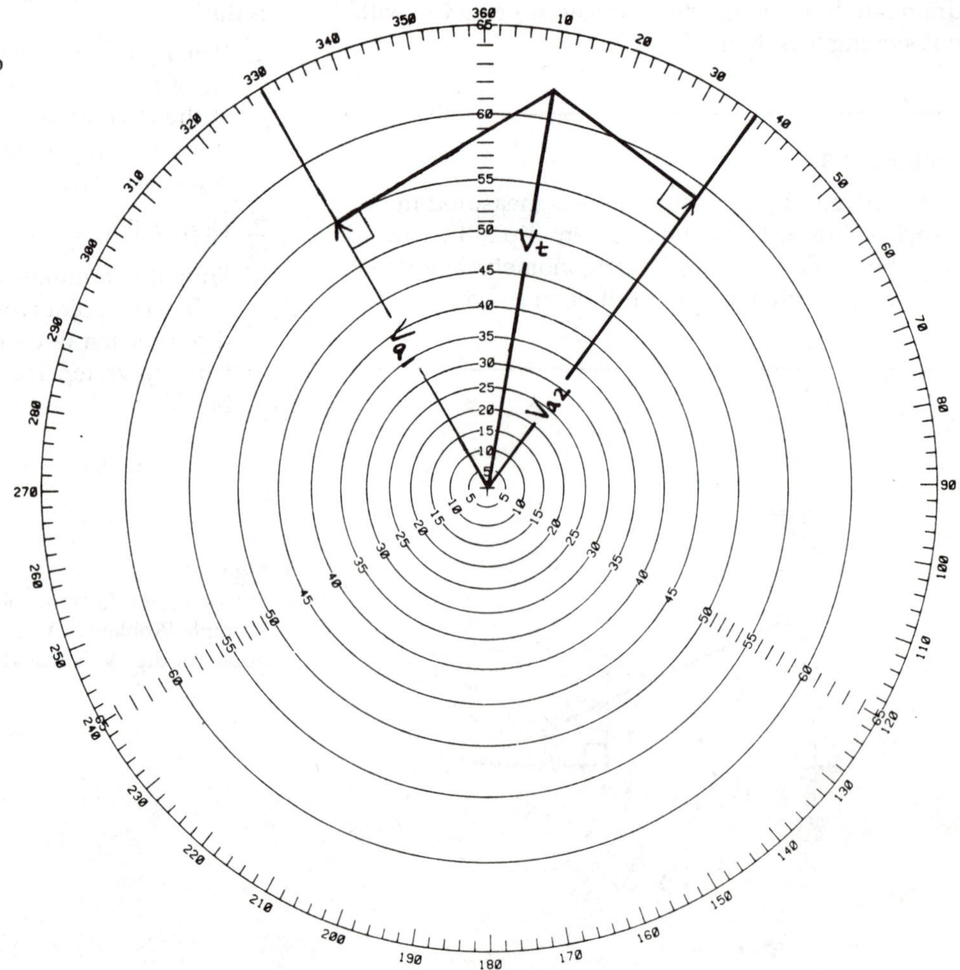

Figure 1-8
Polar tangent diagram solution to example Problem 6. $\mathbf{V}_{a1}$ and $\mathbf{V}_{a2}$ are apparent dips, $\mathbf{V}_t$ is true dip.

Alignment Diagrams

Alignment diagrams (nomograms) usually involve three variables that have a simple mathematical relationship with one another. A straight line connects points on three scales. Figure 1-9 is an alignment diagram for δ, α, and β. If any two of these variables are known, the third may be quickly determined. This technique is particularly convenient for determining apparent dip angles on geologic structure sections that are not perpendicular to strike, as discussed in Chapter 4.

Problem 1-12

Solve Problems 1-2 and 1-9 on the alignment diagram.

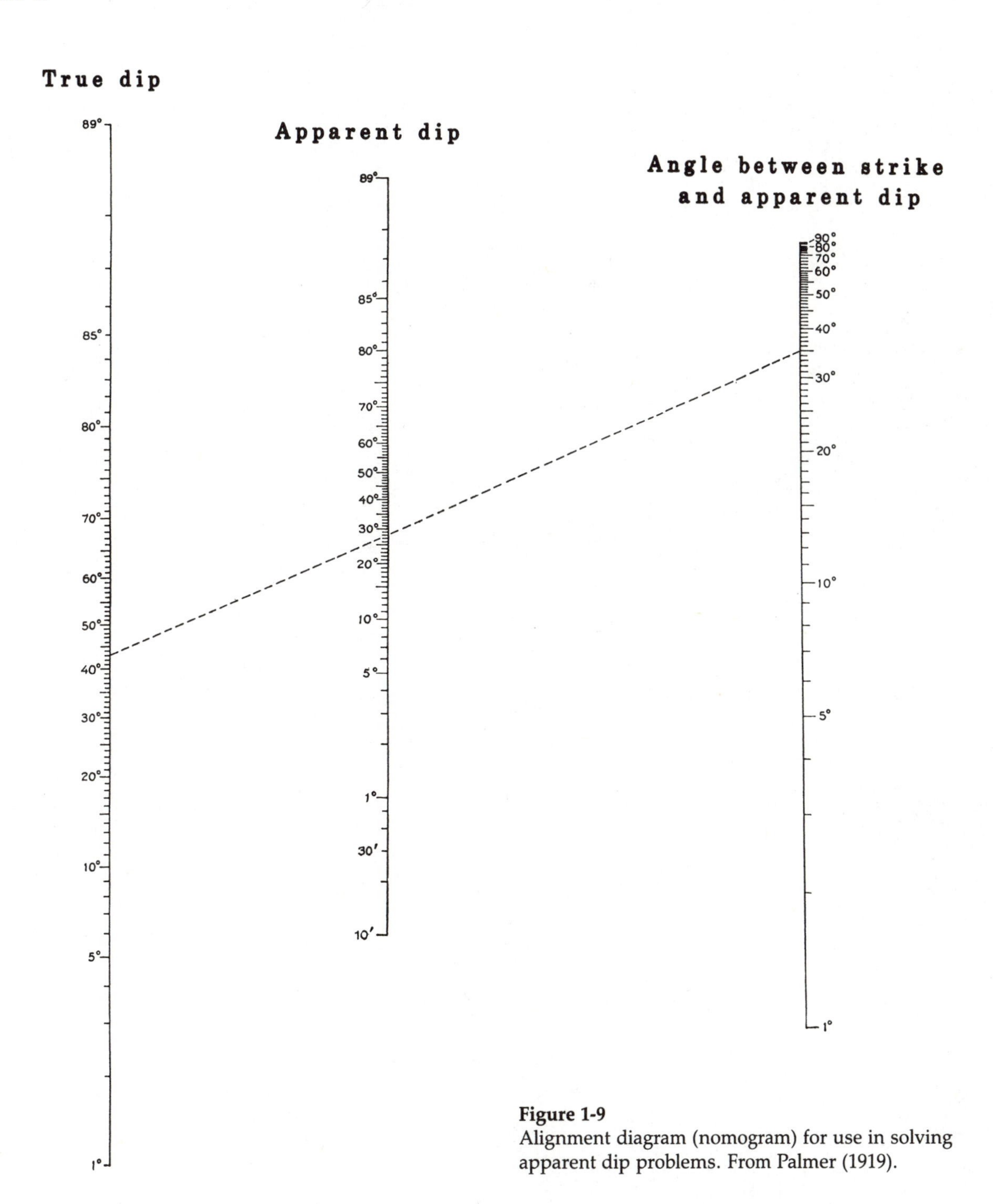

Figure 1-9
Alignment diagram (nomogram) for use in solving apparent dip problems. From Palmer (1919).

Further Reading

Bengtson, C. A. 1980. "Structural uses of tangent diagrams." *Geology,* v. 8, 599–602. An explanation of various uses of tangent diagrams, with several examples.

De Jong, K. A. 1975. "Electronic calculators facilitate solution of problems in structural geology." *Journal of Geological Education,* v. 23, 125–128. A review of the use of pocket calculators in structural geology; includes an extensive bibliography.

Dennison, J. M. 1968. *Analysis of Geologic Structures.* New York: Norton. A structural geology workbook emphasizing the trigonometric approach.

2

Outcrop Patterns and Structure Contours

OBJECTIVES

Determine the general attitude of a plane from its outcrop pattern

Draw structure contour maps

Solve three-point problems

Determine the outcrop patterns of planar and folded layers from attitudes at isolated outcrops

Because the earth's surface is irregular, planar features such as beds, dikes, and faults typically form irregular outcrop patterns. Thus outcrop patterns can serve as clues to the orientations of the planes. Following are seven generalized cases showing the relationships between topography and the outcrop patterns of planes as seen on a map. In Figures 2-1 through 2-7 cover the block diagram (a) and try to visualize the orientation of the bed from its outcrop pattern in map view (b). Note the symbols that indicate attitude.

1. Inclined planes "V" updip as they cross ridges (Fig. 2-1).
2. Vertical planes are not deflected at all by valleys and ridges (Fig. 2-2).
3. Horizontal planes appear parallel to contour lines and "V" upstream (Fig. 2-3).
4. Planes that dip downstream at a steeper gradient than the stream (the usual case) "V" downstream (Fig. 2-4).
5. Planes that dip downstream at the same gradient as the stream appear parallel to the stream bed (Fig. 2-5).
6. Planes that dip downstream at a gentler gradient than the stream "V" upstream (Fig. 2-6).
7. Planes that dip upstream "V" upstream (Fig. 2-7).

After examining Figures 2-1 through 2-7, do problem 2-2 on page 19.

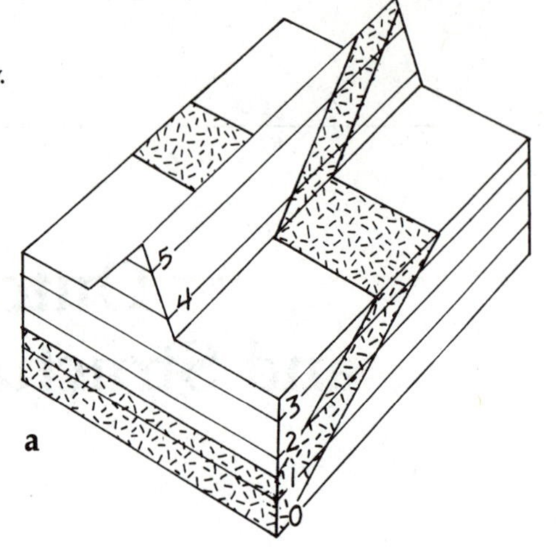

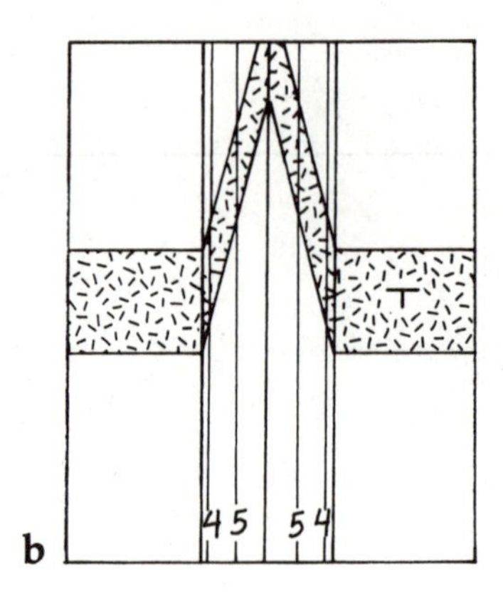

Figure 2-1
Inclined plane crossing a ridge. (a) Block diagram. (b) Map view.

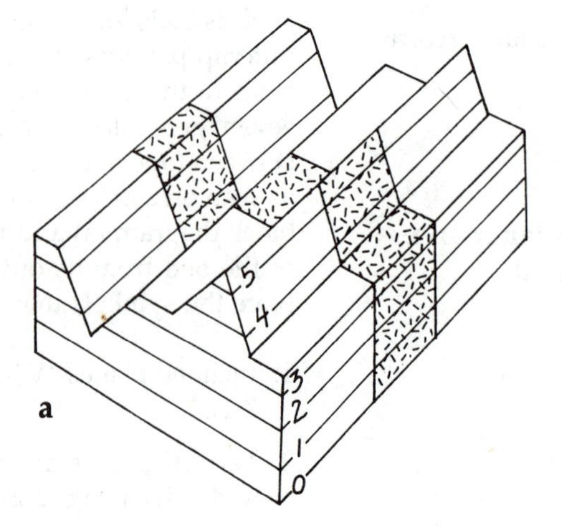

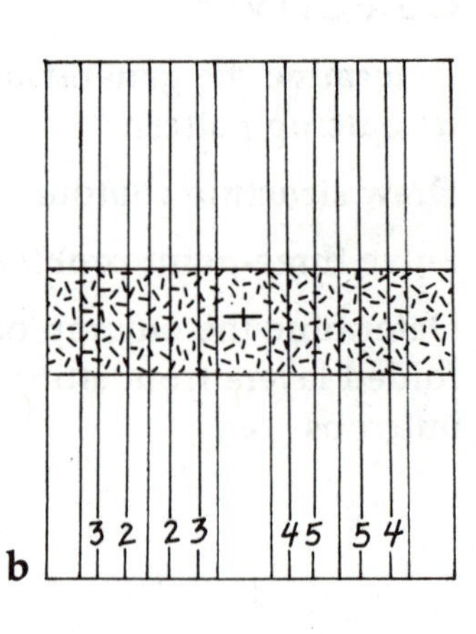

Figure 2-2
Vertical plane crossing a ridge and a valley. (a) Block diagram. (b) Map view.

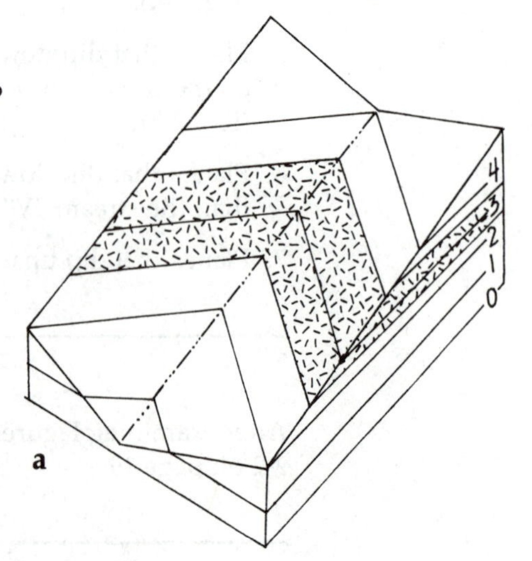

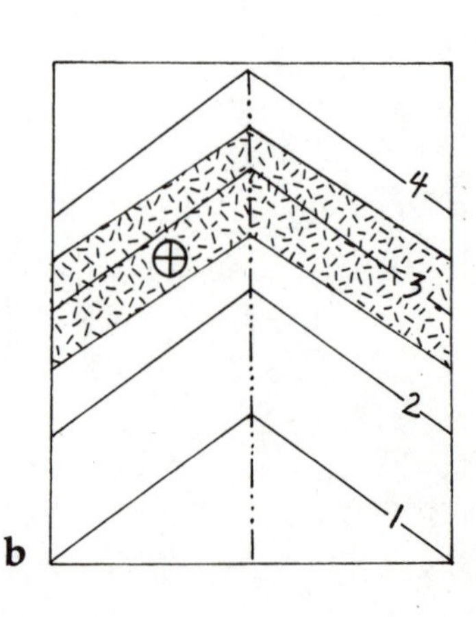

Figure 2-3
Horizontal plane in a stream valley. (a) Block diagram. (b) Map view.

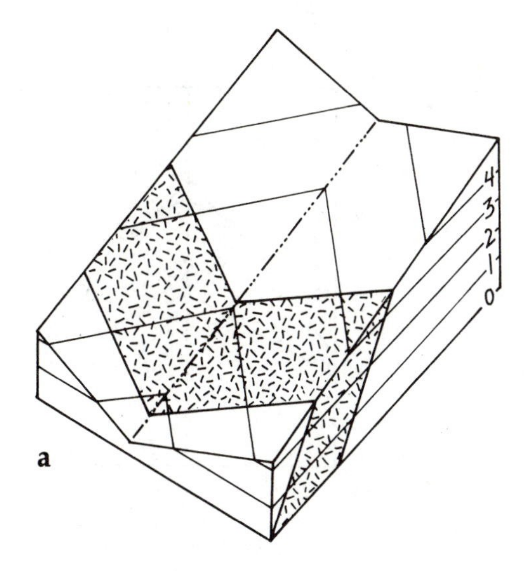

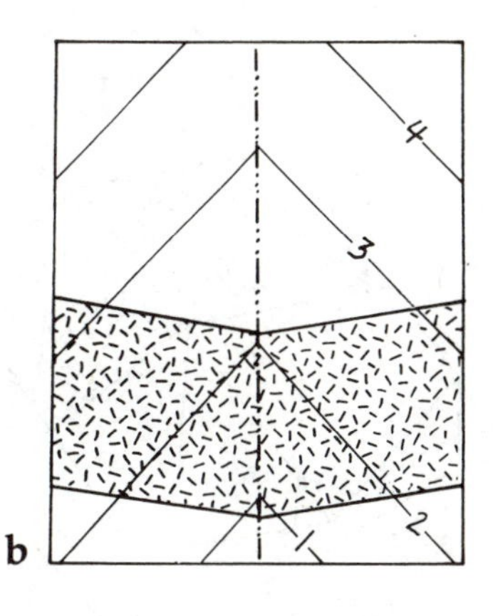

Figure 2-4
Steeply dipping plane dipping downstream. (a) Block diagram. (b) Map view.

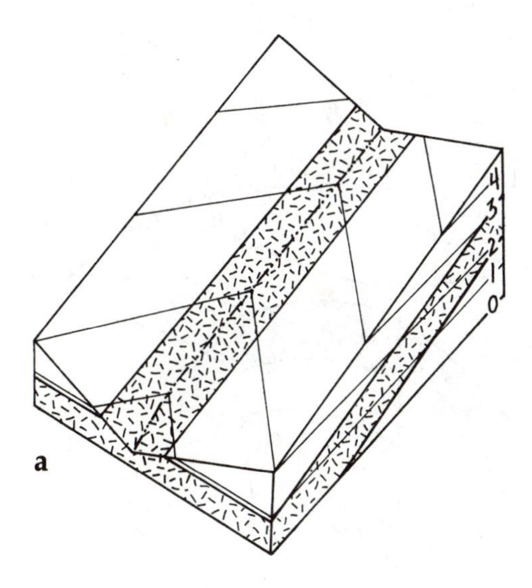

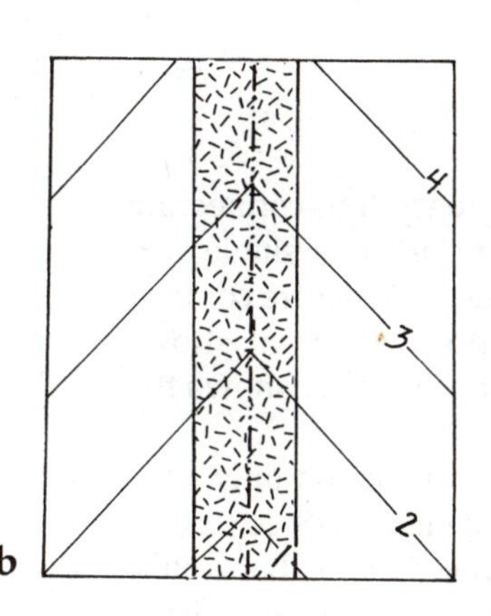

Figure 2-5
Plane dipping parallel to stream gradient. (a) Block diagram. (b) Map view.

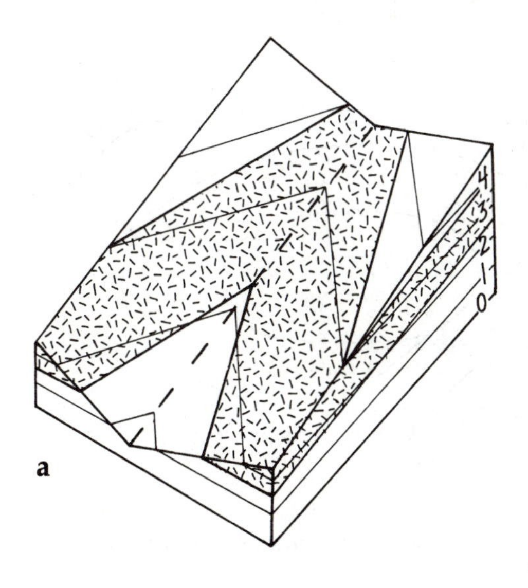

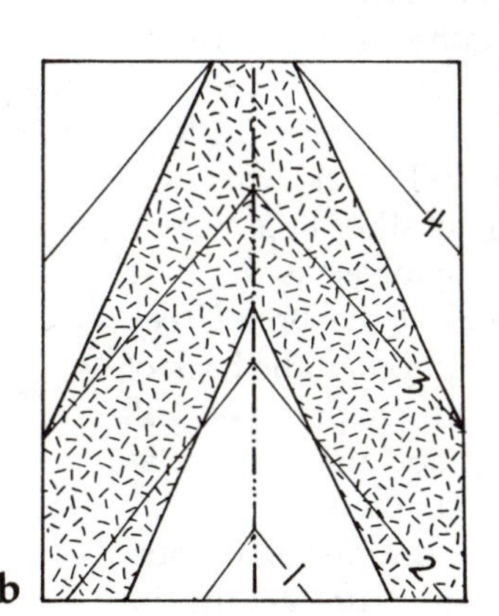

Figure 2-6
Gently dipping plane dipping downstream. (a) Block diagram. (b) Map view.

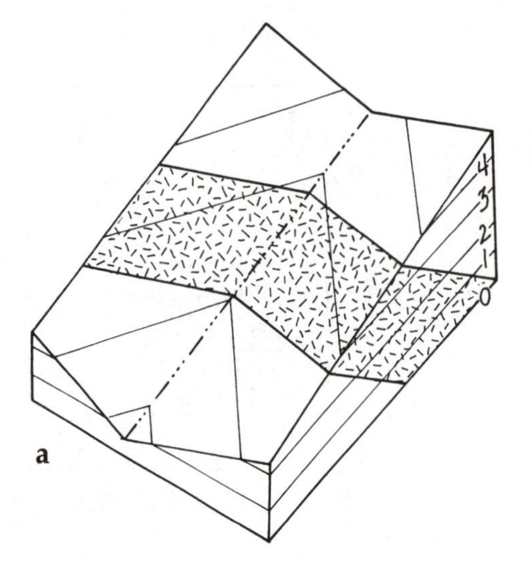

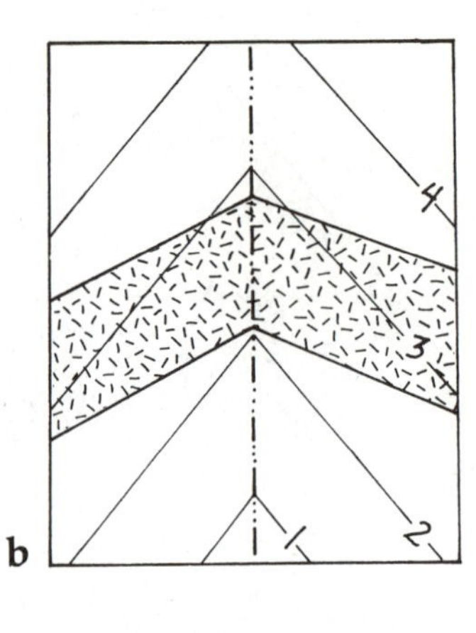

Figure 2-7
Inclined plane dipping upstream. (a) Block diagram. (b) Map view.

Structure Contours

A **structure contour** is an imaginary line connecting points of equal elevation (a contour) on a single surface, such as the top of a formation. Structure contour maps are analogous to topographic maps: one shows the surface of a geologic horizon, the other shows the surface of the earth.

Structure contour maps are most commonly constructed from drill-hole data. See Figure 2-8, for example, which shows a faulted dome. Notice that unlike topographic contours, structure contours sometimes terminate abruptly. Gaps in the map indicate normal faults, and overlaps indicate reverse faults.

Structure contour maps are used extensively in petroleum exploration to identify structural traps. The objective here will be to introduce you to structure contour maps so that you are generally familiar with them and can use them to determine outcrop patterns later in the chapter.

Figure 2-9 is a map showing the elevation (in feet) of the top of a formation in 26 drill holes. This area is in the northeastern corner of the Bree Creek Quadrangle, and the formation involved is the Bree Conglomerate. The geologic map of the Bree Creek Quadrangle may be found on six pages in the back of this book. As explained later in this chapter, you will combine these into one big map and use it often as you work through the following chapters.

There are various techniques for contouring numerical data such as the elevations on Figure 2-9. In the case of geologic structure contours, there are usually not enough data to produce an unequivocal map, so experienced interpretation becomes extremely valuable.

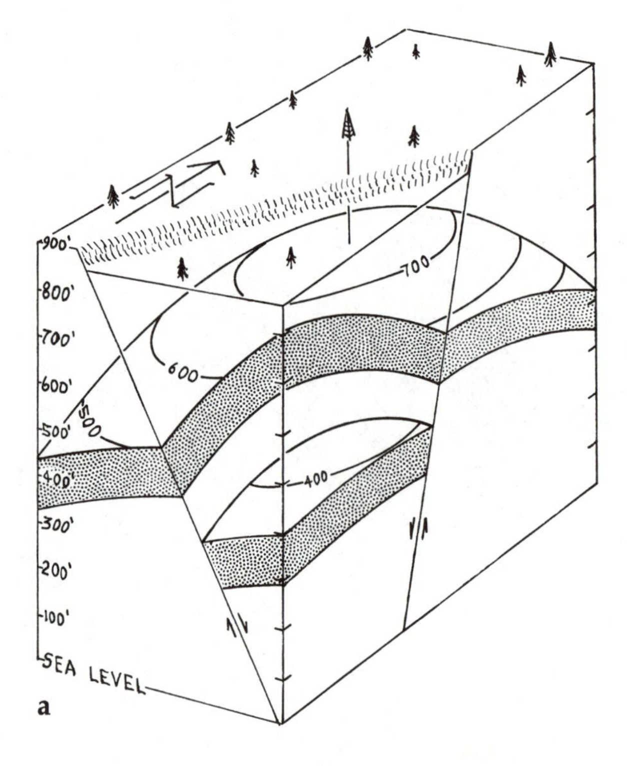

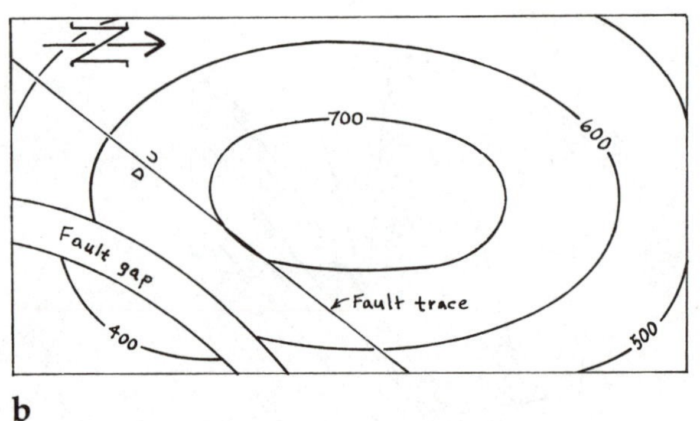

Figure 2-8
Block diagram (a) and structure contour map (b) of a faulted dome.

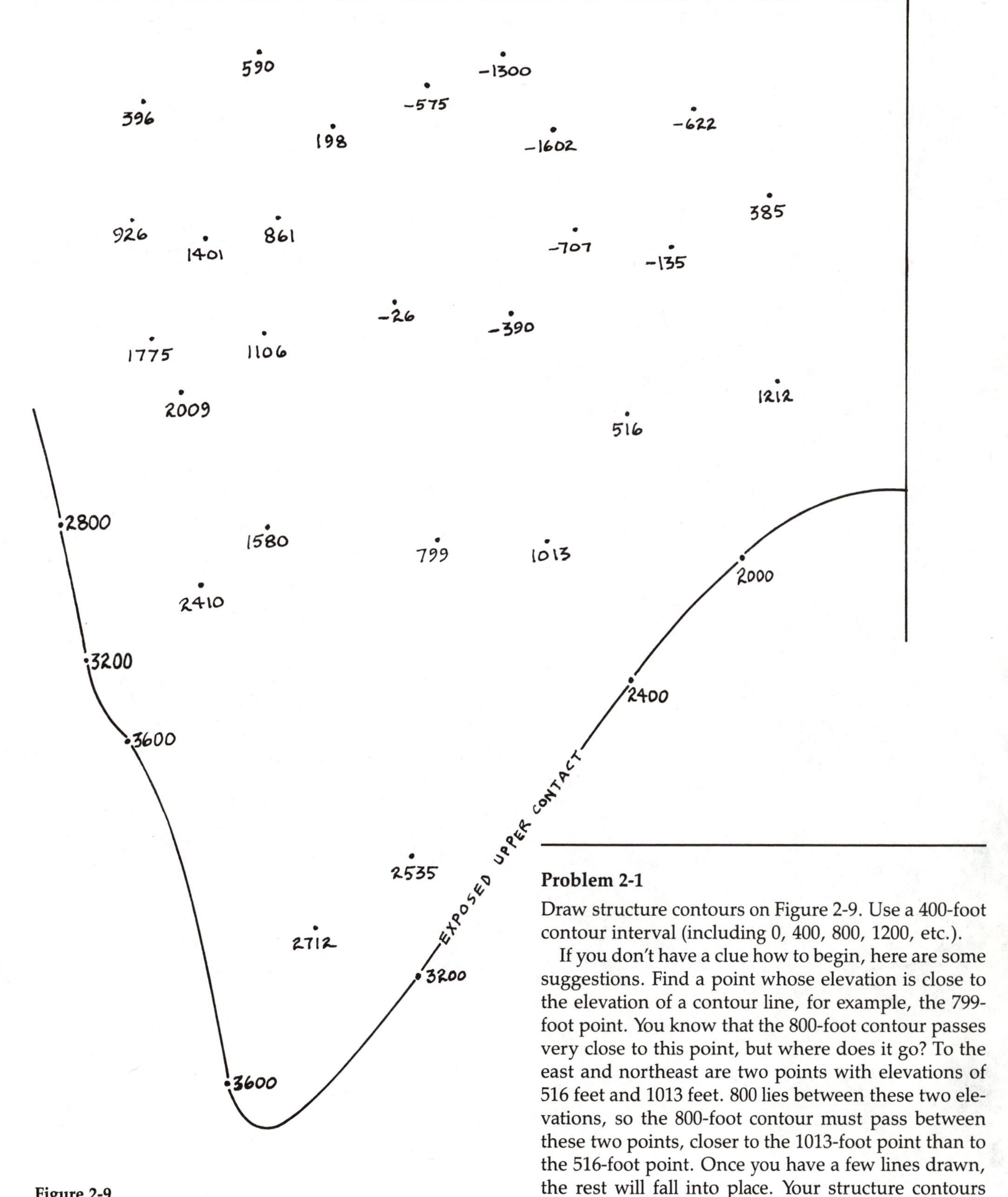

Figure 2-9
Map to accompany Problem 2-1. Elevation of upper surface of Bree Conglomerate in 26 drill holes, northeastern Bree Creek Quadrangle. Elevations are marked along upper contact.

Problem 2-1

Draw structure contours on Figure 2-9. Use a 400-foot contour interval (including 0, 400, 800, 1200, etc.).

If you don't have a clue how to begin, here are some suggestions. Find a point whose elevation is close to the elevation of a contour line, for example, the 799-foot point. You know that the 800-foot contour passes very close to this point, but where does it go? To the east and northeast are two points with elevations of 516 feet and 1013 feet. 800 lies between these two elevations, so the 800-foot contour must pass between these two points, closer to the 1013-foot point than to the 516-foot point. Once you have a few lines drawn, the rest will fall into place. Your structure contours should be smooth, subparallel lines. Use a pencil, because this is a trial-and-error operation.

Notes

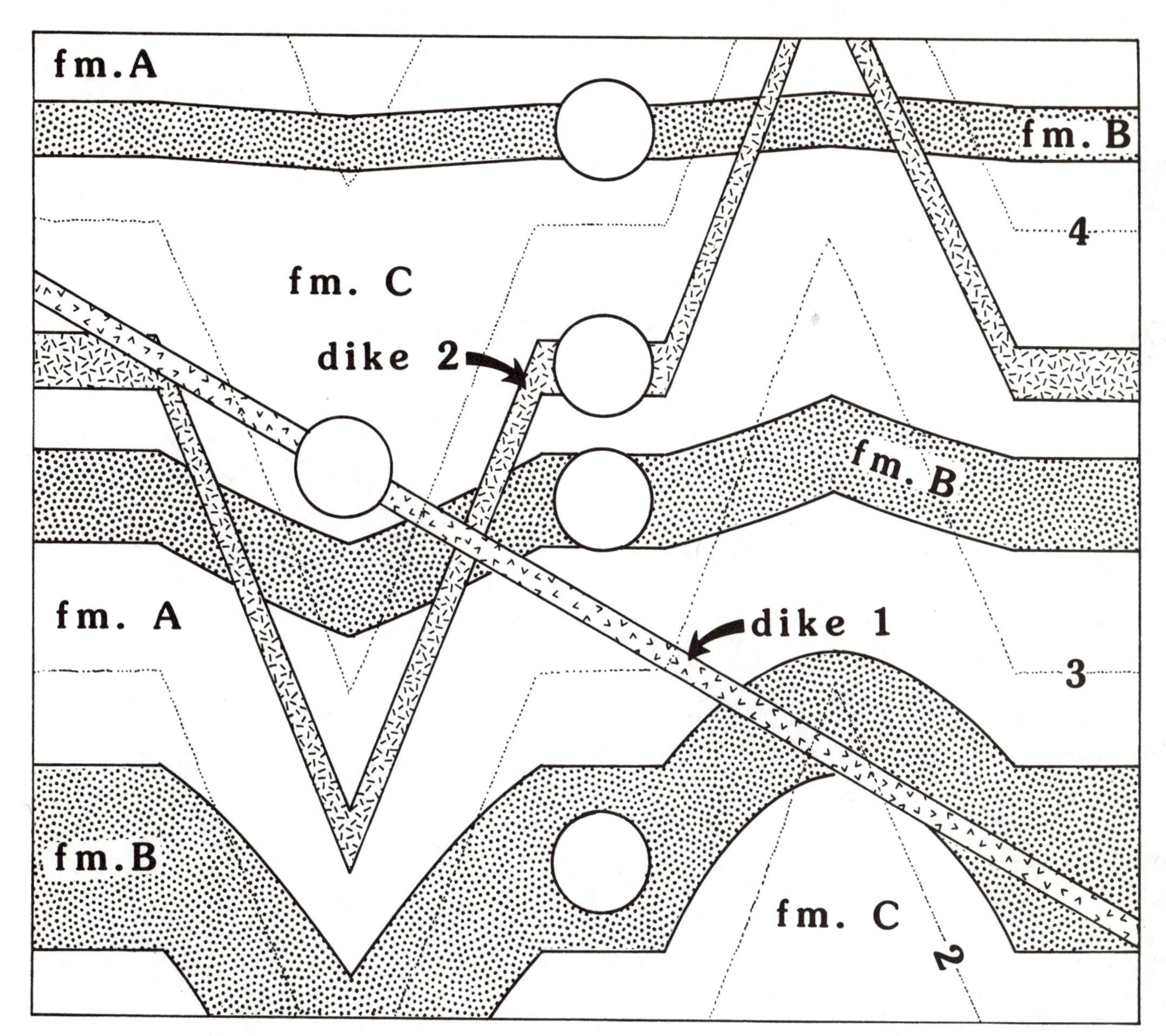

Figure 2-10
Geologic map for use in Problem 2-2. Contours are dotted lines.

Problem 2-2

On the geologic map in Figure 2-10 draw the correct strike and dip symbol in each circle to indicate the attitude of formation B and each dike. To verify your attitude symbols, Figure 2-11 can be photocopied, cut out, and folded to form a block model of this map. (Figure 3-13 shows numerous map symbols.)

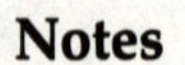

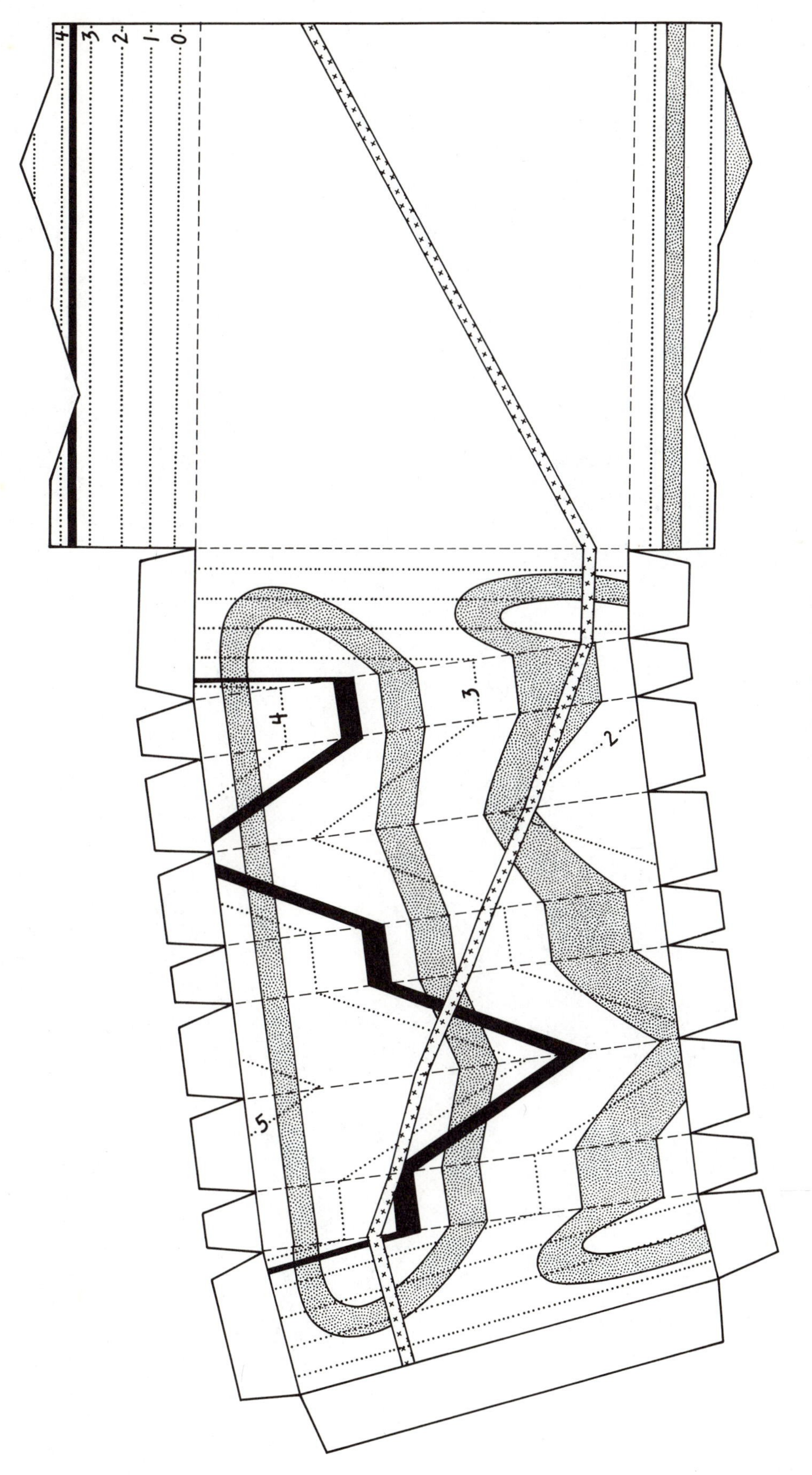

Figure 2-11
Block model to accompany Problem 2-2. Cut out or photocopy the diagram, fold on the dashed lines, and glue the tabs.

Notes

The Three-Point Problem

A classic type of problem in structural geology is called the three-point problem. The elevation and location of three points in a plane are given and the attitude of the plane must be determined. Consider Figure 2-12a, which shows three points (A, B, C) on a topographic map. These three points lie on the top of a sandstone layer. The problem is to determine the attitude of the layer. We will solve this problem two different ways, using first a structure contour approach, then a two-apparent-dip approach.

Solution 1

1. Place a piece of tracing paper over the map, and label the three known points and their elevations. On the tracing paper draw a line connecting the highest of the three points with the lowest. Take the tracing paper off the map, then find the point on this line that is equal in elevation to the intermediate point. In Figure 2-12b point B has an elevation of 160 feet, so the point B′ on the AC line equal in elevation to point B lies 6/10 of the way from point A (200 ft) to point C (100 ft).
2. The bed in question is assumed to be planar, so B′ must lie in the plane. We now have two points, B and B′, of equal elevation lying in the plane of the bed, which define the strike of the plane. The structure contour line B–B′ is drawn, and the strike is measured with a protractor to be N48E (Fig. 2-12c).
3. The direction and amount of dip are determined by drawing a perpendicular line to the strike line from point A, the lowest of the three known outcrop points (Fig. 2-12d). The amount of dip can be determined trigonometrically as shown:

$$\tan \delta = \frac{\text{change in elevation}}{\text{map distance}}$$

$$= \frac{60'}{104'} = .57$$

$$.57 = \tan 30^\circ \quad \delta = 30^\circ$$

Figure 2-12
Solution of a three-point problem using a combination of graphical and trigonometric techniques. (a) Three coplanar points (A, B, and C) on a topographic map. (b) Location of a fourth point, B′, at the same elevation as point B. (c) Line B–B′ defining the strike of the plane. (d) Dip direction line perpendicular to line B–B′.

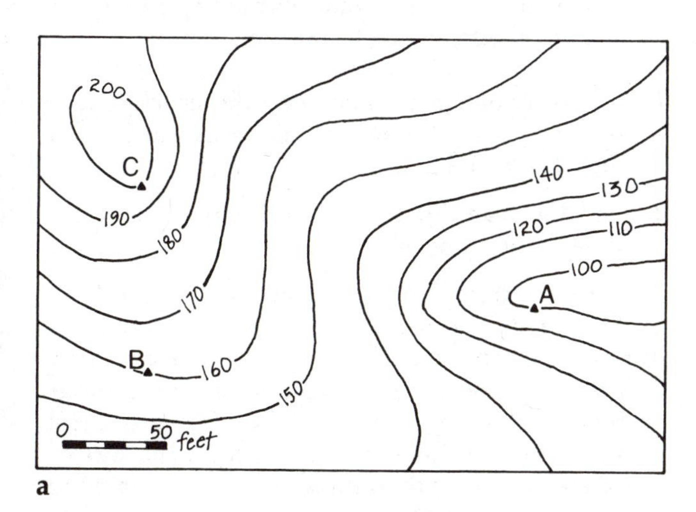

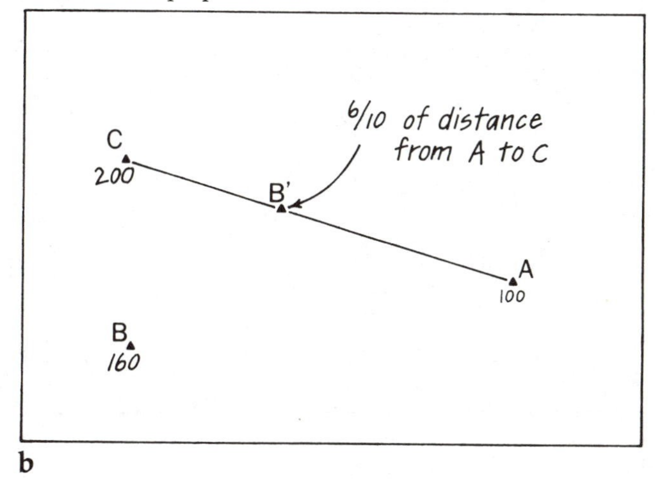

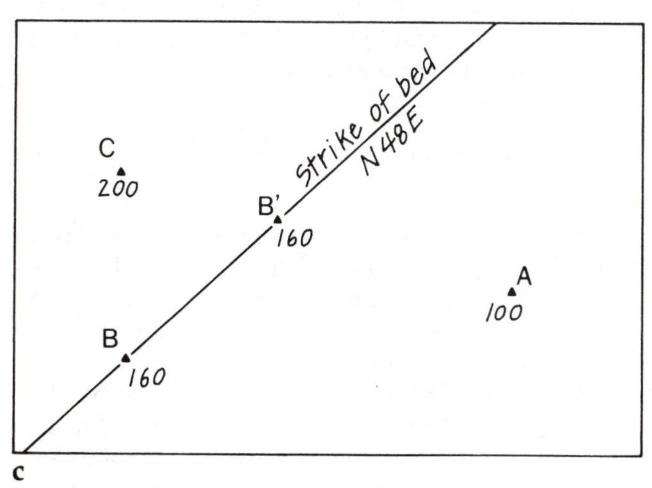

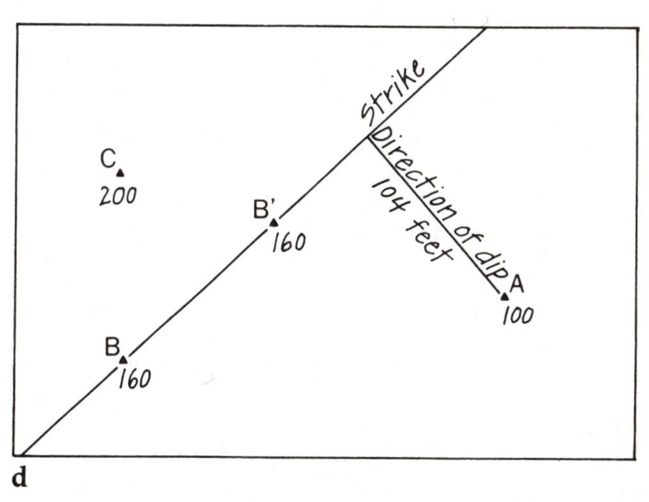

Solution 2

Another approach to solving a three-point problem is to convert it into a two-apparent-dip problem.

1. Draw lines from the lowest of the three points to each of the other two points (Fig. 2-13a). These two lines represent apparent dip directions from B to A and from C to A.

2. Measure the bearing and length of lines CA and BA on the map (Fig. 2-13b), and determine their plunges:

$$\theta_1 = 80^\circ \qquad \theta_2 = 107^\circ$$

$$\tan \alpha_1 = \frac{\text{diff. in elevation}}{\text{map distance}} = \frac{60'}{198'} = .303$$

$$\tan \alpha_2 = \frac{100'}{204'} = .490$$

3. Use equation 4 of Chapter 1 to find the true dip direction, and then equation 3 of Chapter 1 to find the amount of dip.

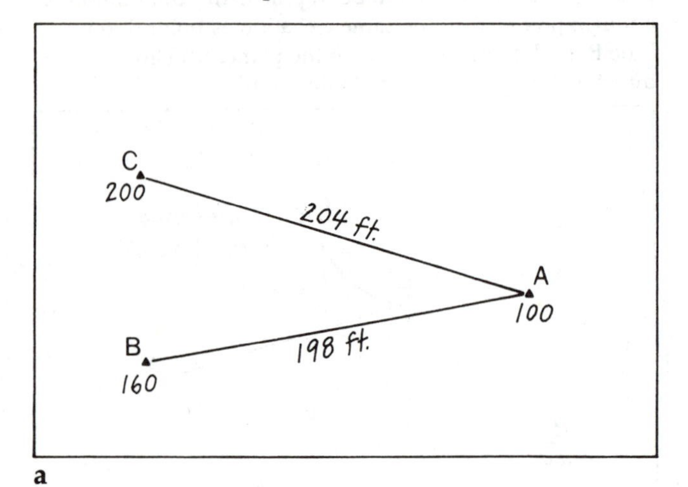

a

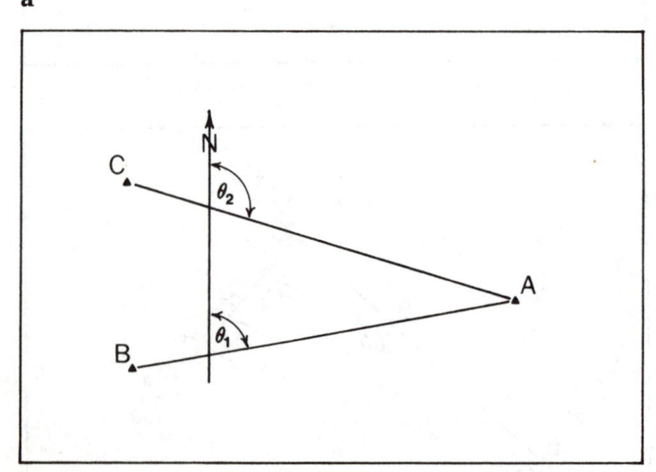

b

Figure 2-13
Three-point problem converted to a two-apparent-dip problem. (a) Three coplanar points. Lines are drawn to the lowest of the three points from the other two points. (b) Apparent dip directions θ_1 and θ_2.

Problem 2-3

Points A, B, and C on Figure 2-14 are oil wells drilled on a flat plain. All tap the same oil-bearing sandstone. The depth of the top of this sandstone in each well is as follows: A = 5115 feet, B = 6135 feet, and C = 5485 feet.

1. Determine the attitude of the sandstone.

2. If a well is drilled at point D, at what depth would it hit the top of the sandstone?

Determining Outcrop Patterns with Structure Contours

Earlier we discussed structure contour maps derived from drill-hole data. Structure contour maps may also be constructed from surface data. Suppose, for example, that an important horizon is exposed in three places on a topographic map, as in Figure 2-12a. If this horizon is planar we can determine its outcrop pattern on the map by the following technique:

1. On a piece of tracing paper draw the structure contour that passes through the middle elevation point (Figs. 2-12b and 2-12c).

2. Find the true dip as described above under the three-point problem.

3. Draw structure contours parallel to the line B–B′ (Fig. 2-12c). In order to determine the outcrop pattern, these structure contours must have a contour interval equal to (or a multiple of) the contour interval on the topographic map. They also must represent the same elevations. Because the surface we are dealing with in this example is assumed to be planar, the structure contours will be a series of straight, equidistant parallel lines. The spacing can be determined trigonometrically:

$$\text{Map distance} = \frac{\text{Contour interval}}{\tan \delta}$$

In this example the spacing turns out to be 17.5 feet in plan view (Fig. 2-15a). Point B is at an elevation of 160 feet, which is conveniently also the elevation of a topographic contour. Points on the bedding plane whose elevations are known (points A and C in this problem) should serve as control points; that is, lay the tracing paper over the map and make sure that the elevations of known outcrop points match their elevations on the structure contour map.

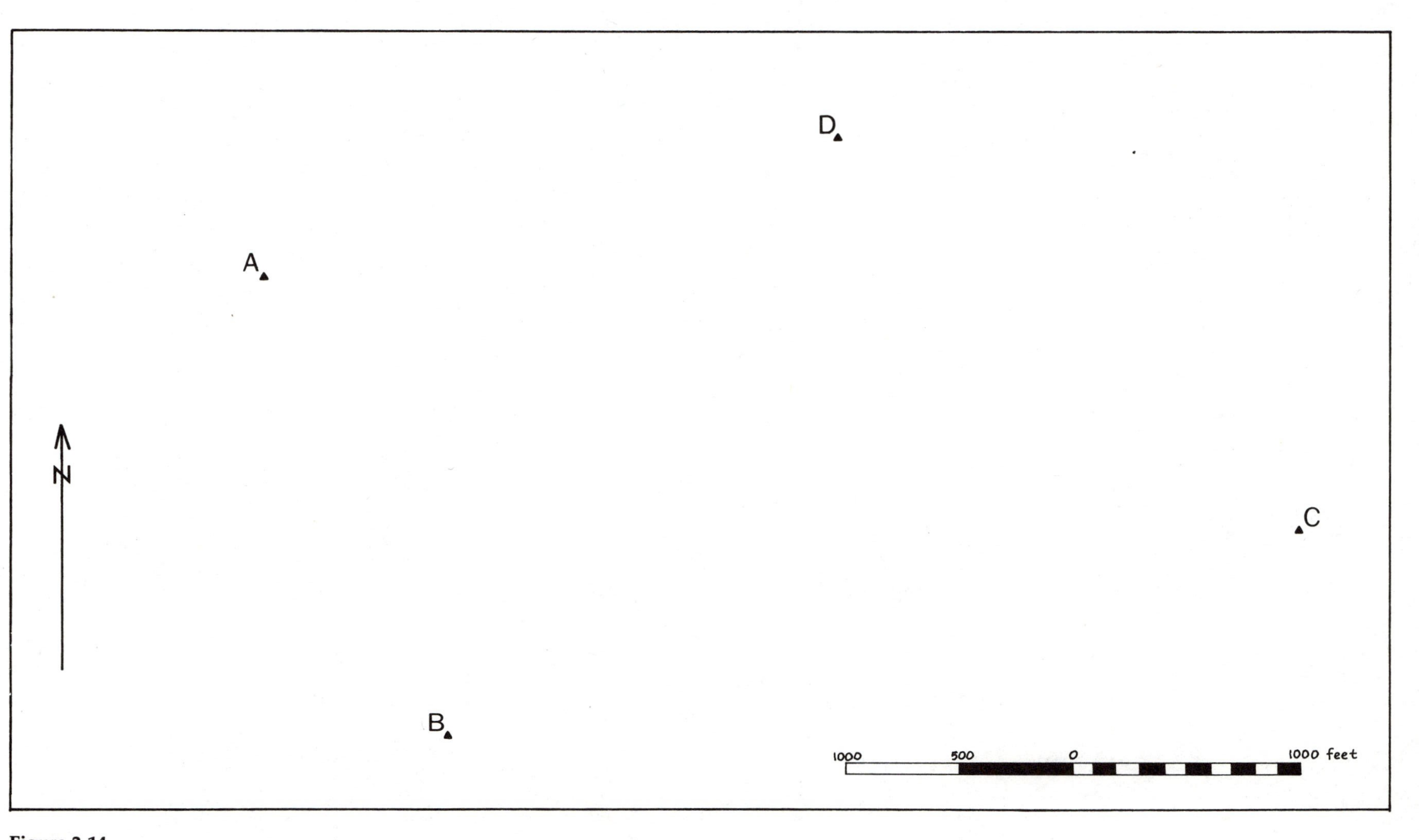

Figure 2-14
Map for use in Problem 2-3.

Notes

If the surface is not quite planar but is changing dip slightly, adjustments can constantly be made on the structure contour map. Figure 2-15b shows the completed structure contour map for this example.

4. Superimpose the structure contour map and the topographic map (Fig. 2-15c). Every point where a structure contour crosses a topographic contour of equal elevation is a surface outcrop point. The outcrop line of the plane is made by placing the structure contour map beneath the topographic map and marking each point where contours of the same elevation cross. A light table may be necessary to see through the topographic map. The points of intersection are connected to display the outcrop pattern on the topographic map (Fig. 2-15d).

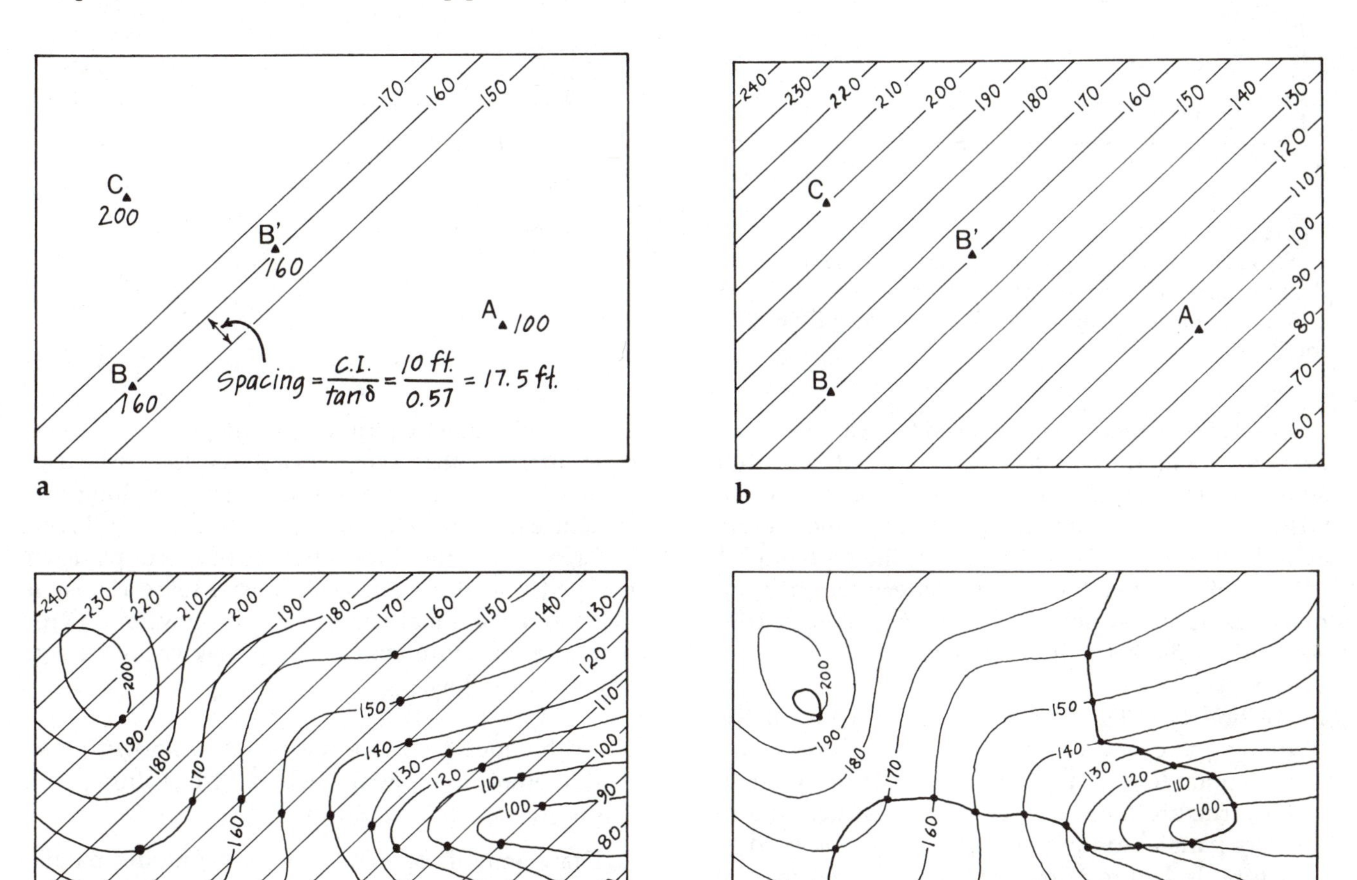

Figure 2-15
Determination of outcrop pattern using structure contours. (a) Three structure contours on a base map (from Fig. 2-12c). (b) Structure contour map. (c) Structure contour map superimposed on a topographic map. (d) Outcrop pattern of a plane on a topographic map.

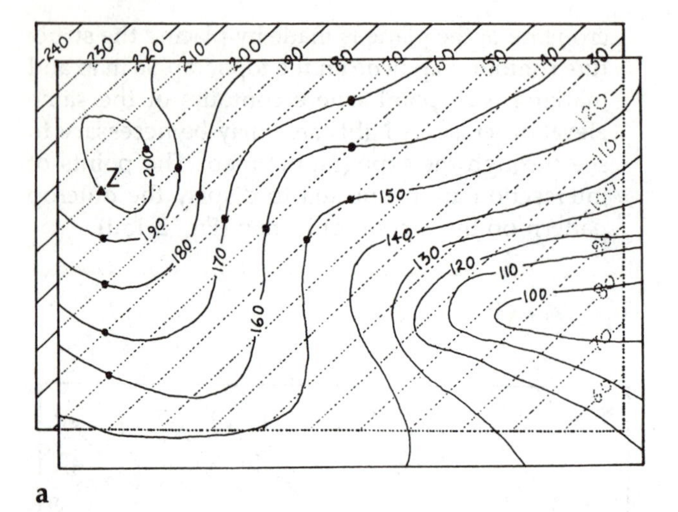

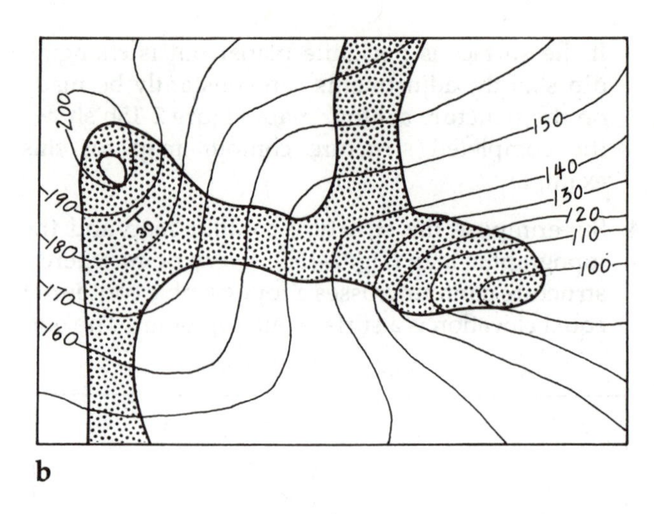

Figure 2-16
(a) Structure contour map shifted to the position of point Z on the bottom of formation whose top is shown in Fig. 2-15 d. (b) Outcrop pattern of rock unit dipping 30° to the southeast.

This same technique can be used to locate a second surface parallel to the first. Suppose that the contact shown in Figure 2-15d is the top of a bed, and we wish to determine the outcrop pattern of the bottom as well. If a single outcrop point on the topographic map is known, then the outcrop pattern can easily be found using the structure contour map already constructed for the bed's upper surface.

1. Position the structure contour map beneath the topographic map such that the bottom surface outcrop point (or points) lie at the proper elevation on the structure contour map. With the structure contours parallel to their former position, proceed as before. In Figure 2-16a, point Z, at an elevation of 200 feet, is a known outcrop point of the bottom of the bed. The structure contour map has been moved so that the 200-foot structure contour passes through point Z, and the predicted outcrop points have been located as before.
2. Once the upper and lower contacts are drawn on the topographic map, the outcrop pattern of the bed can be shaded or colored (Fig. 2-16b).

This technique for locating the intersection of a geologic surface with the surface of the earth may be used even when the surface is not a plane, as long as a structure contour map can be constructed. In Figure 2-17a, for example, three attitudes of a fault plane are mapped, and all are different. If we assume a constant slope and a gradual change in dip between outcrop points, a structure contour map may easily be constructed as follows:

1. Arithmetically interpolate between known elevation points to locate the necessary elevation points on the surface (Fig. 2-17b).
2. Draw smooth parallel structure contours parallel to the strikes at the outcrop points (Fig. 2-17c).
3. Superimpose the structure contour map and the topographic map and mark points where contours of equal elevation intersect (Fig. 2-17d).
4. Connect these intersection points to produce the outcrop map (Fig. 2-17e).

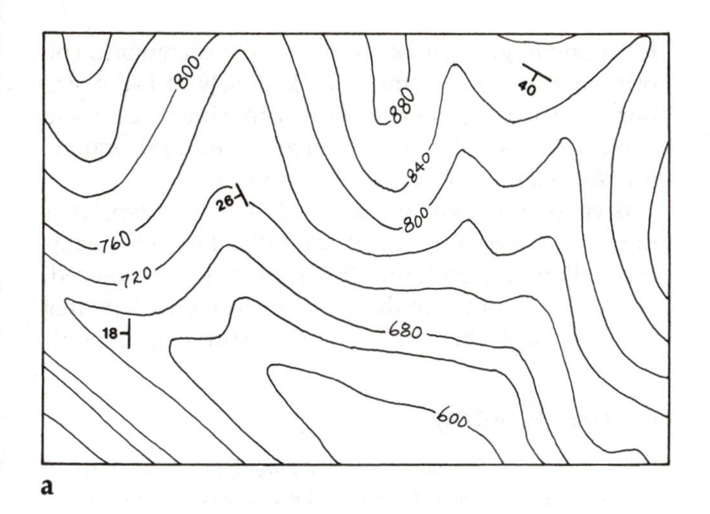

a

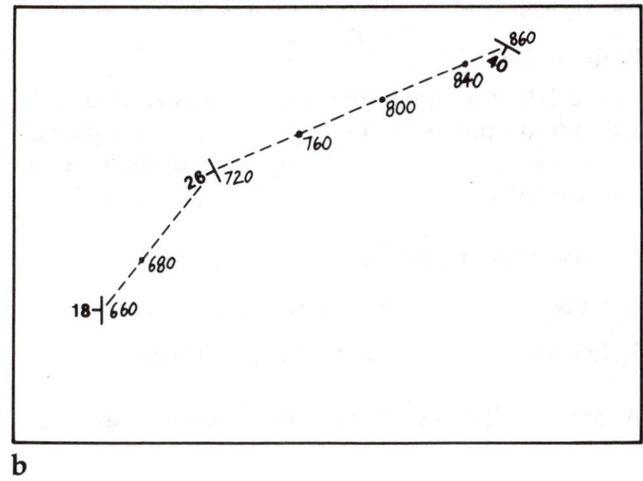

b

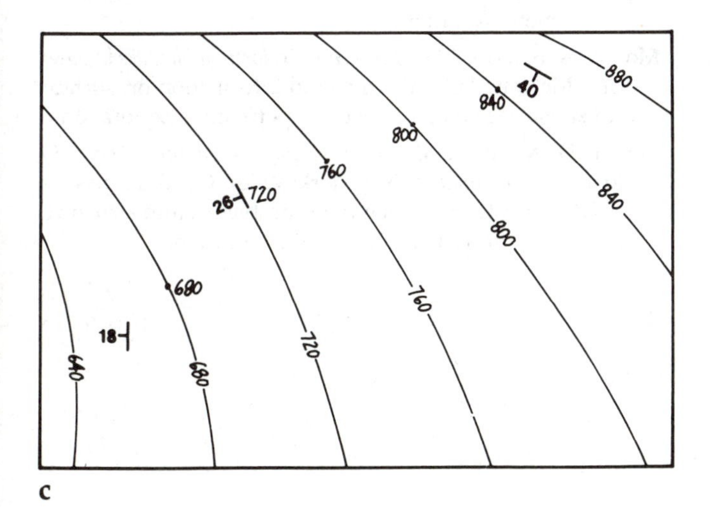

c

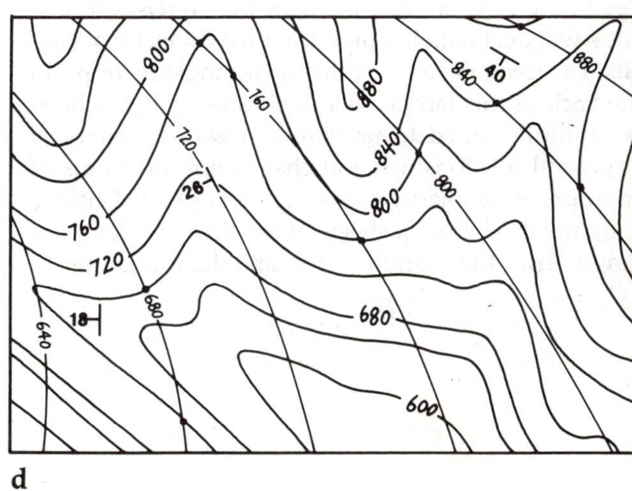

d

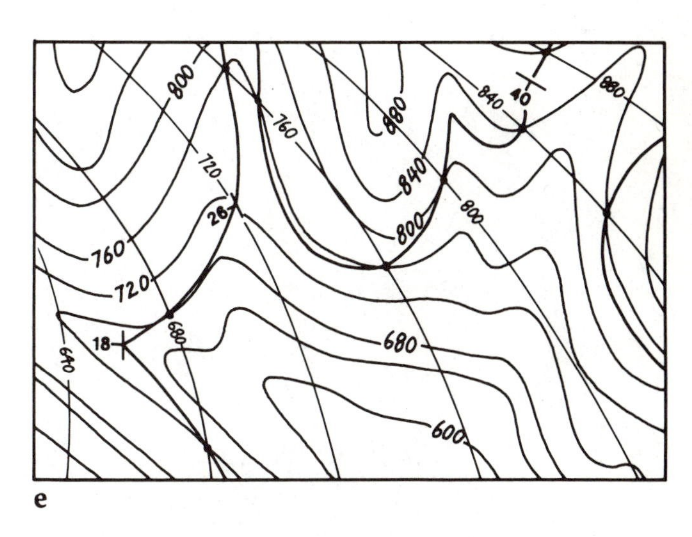

e

Figure 2-17.
Determining outcrop pattern of a gently folded surface using structure contours. (a) Three attitudes of folded horizon on a topographic map. (b) Interpolation of elevations between points of known elevation. (c) Structure contour map of gently folded surface. (d) Structure contour map superimposed on topographic map. (e) Outcrop pattern on topographic map.

Problem 2-4

Figure 2-18 is a topographic map. Points A, B, and C are outcrop points of the upper surface of a planar coal seam. Point Z is an outcrop point of the base of the coal seam.

1. Determine the attitude of the coal seam.
2. Draw the outcrop pattern of the coal seam.
3. Determine the thickness of the coal seam.

Attach any drawings and computations you use.

Bree Creek Quadrangle Map

Beginning in Chapter 3 many of the exercises in this book will deal with the mythical Bree Creek Quadrangle. The geologic map of this quadrangle is found in the back of this lab manual in six sheets. It has been carefully designed to provide you with a variety of structural relationships and challenging problems of appropriate complexity. Before continuing on to Chapter 3, lightly color the six sheets of the Bree Creek Quadrangle, trim the margins, and tape them together to form one large map. More than mere busywork, coloring a map forces you to look closely at the distribution of various rock units. For maximum contrast, avoid using similar colors, such as red and orange, for consistently adjacent rock units.

Because you will be using this map often, it is important that you treat it carefully. Tape the seams carefully on the back (not the front) after it is colored, and then fold it up so that it fits into an envelope that you keep with the rest of your structural equipment.

Further Reading

Bishop, M. S. 1960. *Subsurface Mapping*. New York: Wiley. A good introduction to subsurface mapping, especially structure contour maps.

Dennison, J. M. 1968. *Analysis of Geologic Structures*. New York: Norton. Good discussion of outcrop patterns, three-point problems, and structure contours. Emphasis is on trigonometric solutions.

Moody, G. B. (ed.). 1961. *Petroleum Exploration Handbook*. New York: McGraw-Hill. Much useful information on surface and subsurface maps applied to petroleum exploration.

Ragan, D. M. 1985. *Structural Geology—An Introduction to Geometrical Techniques*. New York: Wiley. Good discussion of outcrop patterns, three-point problems, and structure contours. Emphasis is on graphical solutions.

Figure 2-18
Map for use in Problem 2-4.

3

Interpretation of Geologic Maps

OBJECTIVES

Determine the exact attitude of a plane from its outcrop pattern

Determine stratigraphic thickness from outcrop pattern

Determine the nature of contacts from outcrop patterns and attitudes

Construct a stratigraphic column

Geologic maps are drawn primarily from observations made on the earth's surface, often with reference to aerial photographs. The purposes of a geologic map are to show the surface distributions of rock units, the locations of the interfaces or **contacts** between adjacent rock units, the locations of faults, and the orientations of various planar and linear elements. (Standard geologic symbols are shown in Figure 3-13, p. 42.)

Some aspects of constructing a geologic map, such as the defining of rock units, are quite subjective and are done on the basis of the geologist's interpretations of how certain rocks formed. This being the case, many neatly inked, multicolored maps belie the uncertainty that went into their construction.

Accompanying this manual is a geologic map of the Bree Creek Quadrangle. An important teaching strategy of this book is to have you analyze the map in detail throughout the course, one step at a time, and then to have you synthesize it all into a cohesive structural history. The analysis begins with this chapter; the synthesis will come in Chapter 11.

Determining Exact Attitudes from Outcrop Patterns

Because the strike of a plane is a horizontal line, any line drawn between points of equal elevation on a plane defines the plane's strike. Figure 3-1a is a geologic map with two rock units, formation M and formation X. The contact between these two rock units crosses several contours. To find the strike of the contact a straight line is drawn from the intersection of the contact with the 1920-foot contour on the west side of the map to the intersection of the contact with the 1920 contour on the east side of the map (Fig. 3-1b). The strike of this contact is thus determined to be N79E.

It should be clear to you from the outcrop pattern in Figure 3-1a that the beds dip south. To determine the exact dip, draw a perpendicular line to the strike line from another point of known elevation on the contact. In Figure 3-1c a line has been drawn from the strike line to a point where the contact crosses the 1680 contour. The length of this line h and the change in elevation from the strike line to this point v yield the dip δ with the following equation (Fig. 3-1d):

$$\tan \delta = \frac{v}{h}$$

The solution to this example is:

$$\tan \delta = \frac{v}{h} = \frac{240}{3000} = 0.08$$
$$\delta = 5°$$

This method for determining attitudes from outcrop patterns can only be used if the rocks are not folded.

Figure 3-2 shows the Neogene (Miocene and Pliocene) units of the northeastern block of the Bree Creek Quadrangle. Straight lines have been drawn connecting points of known elevation on the bottom contact of the Rohan Tuff, unit Tr. The strike, measured directly, is N16W, and the dip is 11NE as determined by:

$$\tan \delta = \frac{v}{h} = \frac{400}{2000} = 0.2$$
$$\delta = 11°$$

But what is the attitude of the southern outcrop of the Rohan Tuff? Even though no two points of equal elevation can be found on the bottom contact, notice that the points of known elevation lie on the same straight lines drawn for the northern outcrop. This is strong evidence that the attitude of the southern outcrop of Rohan Tuff is exactly the same as that for the northern one. This kind of reasoning is typical of what must routinely be done when interpreting geologic maps.

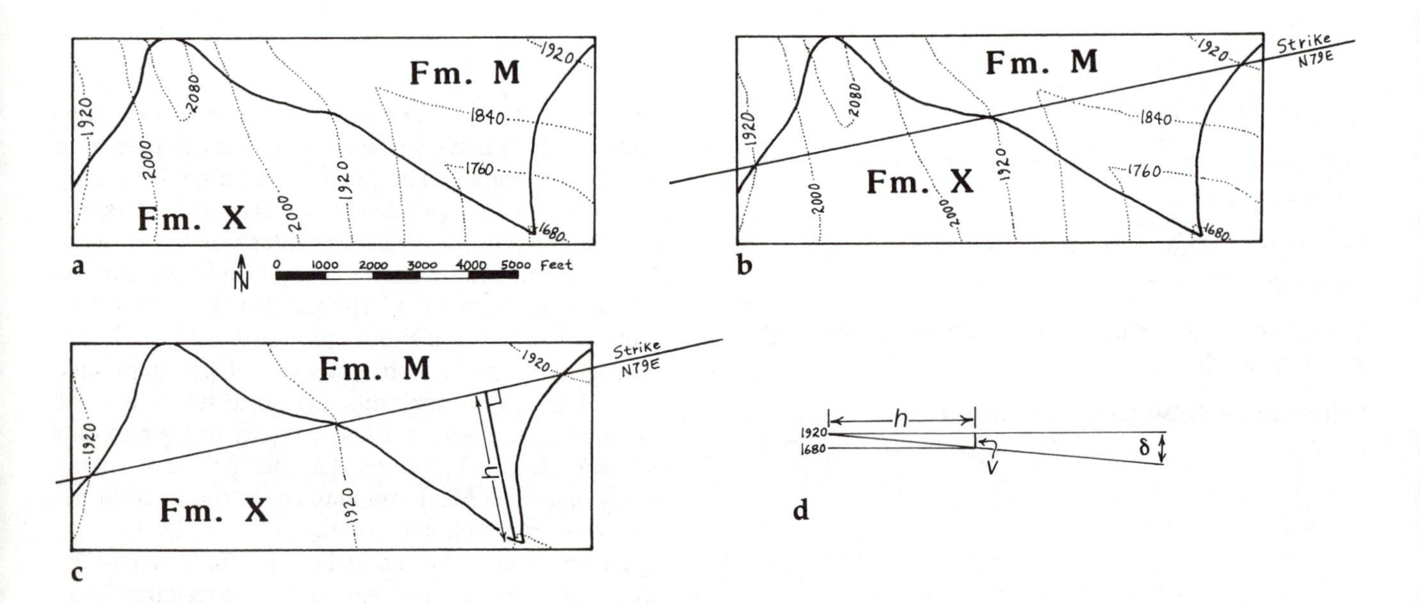

Figure 3-1
Technique for determining the attitude of a plane from its outcrop pattern. (a) Contact between formations X and M. (b) Line connecting points of equal elevation defines strike of plane. (c) Perpendicular is drawn to a point on contact at different elevation. (d) Dip angle δ is found from $\tan \delta = v/h$.

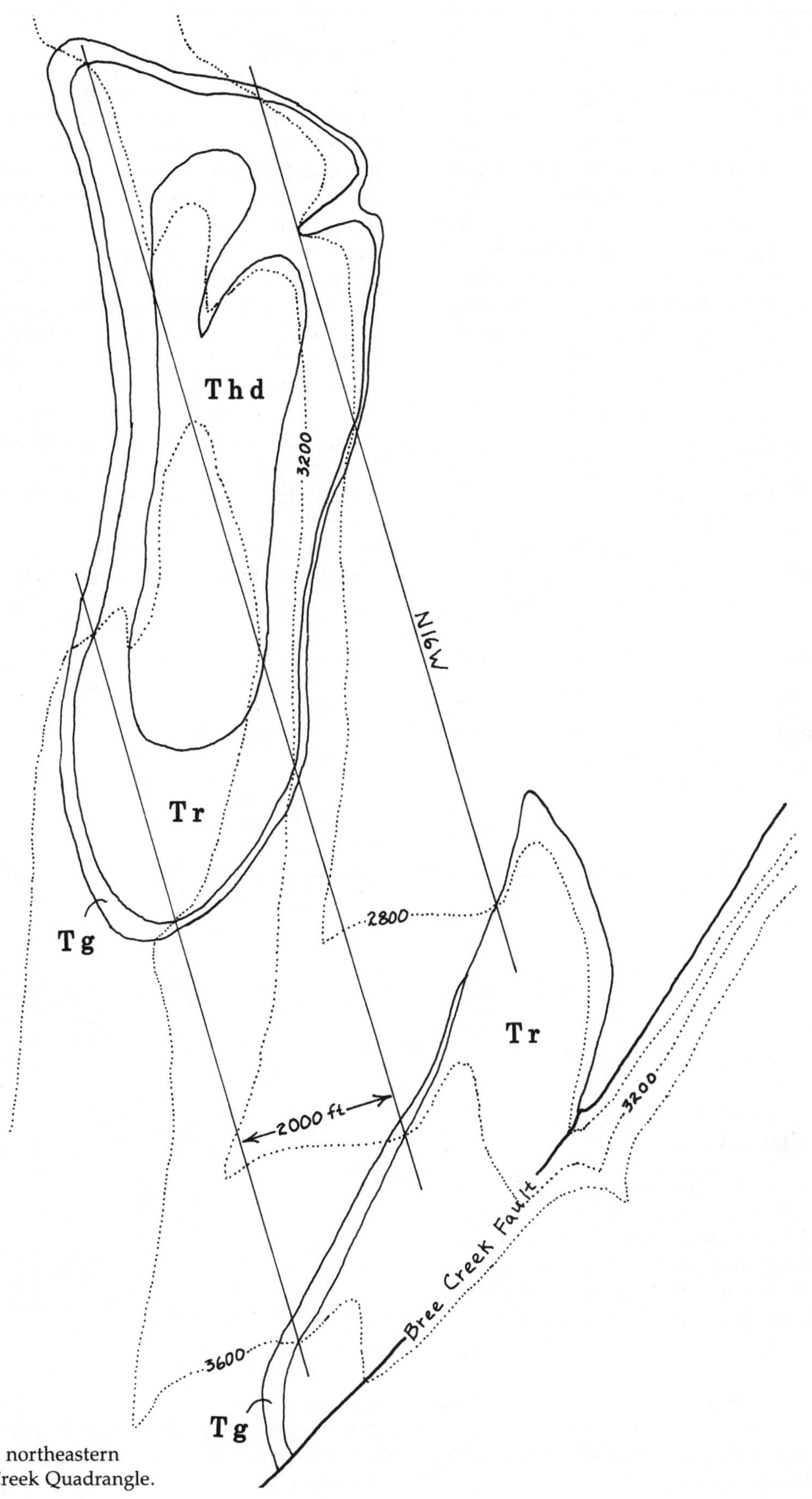

Figure 3-2
Neogene units in northeastern portion of Bree Creek Quadrangle.

Problem 3-1

On the Bree Creek Quadrangle map determine the exact strike and dip of the Miocene and Pliocene units and label the map accordingly with the appropriate symbol. List each attitude in the space below as well as on your map. Use the lower contact of each unit, because the upper contact may have been eroded. After determining the strike and dip of each formation, try to visualize the geology in three dimensions. Make sure that the attitude you determined is in agreement with the outcrop pattern. (Answer sheet for problems 3-1, 3-2, and 3-3 on p. 47)

	Thd	Tr	Tg
Northeastern fault block			
Northern exposures	____	____	____
Southern exposures		____	____
Central fault block			
Northern area		____	____
Galadriel's Ridge		____	____
Southwestern area		____	____
Western fault block			
Gandalf's Knob	____		
Southern exposures	____		

Problem 3-2

The Paleogene (Paleocene through Oligocene) units of the Bree Creek Quadrangle were folded and then eroded nearly flat. Determine the approximate stratigraphic thickness of each of these Paleogene units. To do this, for each unit find a place that is not in the hinge zone of a fold, that is nearly flat, that contains exposures of both the upper and lower contacts, and where the dip is fairly constant. The horizontal distance h must be measured perpendicular to the strike. A good place to measure the Bree Conglomerate, for example, is where Galadriel's Creek crosses it in the southwestern corner of the map. In places where the dip is not completely consistent you may have to use an average dip. For a unit such as the Dimrill Dale Diatomite, whose upper contact is not exposed on the map, indicate that the determined thickness is a minimum thickness. (Because of the large contour interval and the absence of completely flat terrain, thickness determinations will be somewhat variable.)

Tdd	____
Tmm	____
Tm	____
Tts	____
Tb	____
Te	____

Determining Stratigraphic Thickness in Flat Terrain

It is usually possible to determine the approximate stratigraphic thickness of a rock unit from a geologic map if its attitude is known. If a unit is steeply dipping, and if its upper and lower contacts are exposed on flat or nearly flat terrain, then the thickness is determined from the trigonometric relationships shown in Figure 3-3:

$$t = h \sin \delta$$

where t is stratigraphic thickness, h is horizontal thickness or plan width, and δ is dip.

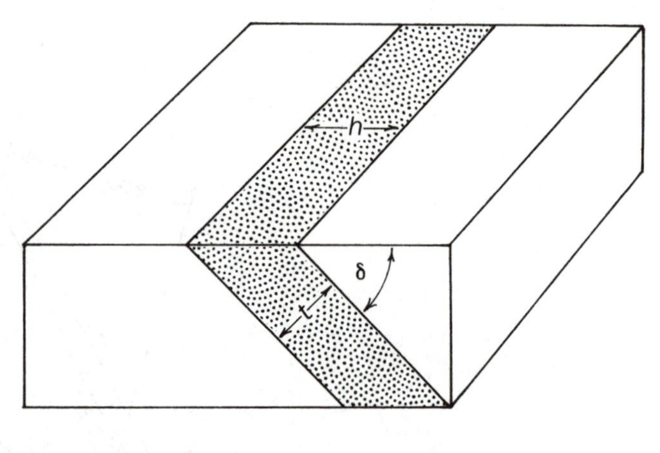

$t = h \sin \delta$

Figure 3-3
Trigonometric relationships used for determining stratigraphic thickness t in flat terrain from dip δ and plan width h.

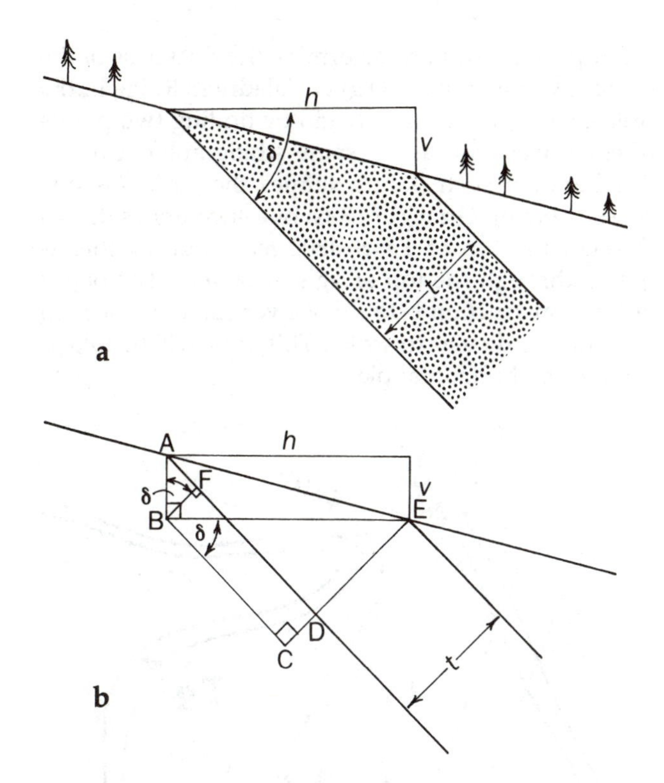

Figure 3-4
Determining stratigraphic thickness t on slopes. (a) Lengths h and v and dip angle δ are needed to derive t. (b) Geometry of derivation.

Determining Stratigraphic Thickness on Slopes

The thickness of layers exposed on slopes may be determined trigonometrically if, in addition to dip δ and plan width of the layer h, the vertical distance v (i.e., difference in elevation) from base to top of the layer is known. Figure 3-4a shows a situation in which the layer and the slope are dipping in the same direction. Relevant angles have been added in Figure 3-4b, from which the following derivation is made:

$$t = \text{DE} = \text{CE} - \text{CD}$$
$$\delta = \text{CBE} = \text{ABF}$$
$$\sin \text{CBE} = \frac{\text{CE}}{\text{BE}} = \frac{\text{CE}}{h}$$
$$\text{CE} = h \sin \text{CBE} = h \sin \delta$$
$$\cos \text{ABF} = \frac{\text{BF}}{\text{AB}} = \frac{\text{BF}}{v}$$
$$\cos \delta = \frac{\text{BF}}{v} = \frac{\text{CD}}{v}$$
$$\text{CD} = v \cos \delta$$
$$t = h \sin \delta - v \cos \delta$$

This relationship applies to situations where bedding dips more steeply than topography and both dip in the same direction (right-hand example of Fig. 3-5). Similar trigonometric derivations can be used to show that in situations where bedding dips more gently than topography and both dip in the same direction (left-hand example of Fig. 3-5), the equation becomes

$$t = v \cos \delta - h \sin \delta,$$

and where bedding and topography dip in opposite directions (middle example of Fig. 3-5) the equation becomes

$$t = h \sin \delta + v \cos \delta.$$

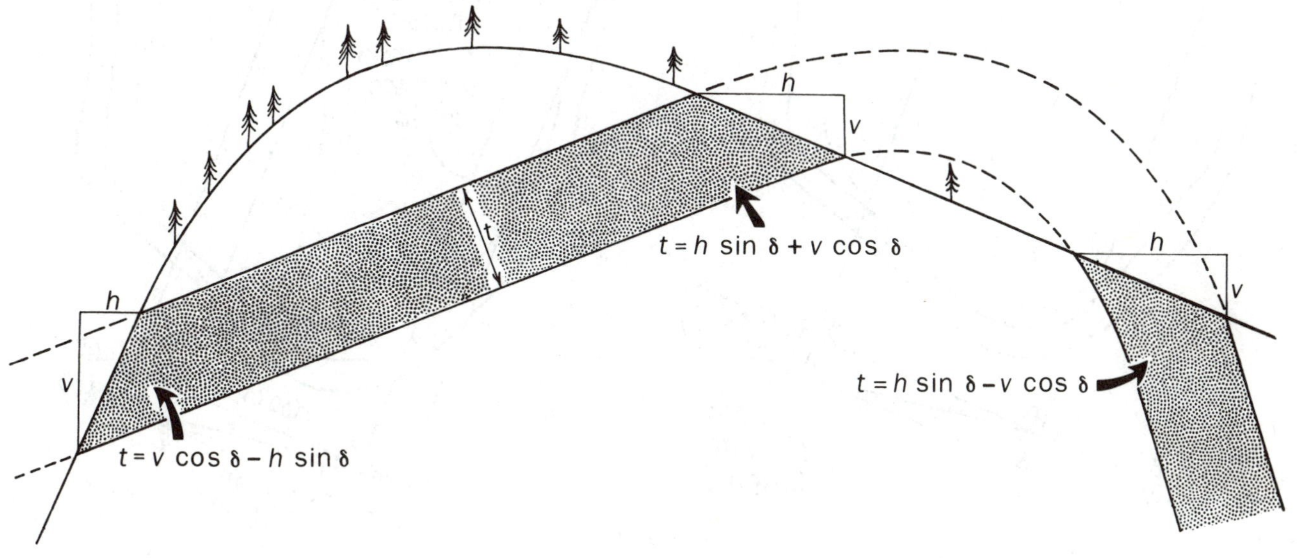

Figure 3-5
Three combinations of sloping topography and dipping layers, with appropriate formula for each.

Determining Stratigraphic Thickness by Orthographic Projection

In some situations the preceding trigonometric techniques for determining stratigraphic thickness cannot be used. On the Bree Creek map, for example, the 400-foot contour interval does not allow the difference in elevation from the base to the top of a unit to be precisely determined. Orthographic projection can be used in such cases, however.

Suppose we want to determine the thickness of the Gondor Conglomerate (Tg) at Galadriel's Ridge in the Bree Creek Quadrangle. Begin by finding two points of equal elevation at the same stratigraphic level. A line between such points defines the strike (as discussed above). On Figure 3-6a one such line is drawn through the top of Tg at 4800 feet, and another is drawn through the top of Tg at 4400 feet. The object of this construction is to draw a vertical cross-section view perpendicular to strike. This view will be folded up into the horizontal plane.

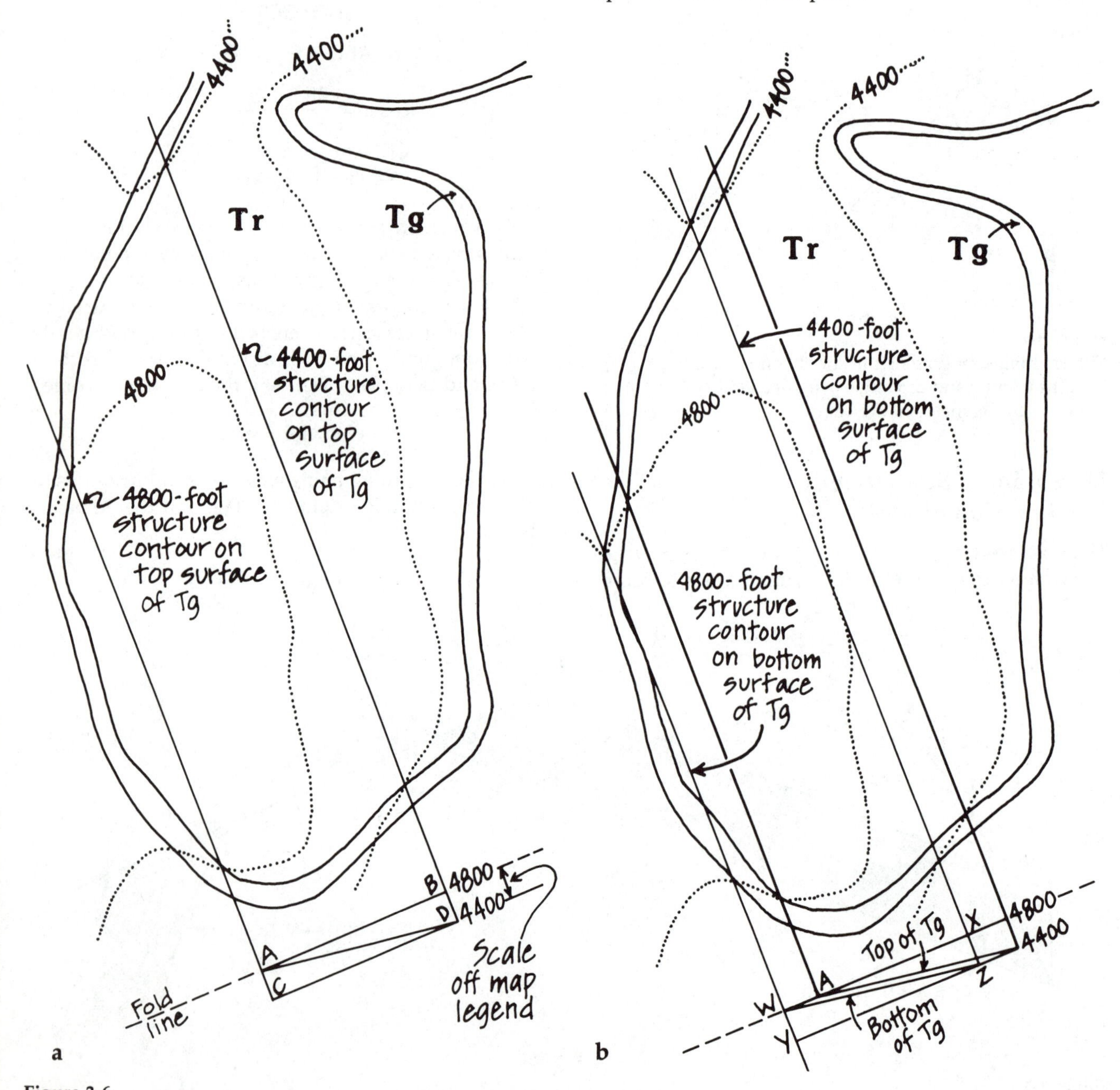

Figure 3-6
Technique for determining stratigraphic thickness by orthographic projection. (a) Plotting top surface. (b) Plotting bottom surface and deriving thickness.

Line AB is drawn perpendicular to the two strike lines (Fig. 3-6a). This will represent the 4800-foot elevation line in the orthographic projection. A second line, CD, is now drawn perpendicular to the two strike lines. Line CD represents the 4400-foot elevation line. The distance between lines AB and CD is taken directly off the map legend. Now we draw line AD, which represents the eastward-dipping top of Tg in orthographic projection.

Repeating this same procedure with the bottom contact of Tg results in points W, X, Y, and Z (Fig. 3-6b). Line WZ represents the base of Tg in orthographic projection, and the thickness can be measured directly off the diagram. The precision is limited primarily by the scale of the map. In this example Tg can be measured to be about 100 feet thick.

Problem 3-3

Determine the approximate thickness of the Neogene units in the areas indicated. (Both the Helm's Deep Sandstone and the Rohan Tuff have quite variable thicknesses. The Rohan Tuff was tilted and partly eroded prior to the deposition of the Helm's Deep Sandstone, and the Helm's Deep Sandstone has no upper contact. In both cases determine the range of thicknesses.)

	Thd	Tr	Tg
Gollum Ridge	______	______	
Gandalf's Knob	______		
Galadriel's Ridge		______	______
Gollum Creek			______
Mirkwood Creek			______
N. of Edoras Creek	______		

Determining the Nature of Contacts

A contact is the surface between two contiguous rock units. There are three basic types of contacts: (1) depositional, (2) fault, and (3) intrusive. It is important to be able to interpret the nature of contacts from geologic maps whenever possible. Following are a few map characteristics of each type of contact.

Where sedimentary or volcanic rocks have been deposited on top of other rocks, the contact is said to be **depositional.** If adjacent rock units have attitudes parallel to one another and there is no evidence of erosion on the contact, then the contact is a **conformable** depositional contact. On the map, conformable contacts display no abrupt change in attitude across the contact. In Figure 3-7, for example, although the dips in formation X are steeper than those in formation Y, there is a gradual steepening across the contact. A cross-section view is shown below the map view.

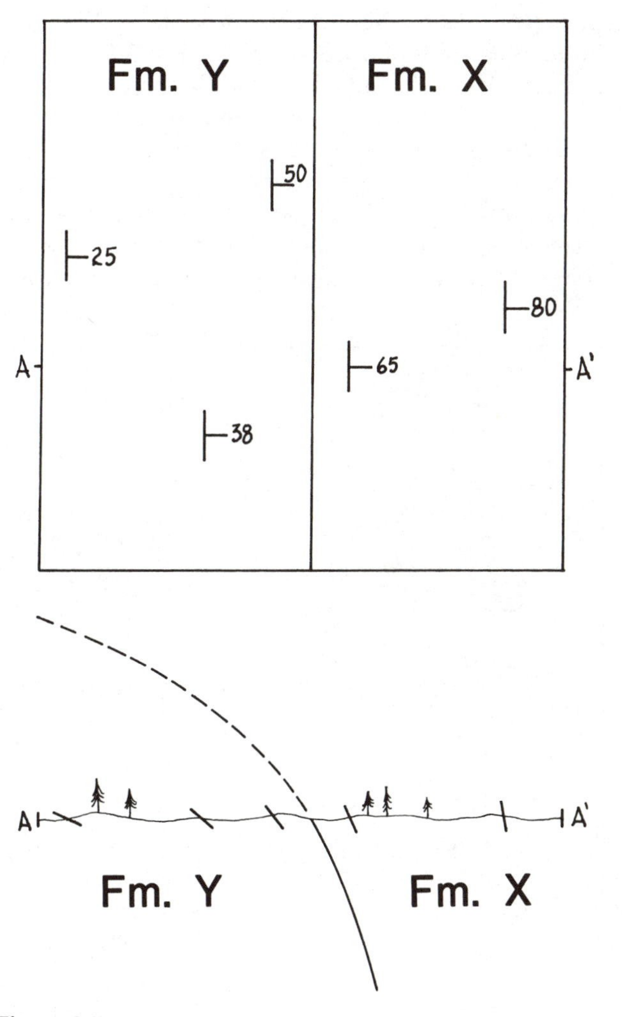

Figure 3-7
Conformable depositional contact. Map view above and vertical structure section below.

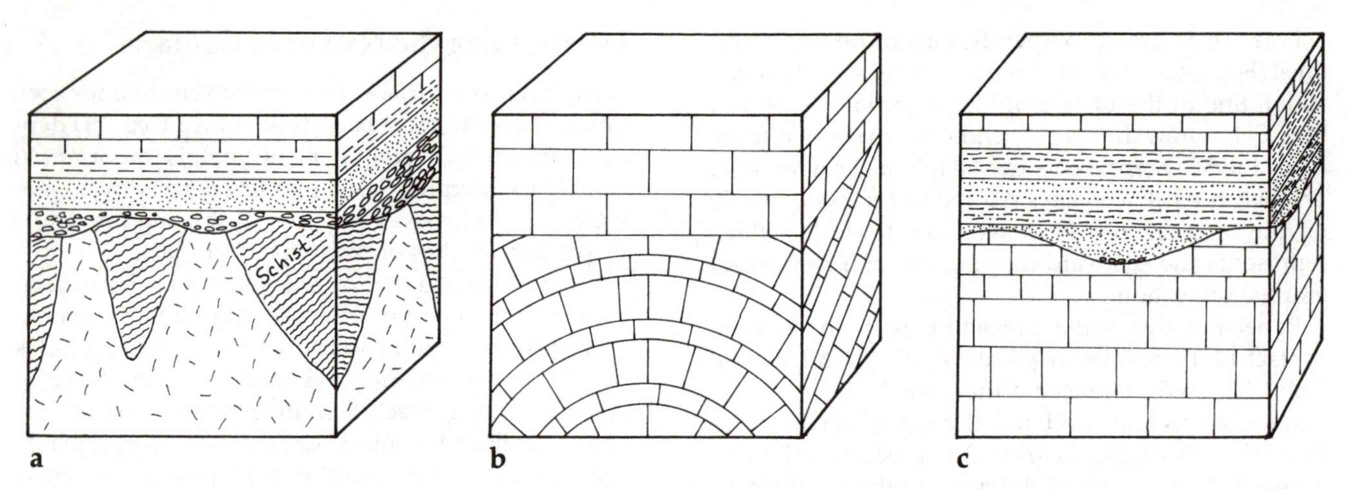

Figure 3-8
Three types of unconformities: nonconformity (a), angular unconformity (b), and disconformity (c).

If a demonstrable surface of erosion or nondeposition separates two rock units then the contact is an **unconformity**—a buried erosion surface. There are three basic types of unconformities (Fig. 3-8): (1) nonconformities (sediments deposited on crystalline rock), (2) angular unconformities (sediments deposited on deformed and eroded older sediments), and (3) disconformities (sediments deposited on eroded but undeformed older sediments). Notice that a disconformity would be indistinguishable from a conformable contact on a geologic map because in both cases the beds are parallel across the contact. Disconformities can only be recognized in the field. In angular unconformities the layers overlying the unconformity are always parallel to the contact, while those beneath it are not (Fig. 3-9b).

Fault contacts are best diagnosed in the field on the basis of fault gouge, slickensides, offset beds, and geomorphic features. On geologic maps faults are often conspicuous because of the rock units that are truncated. Figure 3-10 shows a contact that can only be a fault; if it were an unconformity one of the units would strike parallel to the contact.

Intrusive contacts are obvious where the intrusive rocks have clearly been injected into the country rock. As in the case of faults, this is best determined in the

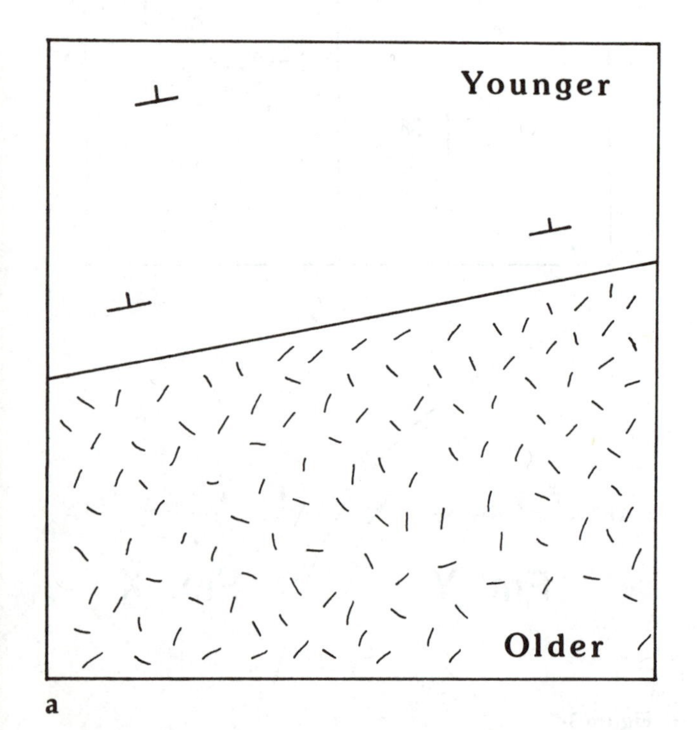

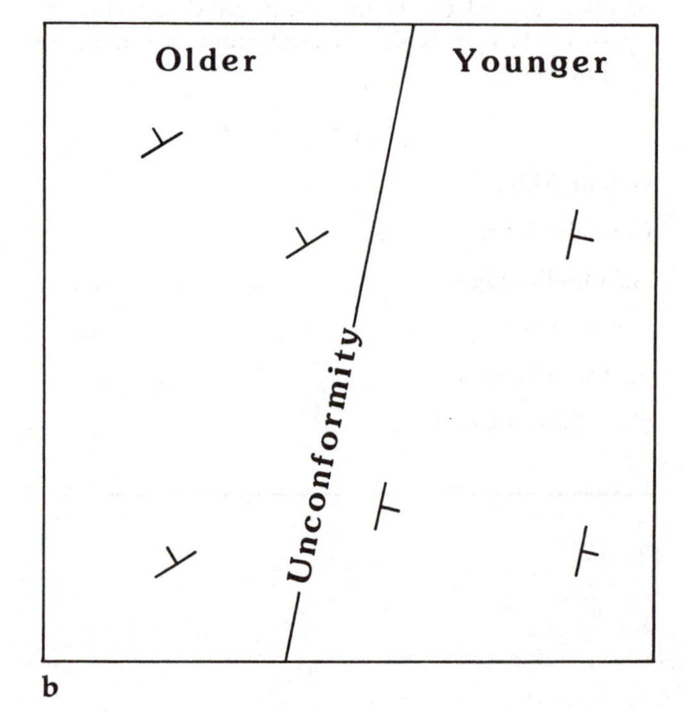

Figure 3-9
Nonconformity (a) and angular unconformity (b) in map view.

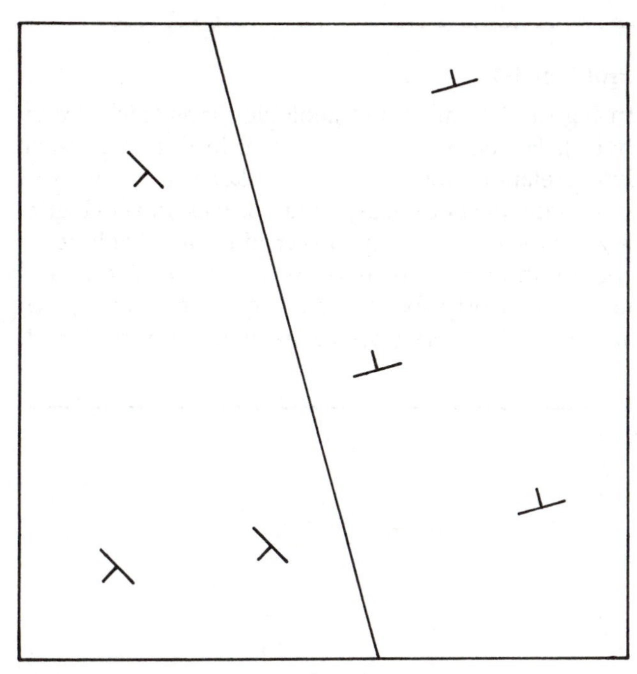

Figure 3-10
Fault contact in map view.

field. Figure 3-11 shows an unequivocal intrusive contact, but sometimes intrusive contacts are not so jagged and cannot be easily distinguished from faults. Intrusions such as sills may even be parallel to the bedding of the country rock, making the contact appear to be a nonconformity.

While the nature of a contact may not always be clear in plan view, a geologist drawing a structure section (cross-section view) must show the nature of the contact. In Figure 3-12a a geologic map is shown with two possible structure sections. Figure 3-12b interprets the contact between the gabbro and formation M as an unconformity, and Figure 3-12c interprets the same contact as a fault. The fact that the strike of the beds in formation M exactly parallels the contact makes the unconformity the preferred interpretation. A fault, or even an intrusive contact, cannot be ruled out, however, without examining the contact in the field.

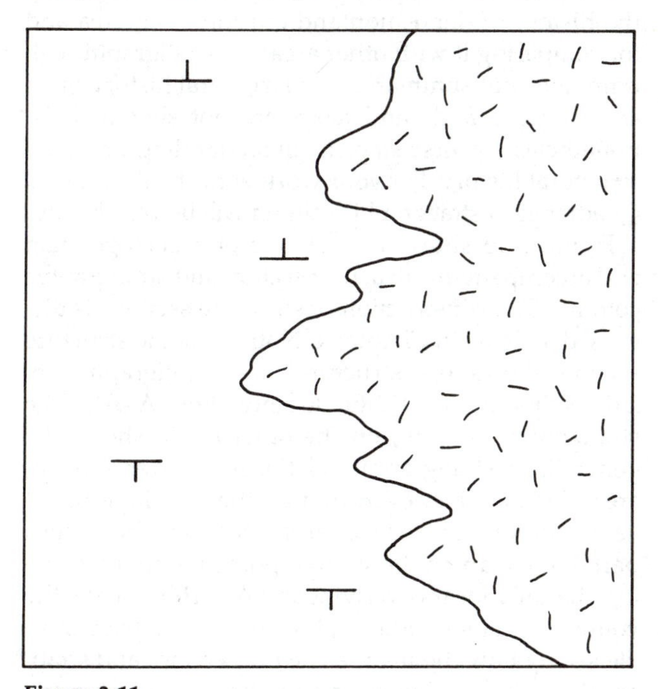

Figure 3-11
Intrusive contact in map view.

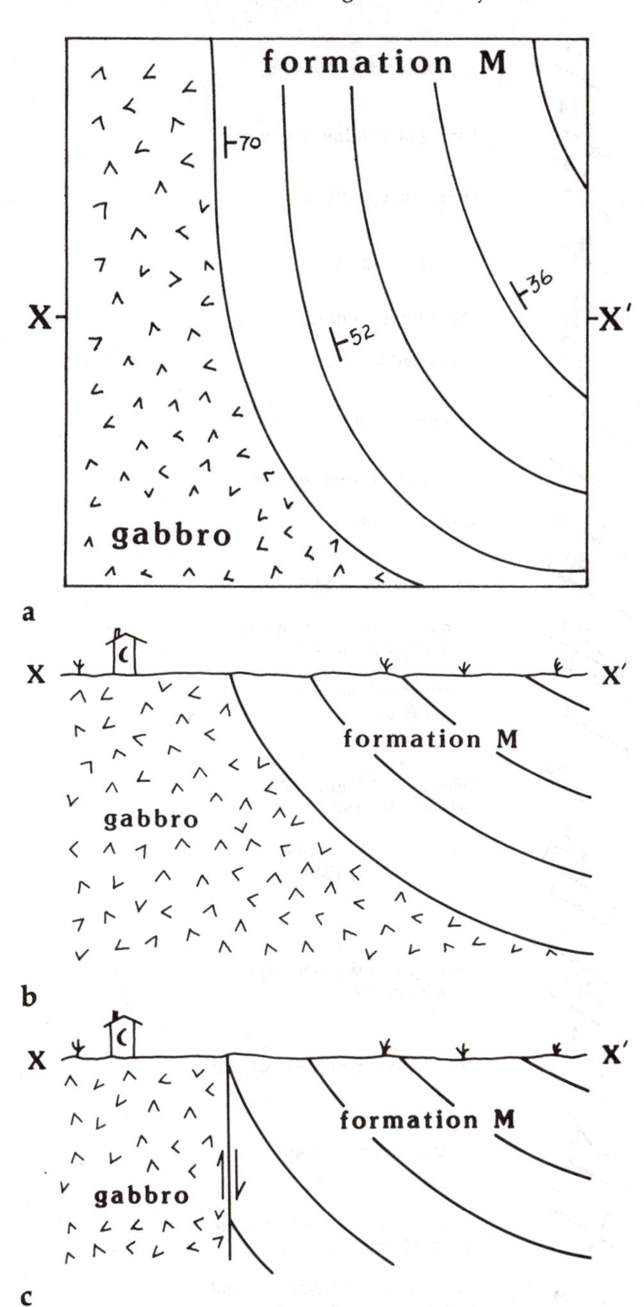

Figure 3-12
Geologic map (a) with two possible structure section interpretations, unconformity (b), and fault (c).

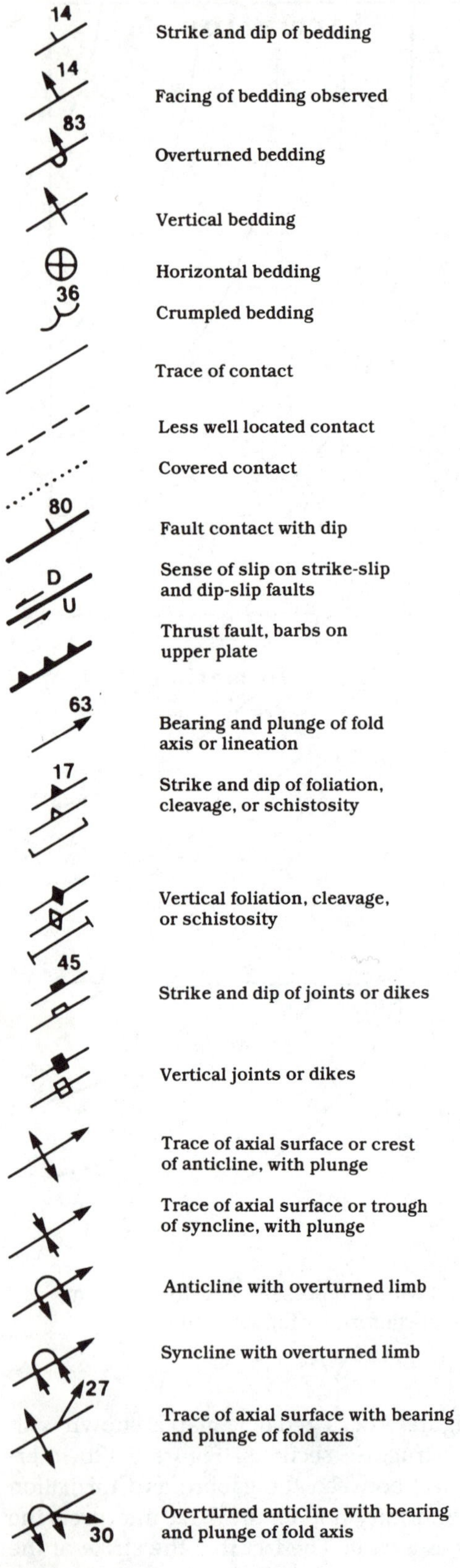

Strike and dip of bedding

Facing of bedding observed

Overturned bedding

Vertical bedding

Horizontal bedding

Crumpled bedding

Trace of contact

Less well located contact

Covered contact

Fault contact with dip

Sense of slip on strike-slip and dip-slip faults

Thrust fault, barbs on upper plate

Bearing and plunge of fold axis or lineation

Strike and dip of foliation, cleavage, or schistosity

Vertical foliation, cleavage, or schistosity

Strike and dip of joints or dikes

Vertical joints or dikes

Trace of axial surface or crest of anticline, with plunge

Trace of axial surface or trough of syncline, with plunge

Anticline with overturned limb

Syncline with overturned limb

Trace of axial surface with bearing and plunge of fold axis

Overturned anticline with bearing and plunge of fold axis

Figure 3-13
Common symbols used on geologic maps.

Problem 3-4

In Figure 3-14 are three geologic maps (each drawn twice). For each map there are at least two possible interpretations for the contact. Sketch structure sections showing both possibilities, as was done in Figure 3-12. Indicate the one you consider most likely to be correct and give your reasons. In your structure sections don't worry about calculating and measuring each apparent dip, merely approximate the dips freehand.

Constructing a Stratigraphic Column

A stratigraphic column is a thumbnail sketch of the geology of an area showing relationships and thicknesses. It is an extremely useful tool for summarizing the history of deposition and erosion of an area and for comparing it with other areas. A stratigraphic column does not summarize the structural history of an area because folds and faults are not shown. It is, nonetheless, a first step in understanding an area's structural history. For your work with the Bree Creek Quadrangle a stratigraphic column will be very handy.

Figure 3-15 shows an example of a geologic map and accompanying structure section and stratigraphic column. The construction of structure sections is discussed in detail in Chapter 4. Notice that the structure section shows the structural and stratigraphic relationships in a specific locality, line A–A′. The stratigraphic column, on the other hand, shows the generalized stratigraphic relationships over a larger area. For example, even though the Antelope Basalt lies unconformably on the Jerome Schist in the eastern part of the map, in the stratigraphic column the Antelope Basalt appears overlying the Waterford Shale, the youngest unit it overlies in the area. If you have trouble seeing how the map, structure section, and stratigraphic column relate to one another, try coloring one or two of the units on all three diagrams.

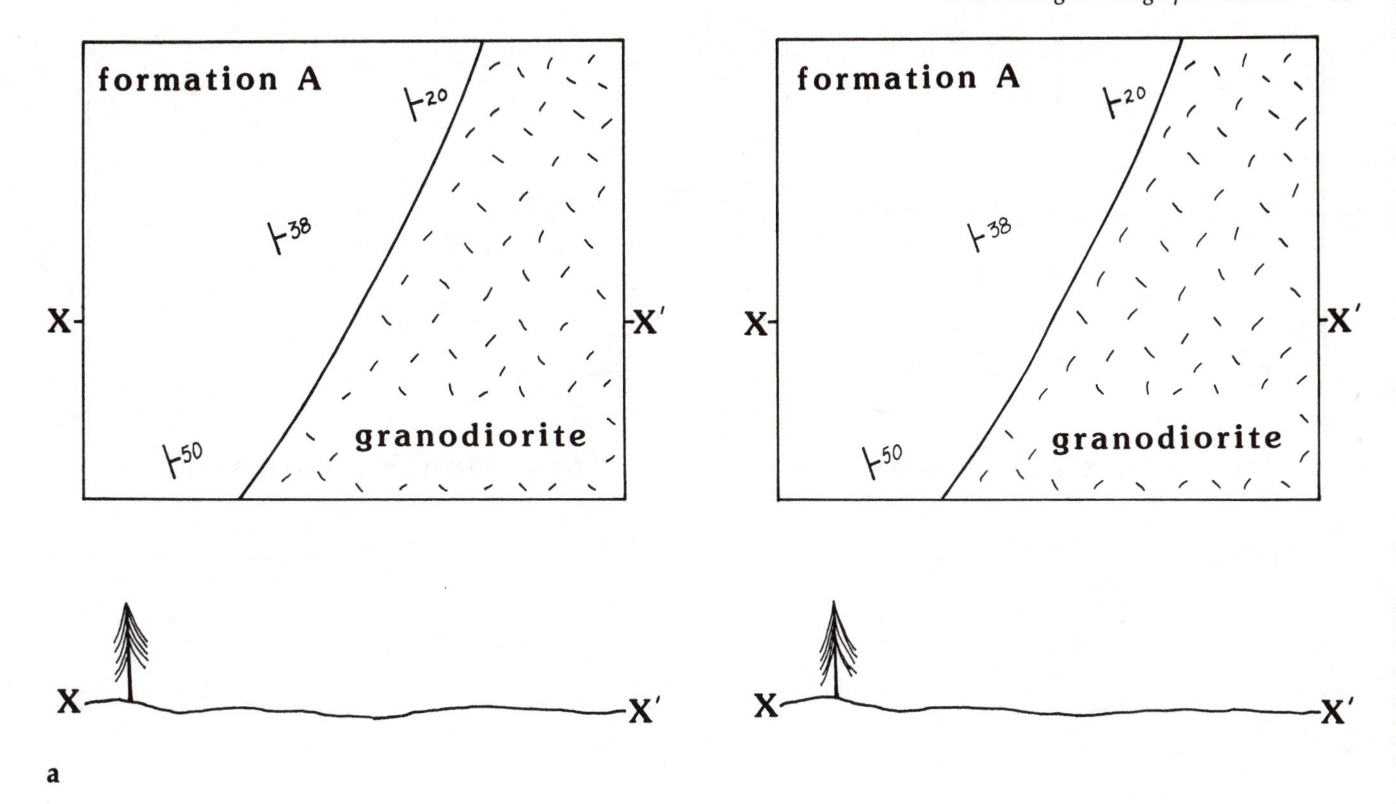

Figure 3-14
Three geologic maps for use in Problem 3-4 (continued).

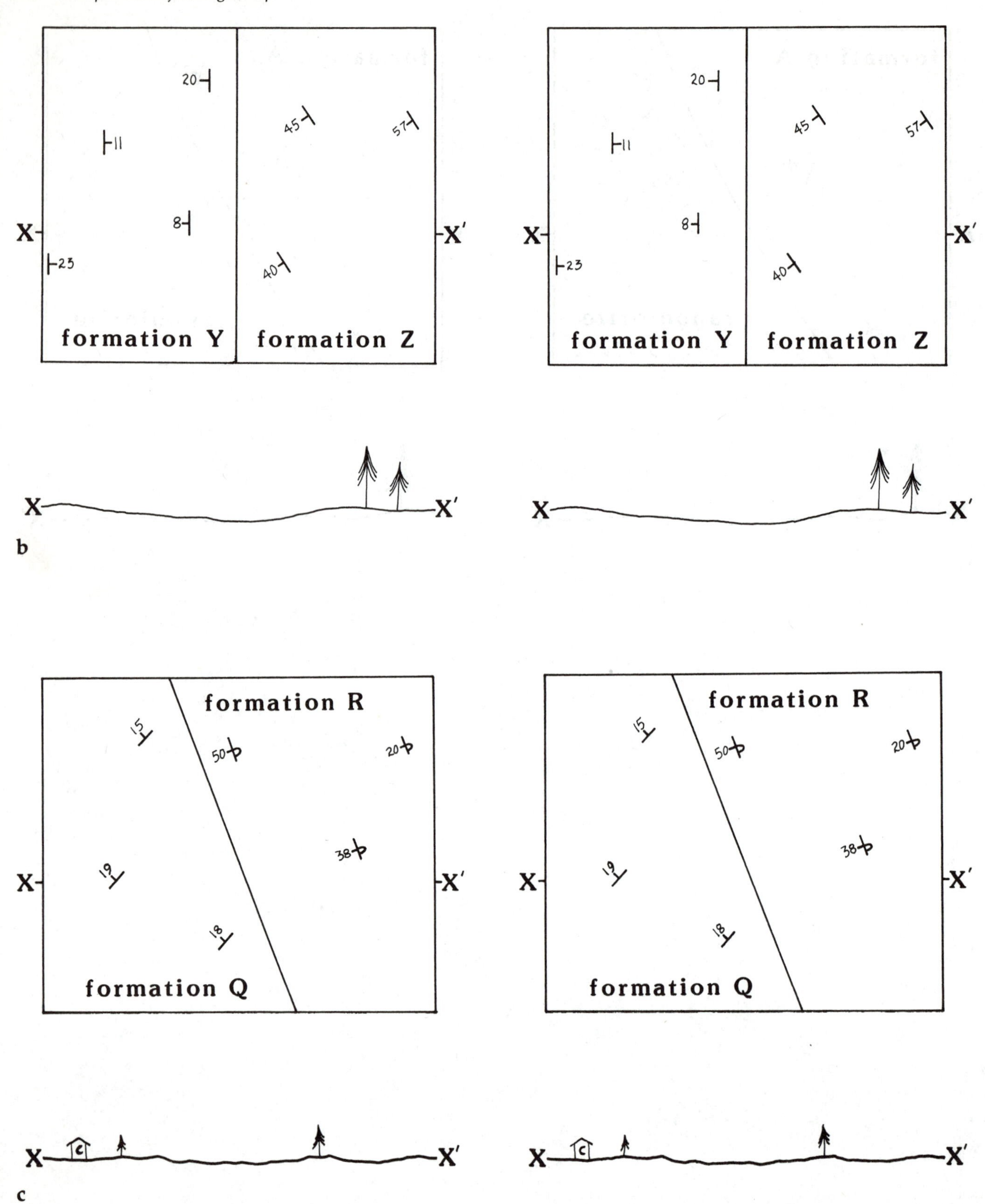

Figure 3-14
(continued)

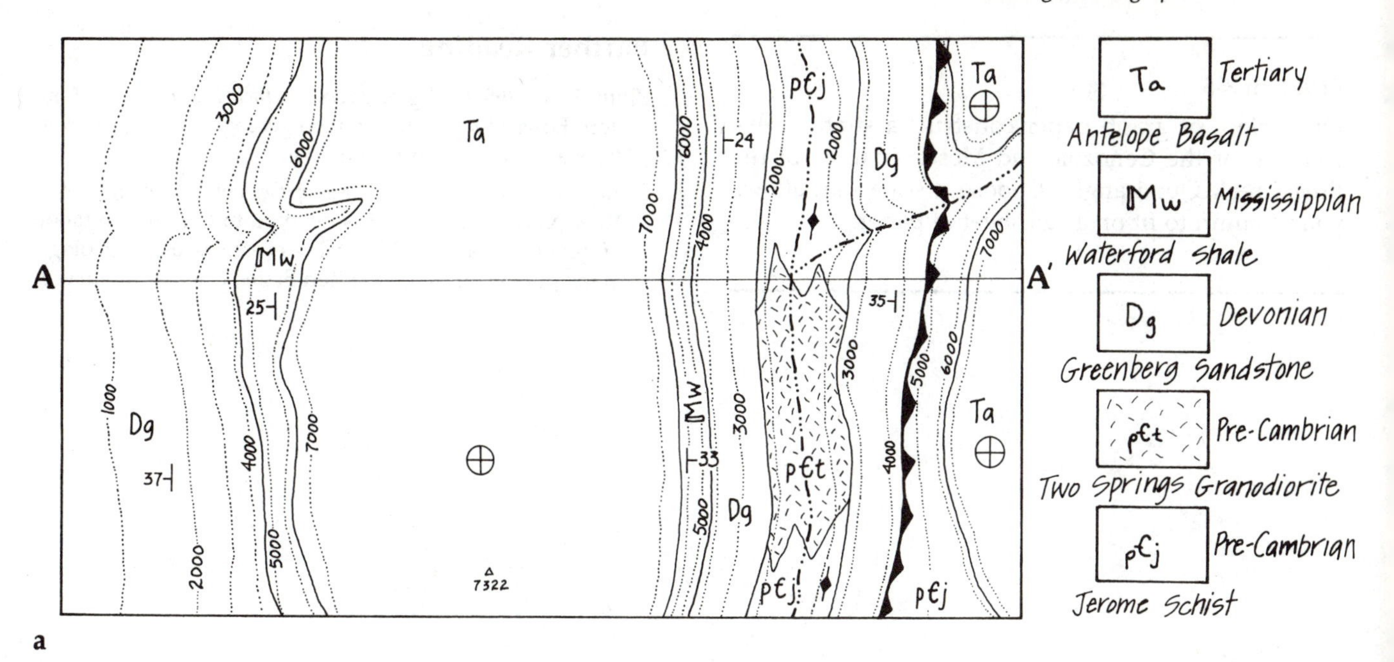

a

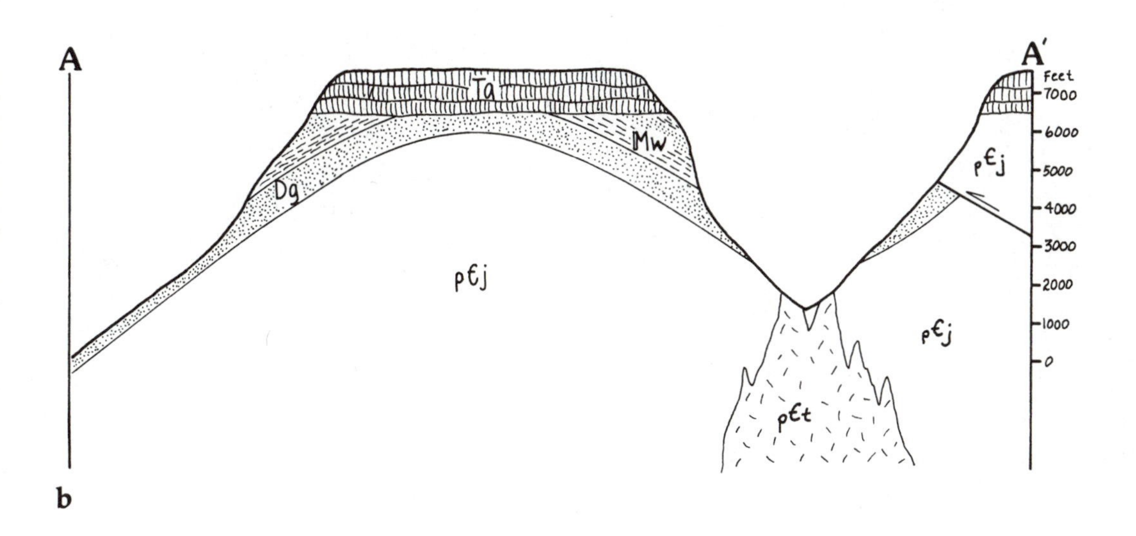

b

Tertiary	Antelope Basalt (1100 feet)
Mississippian	Waterford Shale (min. 1400 feet)
Devonian	Greenberg Sandstone (850 feet)
Pre-Cambrian	Jerome Schist
	Two Springs Granodiorite

Feet: 7000, 6000, 5000, 4000, 3000, 2000, 1000, 0

Unconformity

Unconformity

c

Figure 3-15
Geologic map (a) with corresponding structure section (b) and stratigraphic column (c).

Problem 3-5

On a piece of graph paper construct a stratigraphic column for the Cenozoic and Mesozoic units of the Bree Creek Quadrangle. Choose a scale that allows your column to fit on a full sheet of paper.

Further Reading

Blyth, F. G. 1965. *Geological Maps and their Interpretation*. London: Edward Arnold. Eighteen geologic maps with exercises and notes on interpretation.

Dennison, J. M. 1968. *Analysis of Geologic Structures*. New York: Norton. Chapter 7 is devoted to the interpretation of geologic maps and contains exercises using geologic maps published by the United States Geological Survey.

Name ______________________________

Section ______________________________

Answer Sheet for Problems 3-1, 3-2, and 3-3

Problem 3-1

	Thd	Tr	Tg
Northeastern fault block			
Northern exposures	____	____	____
Southern exposures		____	____
Central fault block			
Northern area		____	____
Galadriel's Ridge		____	____
Southwestern area		____	____
Western fault block			
Gandalf's Knob	____		
Southern exposures	____		

Problem 3-2

Tdd ____

Tmm ____

Tm ____

Tts ____

Tb ____

Te ____

Problem 3-3

	Thd	Tr	Tg
Gollum Ridge	____	____	
Gandalf's Knob	____		
Galadriel's Ridge		____	____
Gollum Creek			____
Mirkwood Creek			____
N. of Edoras Creek	____		

Notes

4

Geologic Structure Sections

OBJECTIVE

Draw geologic structure sections through folded and faulted terrain

A geologic map only shows the geology on the earth's surface. In order to provide a third dimension, it is standard practice to draw one or more **vertical structure sections.** These are vertical cross-sections of the earth showing rock units, folds, and faults. By convention, structure sections are usually drawn with the west on the left. Structure sections oriented exactly north-south are usually drawn with the north on the left.

At best, structure sections are drawn using well logs and geophysical data to supplement the surface information; most structure sections, however, are based solely on the geologic map and the geologist's best guess about how the rocks have been deformed. As such, structure sections must be regarded as interpretations that are subject to change with the appearance of new information. By way of example, examine the geologic map in Figure 4-1 and the two structure sections based on it.

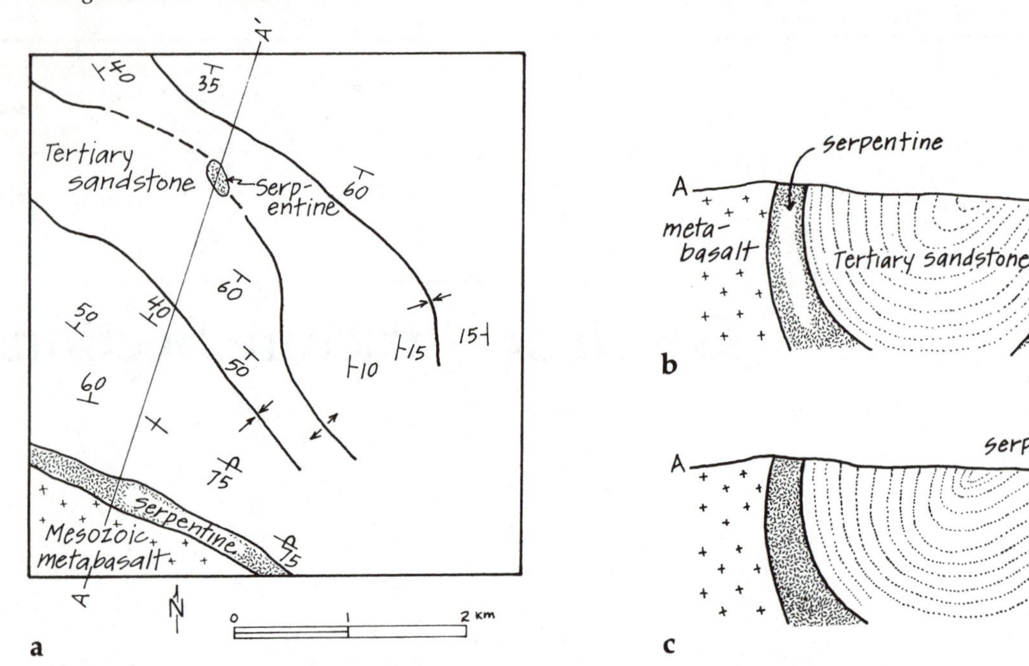

Figure 4-1
Geologic map with two interpretations of structure section A–A′ (generalized from Dibblee, 1966). (a) Geologic map. (b) Original structure section interpreting northern serpentine outcrop as core of an anticline. (c) Revised structure section interpreting northern serpentine outcrop as a landslide block.

The map shows three groups of rocks: Mesozoic metabasalt, serpentine, and Tertiary sandstone. The serpentine occurs as a continuous band between the metabasalt and the sandstone in the southwestern part of the map and as a small patch within the sandstone in the northern part of the map. The original interpretation (Fig. 4-1b) accounts for the northern outcrop of serpentine as occurring in the core of a partially eroded anticline. Further field work has shown, however, that the northern patch of serpentine is more probably a large landslide block that long ago slid off the southern serpentine mass (Fig. 4-1c). Far from being a trivial difference, these two interpretations imply rather different styles of folding as well as predicting completely different stability and permeability characteristics for the entire length of the anticlinal axial trace.

When you are drawing structure sections remember that it should be geometrically possible to unfold the folds and recover the fault slip in order to reconstruct an earlier, less deformed or undeformed state. In other words, your structure section should be **retrodeformable.** Structure sections in which great care is taken concerning retrodeformation are called **balanced structure sections.** The construction and retrodeformation of balanced structure sections in complex regions is beyond the scope of this book (see Suppe, 1985, p. 57–70). If you make sure that sedimentary units maintain a constant thickness (unless you have evidence to the contrary) and that the hanging walls of faults match the footwalls, you will be on the right track.

Drawing a Topographic Profile

The first step in constructing a geologic structure section is drawing a topographic profile along the line of section. Topographic profiles show the relief at the earth's surface along the top of the structure section. Problems 4-1 through 4-3 in this chapter are relatively simple structure sections in which the topographic profile is provided. Problem 4-4 involves the construction of two structure sections on the Bree Creek Quadrangle map, for which you will draw the topographic profiles yourself.

The technique for drawing a topographic profile is as follows:

1. Draw the section line on the map (Fig. 4-2a).
2. Lay the edge of a piece of paper along the section line, and mark and label on the paper each contour, stream, and ridge crest (Fig. 4-2b).

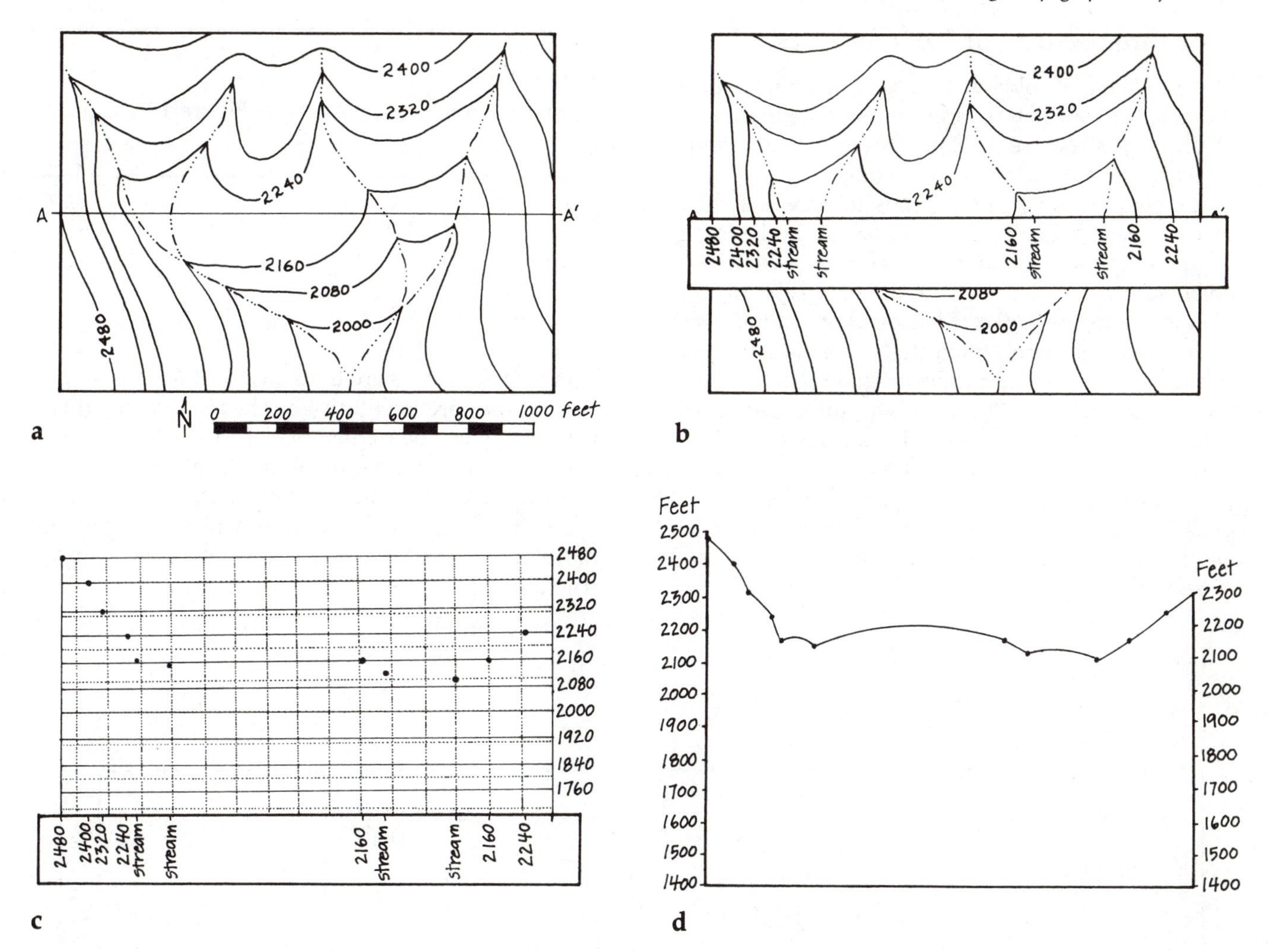

Figure 4-2
Drawing a topographic profile. (a) Draw section line. (b) Transfer contour crossings, streams, and other features to paper. (c) Plot points on paper onto graph paper. (d) Connect points.

3. Scale off and label the appropriate elevations on a piece of graph paper (Fig. 4-2c). Graph paper with 10 or 20 squares per inch is ideal for 7.5-minute quadrangle maps (and the Bree Creek Quadrangle) because the scale is 1 inch to 2000 feet. Notice that the map scales on Figures 4-2a and 4-2b are the same as the vertical scale for Figures 4-2c and 4-2d. It is very important that the vertical and horizontal scales be the same on almost all structure sections. This is a very common oversight. If the scale of the structure section is not the same as the scale of the map then the dips cannot be drawn at their nominal angle. In the rare case where the vertical scale must be exaggerated to emphasize the topography, the graph in Appendix D must be used to find the corrected dips, and the amount of vertical exaggeration must be clearly labeled on the structure section.
4. Lay the labeled paper on the graph paper and transfer each contour, stream, and ridge crest point to the proper elevation on the graph paper (Fig. 4-2c).
5. Connect the points (Fig. 4-2d).

Structure Sections of Folded Layers

The geometry of folds is discussed in Chapter 6. In this chapter we will be concerned with the mechanics of drawing structure sections through folded beds, not with the mechanics of the folding.

The simplest structure sections to draw are those that are perpendicular to the strike of the bedding. Figure 4-3 shows a geologic map with all beds striking north-south. Section A–A′ is drawn east-west perpendicular to the strike. Each bedding attitude and each contact is merely projected parallel to the fold axis to the topographic profile oriented parallel to the section line. On the topographic profile each measured dip is drawn with the aid of a protractor. Using these dip lines on the topographic profile as guides, contacts are drawn as smooth, parallel lines. Dashed lines are used to show eroded structures. As much depth below the earth's surface as the data allow should be shown.

Problem 4-1

Draw structure section A–A′ on Figure 4-7 (page 57).

In very few cases are the strikes of the beds all parallel, as they are in Figure 4-3. The section line, therefore, rarely can be perpendicular to all of the strikes. When the section line intersects the strike of a plane at an angle other than 90°, the dip of the plane as it appears in the structure section will be an apparent dip. Recall that apparent dips are always less than the true dip.

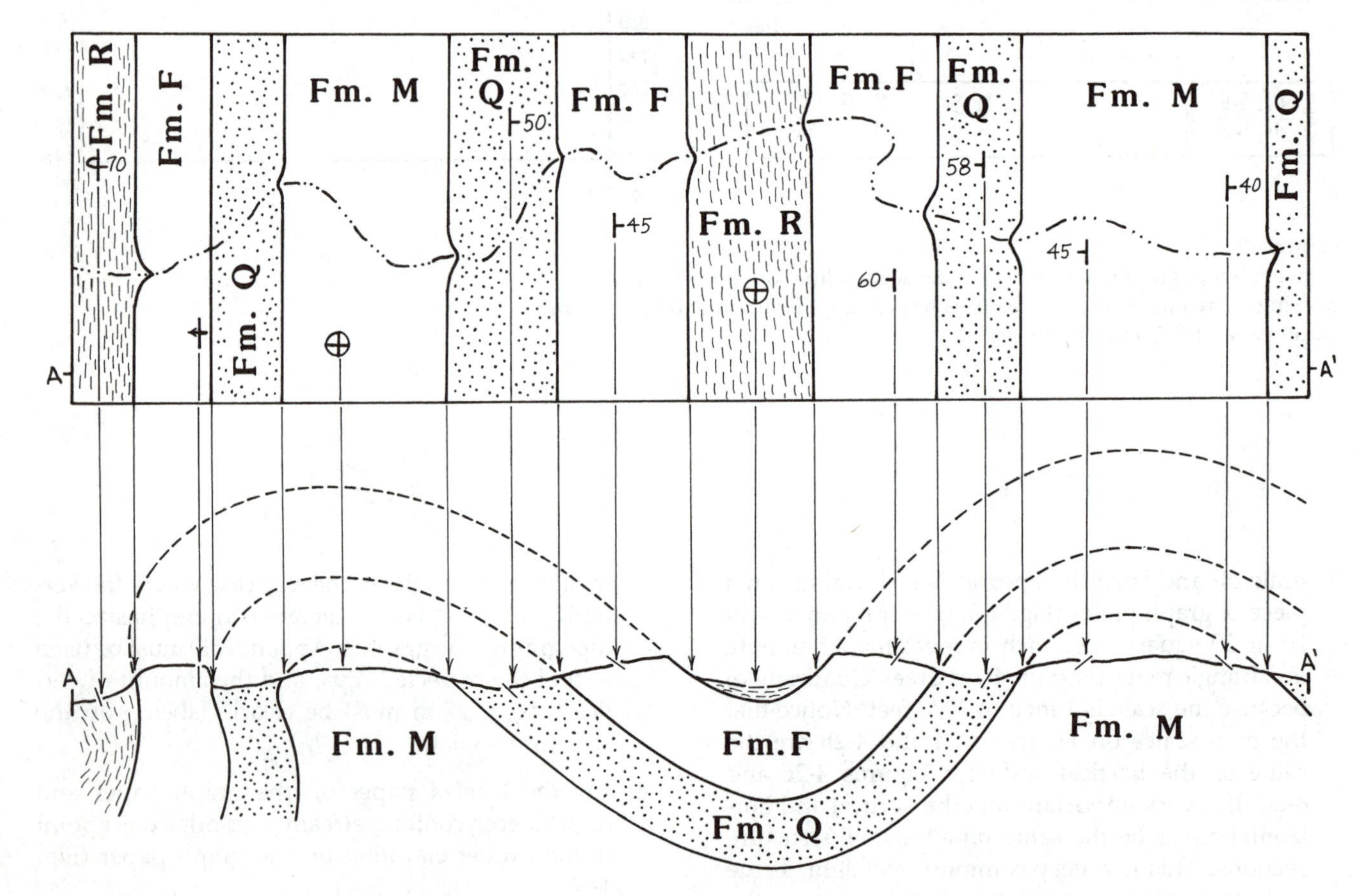

Figure 4-3
Drawing a structure section perpendicular to strike of bedding. Arrows show transfer of attitudes from map to section.

The quickest way to determine the correct apparent dip to draw on the structure section is to use the nomogram in Figure 1-9. Figure 4-4 shows a geologic map in which the strike of formation B has been projected along strike to line X–X′ and then perpendicular to X–X′ to the topographic profile. The angle between the strike and the section line is 35°, the true dip is 43°, and the apparent dip is revealed by the alignment diagram to be 28°, which is the angle drawn on the structure section.

Some rock units have highly variable strikes, and judgment must be exercised in projecting attitudes to the section line. Attitudes close to the section line should be used whenever possible. If the dip is variable, the dip of the contact may have to be taken as the mean of the dips near the section line. The attitudes should be projected parallel to the fold axis, which in the case of plunging folds will not be parallel to the contacts. In all cases, it is important to study the entire geologic map to aid in the construction of structure sections. Critical field relationships that must appear on your structure section may not be exposed along the line of section.

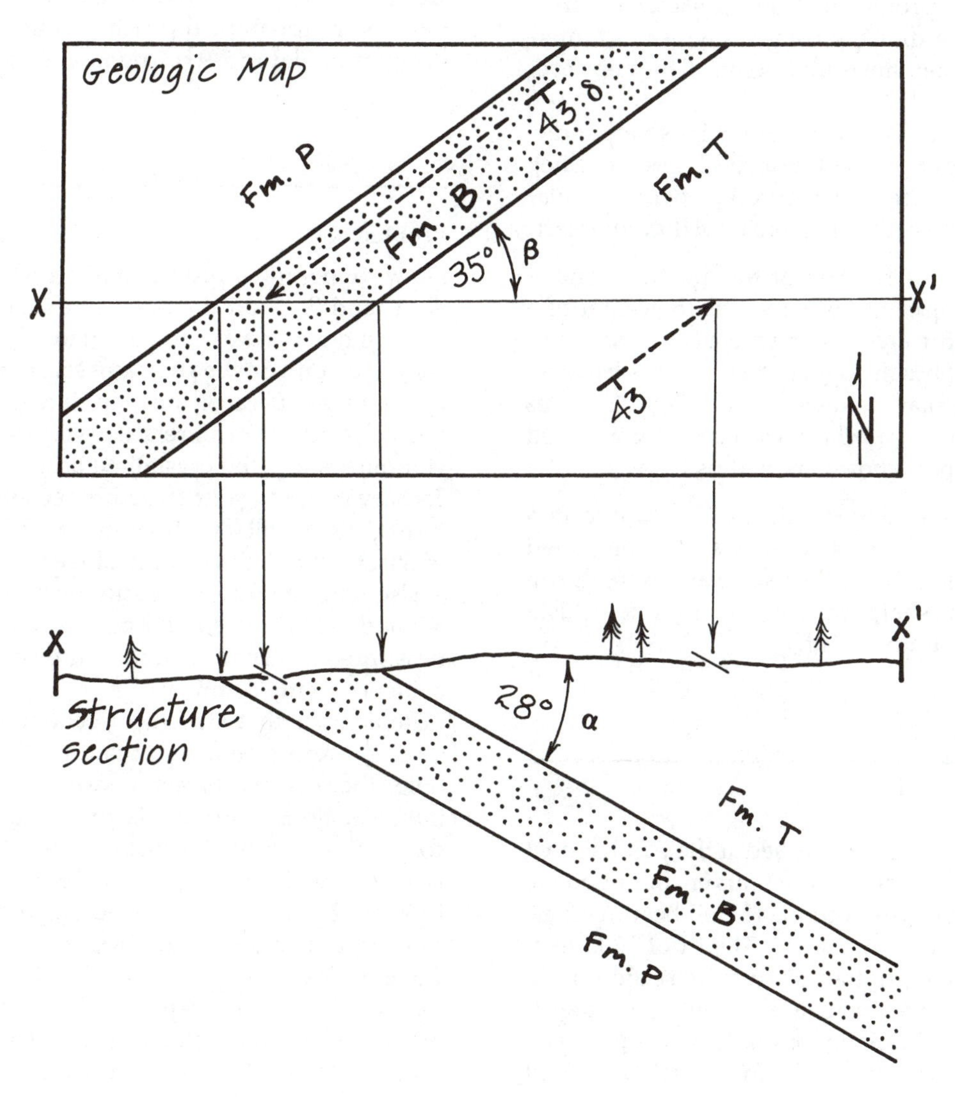

Figure 4-4
Geologic map and corresponding structure section drawn at an angle to strike. Dip from map becomes apparent dip in section.

The Arc Method

A more precise, but not necessarily more accurate, technique than freehand sketching for drawing structure sections is called the arc (or Busk) method. It has proved to be particularly useful in terranes of basins and domes where beds have been "flexure" folded and retain a constant thickness. Such folds are sometimes called **concentric** folds for reasons that will become clear.

The arc method is based on the following two premises: (1) the transition from one dip to the next is smooth, and (2) bed thickness is constant. "Room problems" (loss of volume) at the cusps of folds are completely ignored, which is why this technique is only appropriate for gently folded layers.

Consider the map and topographic profile in Figure 4-5a. Each attitude on the map has been projected to the topographic profile. Instead of sketching freehand, however, a drawing compass is used to interpolate dips between measured points. The steps are:

1. With the aid of a protractor draw lines perpendicular to each dip on the topographic profile. Such lines have been drawn in Figure 4-5b perpendicular to dips a, b, and c. Extend them until they intersect.
2. Each point of intersection of the lines perpendicular to two adjacent dips serves as the center of a set of concentric arcs drawn with a compass. Point 1 on Figure 4-5b is the center of a set of arcs between the perpendiculars to dips a and b. Point 2 serves as the center from which each arc is continued between the perpendiculars to dips b and c.
3. The process is continued until the structure section is completed. Figure 4-5c shows the completed structure section. Notice that some arcs were drawn with unlikely sharp corners in order for thicknesses to remain constant.

Problem 4-2

An exploratory oil well was drilled at the point shown on Figure 4-8 (page 59), and the units encountered are shown on the structure section. The oil-bearing Eagle Bluff Limestone was struck at a depth of 7200 feet. Using the arc method, draw a structure section. Indicate on the map where you, as a consulting geologist, would recommend drilling for oil. How deep do you predict the well will have to be to hit the Eagle Bluff Limestone?

Structure Sections of Intrusive Bodies

Tabular intrusive bodies, such as dikes and sills, present no special problem. Irregular plutons, however, are problematical because in the absence of drill-hole or geophysical data it is impossible to know the shape of the body in the subsurface. Such plutons are usually drawn somewhat schematically in structure sections, displaying the presumed nature of the body without pretending to show its exact shape. For an example, see Figure 4-6.

Problem 4-3

Draw structure section A–A′ on Figure 4-9. Use freehand sketching rather than the arc method. Determine each apparent dip using either the alignment diagram in Figure 1-9 or trigonometry.

Problem 4-4

Draw topographic profiles and structure sections for A–A′ and B–B′ on the Bree Creek Quadrangle map. Draw them as neatly and accurately as possible, and color each unit on the structure sections as it is colored on your map. Because the map shows that the Tertiary section is sitting on a Cretaceous crystalline basement, you must show the crystalline basement beneath the Tertiary rocks on your structure sections. These structure sections will later become part of your synthesis of the structural history of the Bree Creek Quadrangle.

Use your thickness measurements from Problems 3-2 and 3-4. Units should maintain a constant thickness in your structure section unless you have good evidence to the contrary.

Remember that structure sections involve a great deal of interpretation and that, until someone drills a hole, there is no correct answer. As in all scientific interpretations, the best is the simplest one that is compatible with the available data. In the northeastern part of the map area, someone did drill holes. Problem 2-1 (Fig. 2-9) involved the drawing of a structure contour map on the upper surface of the Bree Conglomerate. Use your completed structure contour map to indicate the depth of the Bree Conglomerate in the eastern half of the structure section A–A′.

Be sure that your structure sections have all of the features listed below.

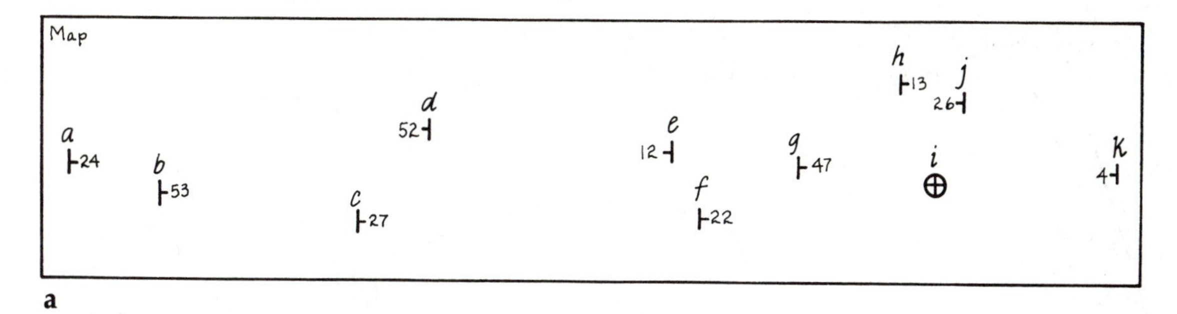

a

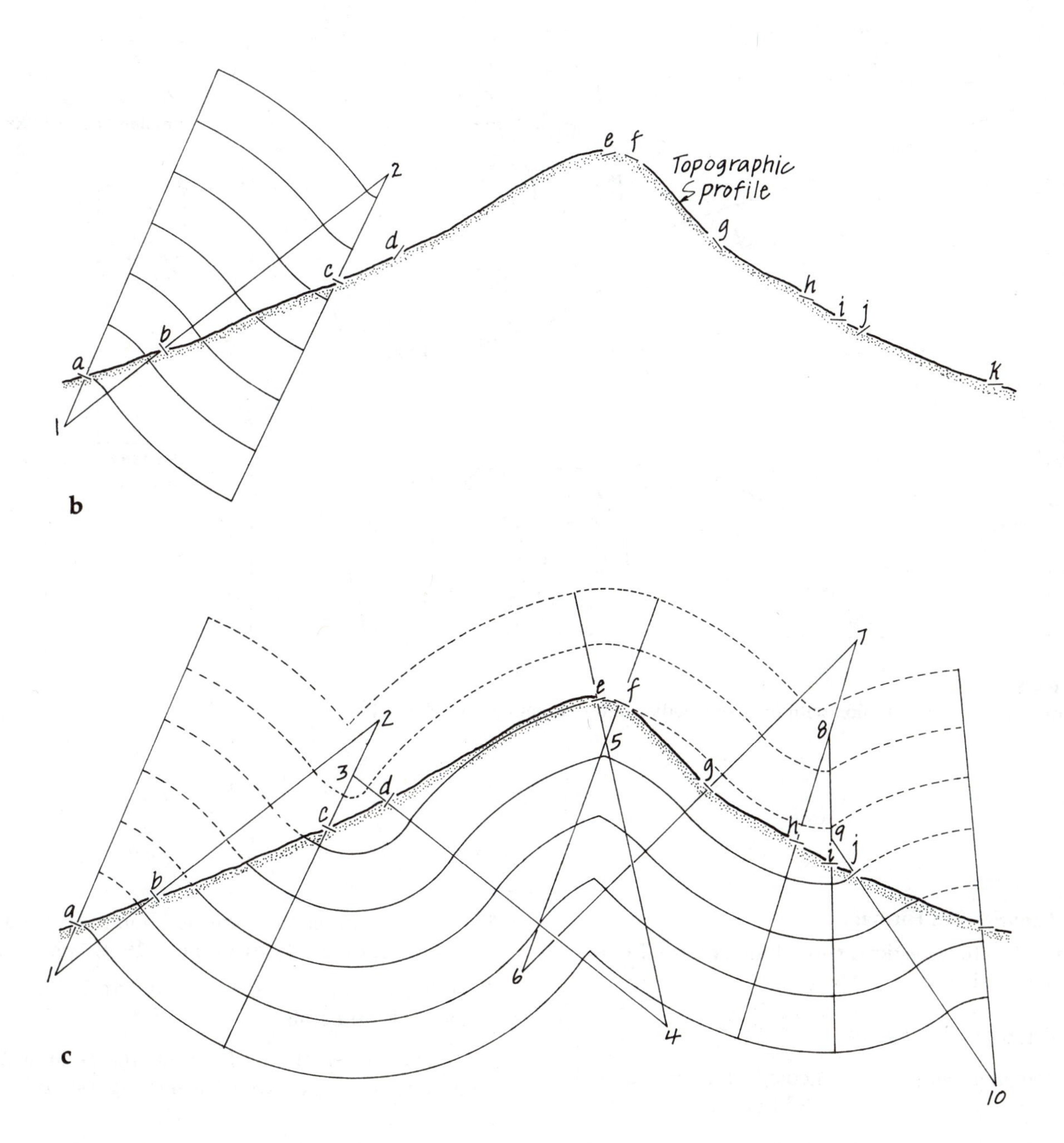

b

c

Figure 4-5
Arc method of drawing structure sections of folded rock layers. (a) Geologic map. (b) Topographic profile with beginning of structure section. (c) Completed structure section.

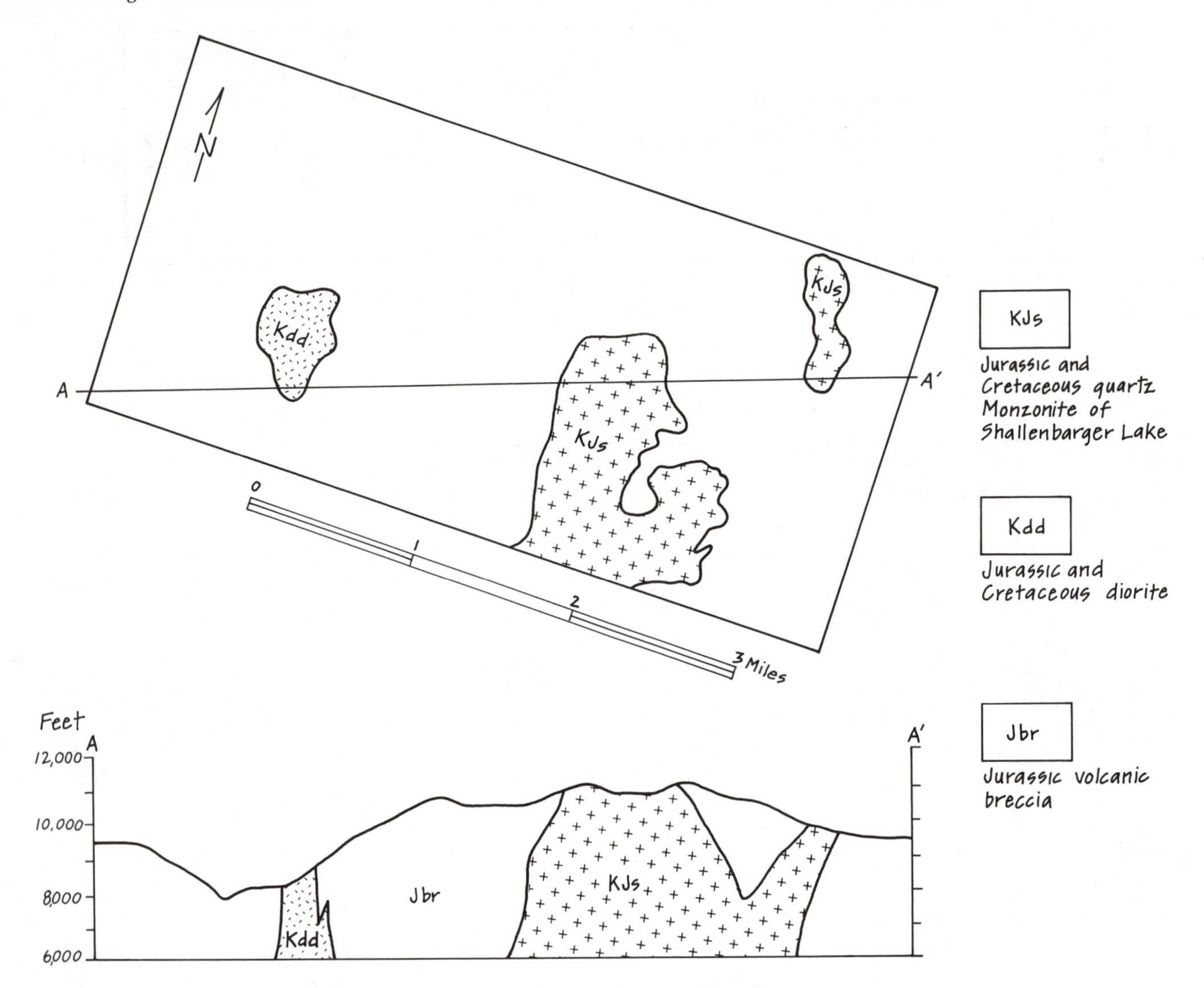

Figure 4-6
Example of a structure section with intrusive bodies (after Huber and Rinehart, 1965).

Structure Section Format

Formal structure sections should include the following characteristics:

1. A descriptive title.
2. Named geographic and geologic features such as rivers, peaks, faults, and folds should be labeled.
3. The section should be bordered with vertical lines on which elevations are labeled.
4. All rock units should be labeled with appropriate symbols.
5. Standard lithologic patterns should be used to indicate rock type (see Compton, 1985: Appendix 8).
6. A legend should be included which identifies symbols and scale.
7. Vertical exaggeration, if any, should be indicated; if none, indicate "No vertical exaggeration."
8. Contacts should be thin dark lines.
9. Construction lines should be erased.
10. Rock units should be colored as they are on the map.

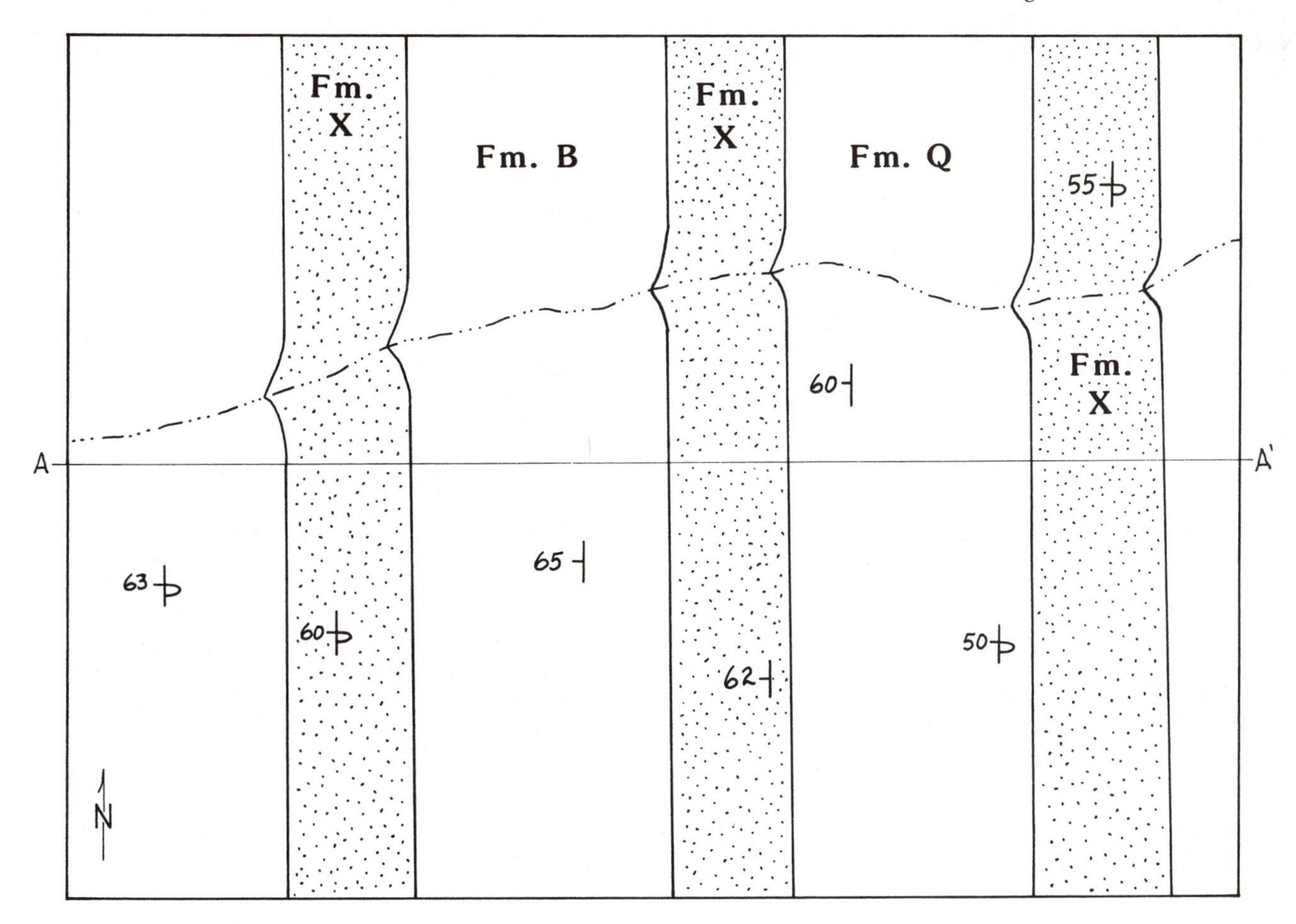

Figure 4-7
Map and topographic profile for Problem 4-1.

Further Reading

Ragan, D. M. 1985 (3rd ed.). *Structural Geology—An Introduction to Geometrical Techniques.* New York: Wiley. Chapter 19 is devoted to maps and cross-sections.

Suppe, John. 1985. *Principles of Structural Geology.* Englewood Cliffs, NJ: Prentice-Hall. Pages 57–70 contain a discussion of retrodeformable or balanced cross-sections, which allow the structure section to be undeformed for the purpose of analyzing earlier, less deformed states.

Notes

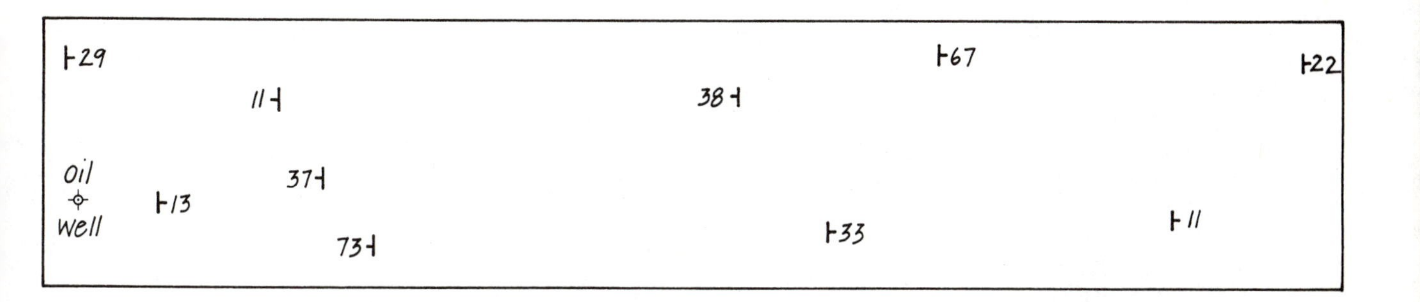

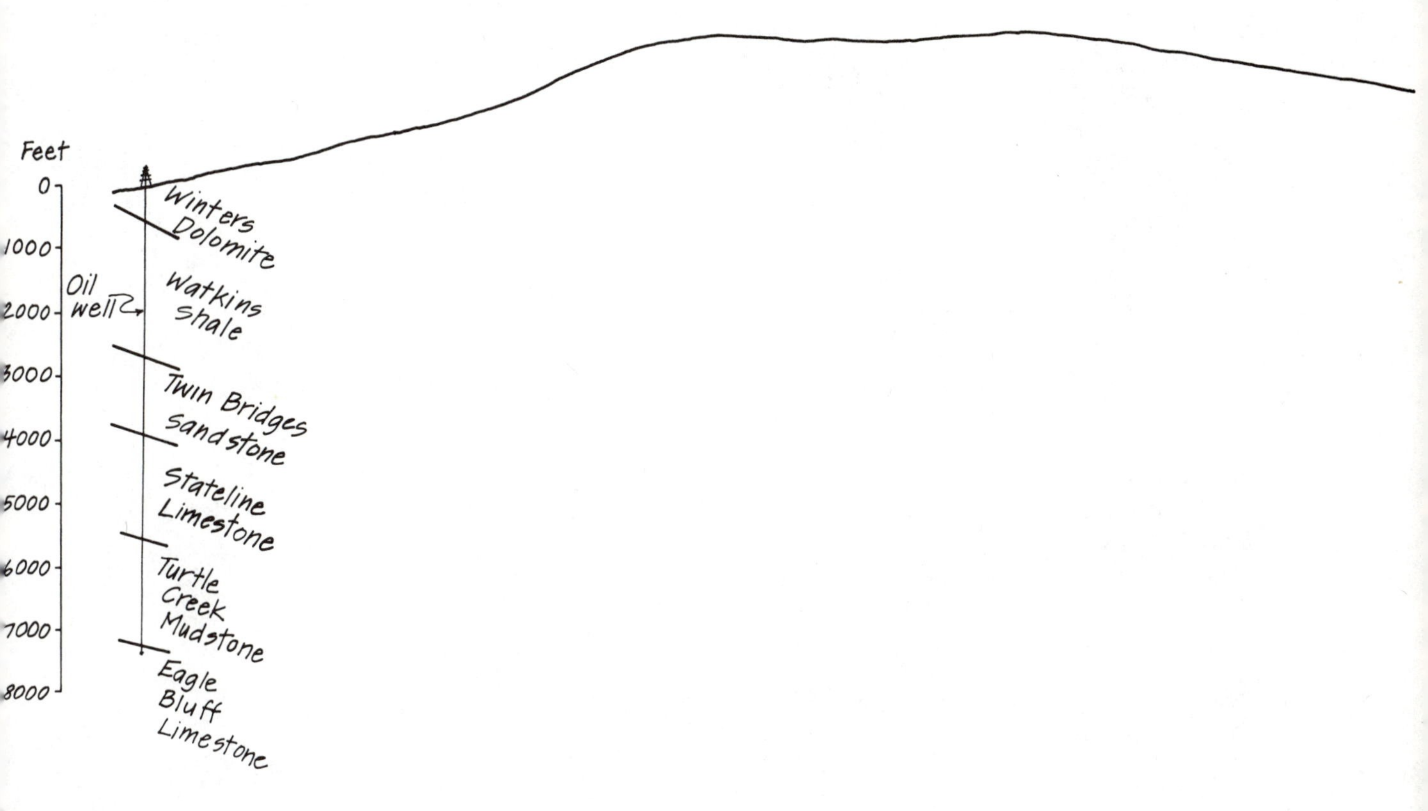

Figure 4-8
Map, topographic profile, and well log for Problem 4-2.

Notes

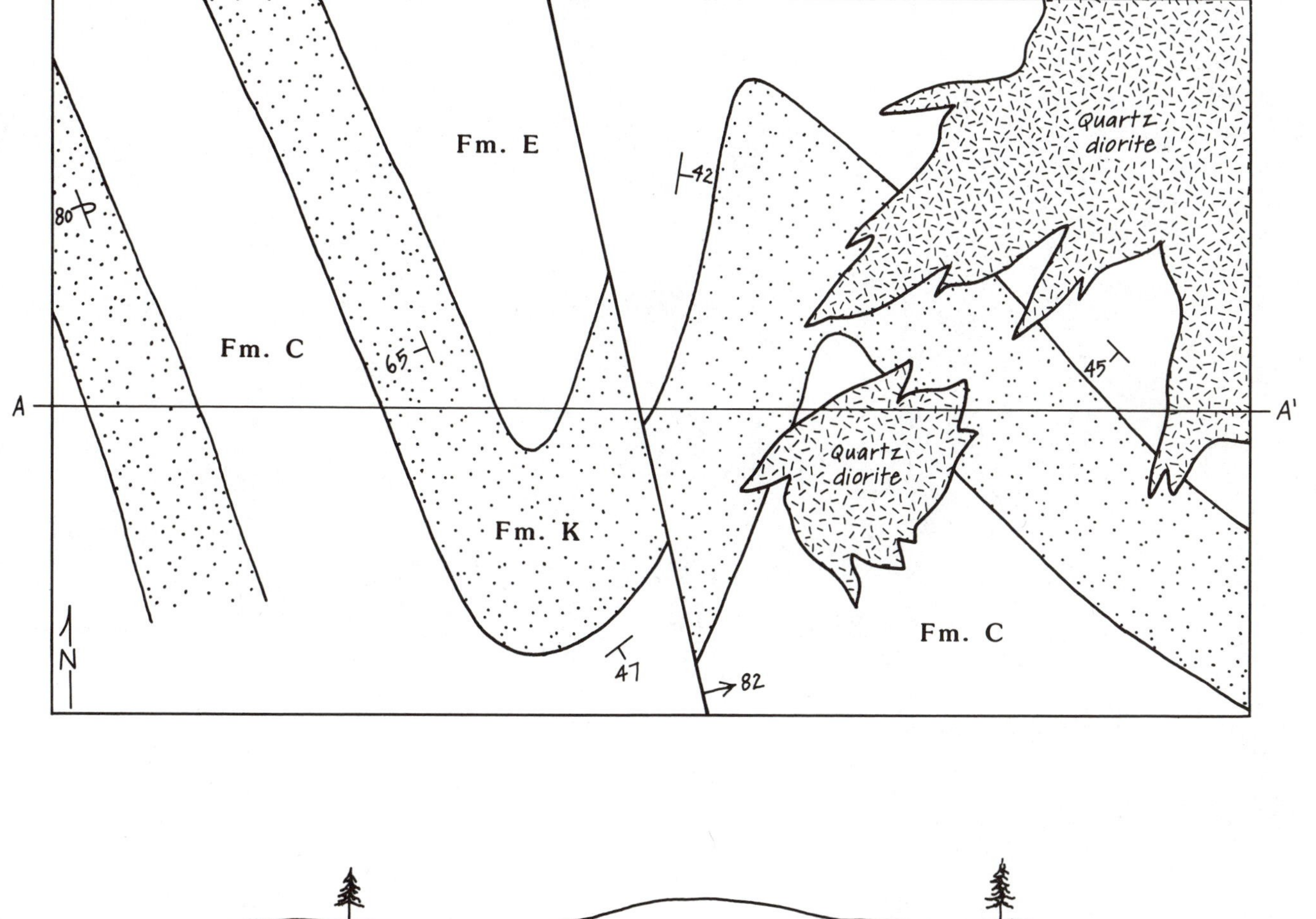

Figure 4-9
Map and topographic profile for Problem 4-3.

Notes

5

Stereographic Projection

OBJECTIVE

Use stereographic projection to determine the attitudes of lines and planes in various situations

An extremely useful technique for solving many structural problems is stereographic projection. This involves the plotting of planes and lines on a circular grid or net. Two types of nets are in common use. The net in Figure 5-1a is called a stereographic net. It is also called a Wulff net, after G. V. Wulff, who adapted the net to crystallographic use. The net in Figure 5-1b is called a Lambert equal-area net, or Schmidt net.

The two nets are constructed somewhat differently; the key difference is that all of the grid squares on the equal-area net are the same size, while on the stereographic net they are smaller toward the center. The situation is similar to map projections of the earth. Some projections sacrifice accuracy of area to preserve spatial relationships, while others do the opposite. In preserving area, the equal-area net does not preserve angular relationships. The construction of the equal-area net does allow the correct measurement of angles, however, and this net may reliably be used even when angular relationships are involved. Because structural geologists are commonly concerned with the relative density of their data, a matter requiring accurate area projection, many use equal-area nets exclusively. In this book, only the equal-area net will be used.

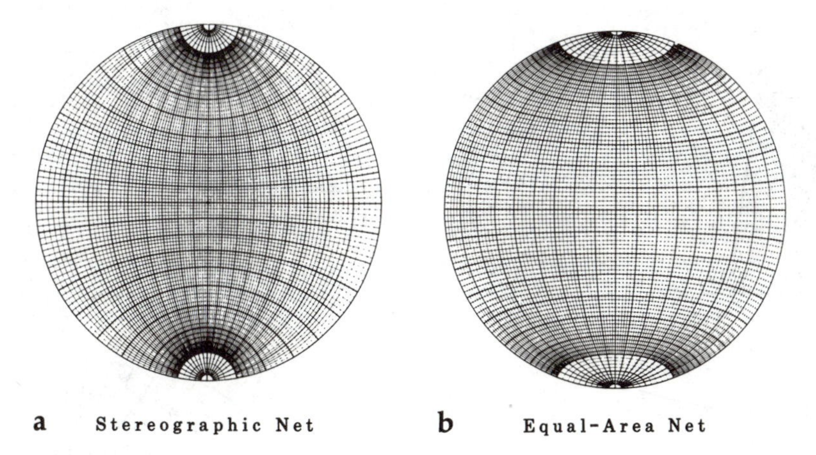

Figure 5-1
Nets used for stereographic projection. (a) Stereographic net or Wulff net. (b) Lambert equal-area net or Schmidt net.

The equal-area net is arranged rather like a globe of the earth, with north-south lines that are analogous to meridians of longitude and east-west lines that are analogous to parallels of latitude. The north-south lines are called **great circles** and the east-west lines are called **small circles.** The perimeter of the net is called the **primitive circle** (Fig. 5-2); here "primitive" has the mathematical sense of "fundamental."

Unlike crystallographers, who use the net as if it were an upper hemisphere, structural geologists use it as a lower hemisphere. To visualize how elements are projected onto the net, imagine looking down into a large bowl in which a cardboard half-circle has been snugly fitted at an angle. The exposed diameter of the half-circle is a straight line, and the curved part of the half-circle describes a curve on the bottom of the bowl. Figure 5-3a is an oblique view of a plane that strikes north-south and dips 50W. It is shown intersecting the lower hemisphere. Figure 5-3b is an equal-area projection of the same plane. Notice that the dip of the plane, 50° in this case, is measured from the perimeter of the net. The great circles on the net represent a set of planes having the same strike and all possible dips. The primitive circle represents a horizontal plane.

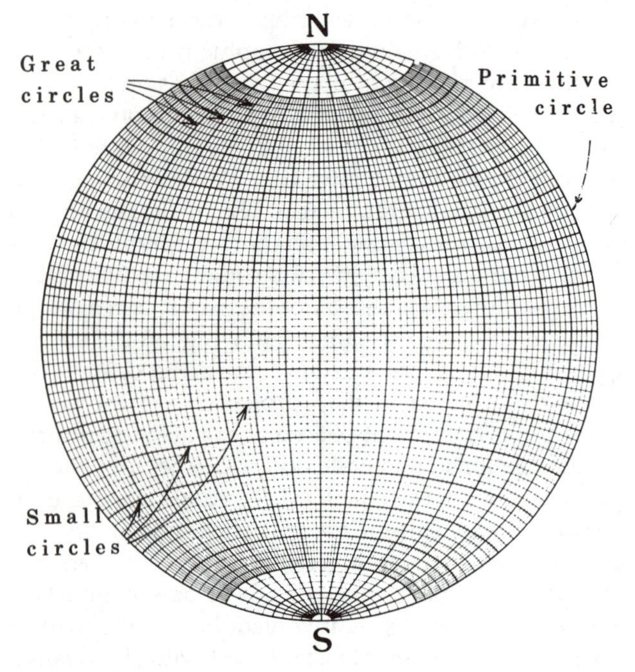

Figure 5-2
Main elements of the stereonet.

Stereographic projection is done by placing a piece of tracing paper over the net and rotating it about a thumbtack through the center of the net. Although north and south poles are labeled in Figures 5-2 and 5-3, geographical coordinates are actually attached to the tracing paper, not to the net. By means of rotating the tracing paper, a great circle corresponding to any plane can be drawn. Similarly, the straight line corresponding to the equator on Figures 5-2 and 5-3 will be referred to in the following examples as the "east-west line," even though it has no fixed geographical orientation.

At the back of this book is an equal-area net for use in this and succeeding chapters. Because your net will be heavily used, it is a good idea to tape it to a piece of thin cardboard to protect it and to ensure that the thumbtack hole does not get larger.

Following are several examples of stereographic projection. Work through each of them on your own net.

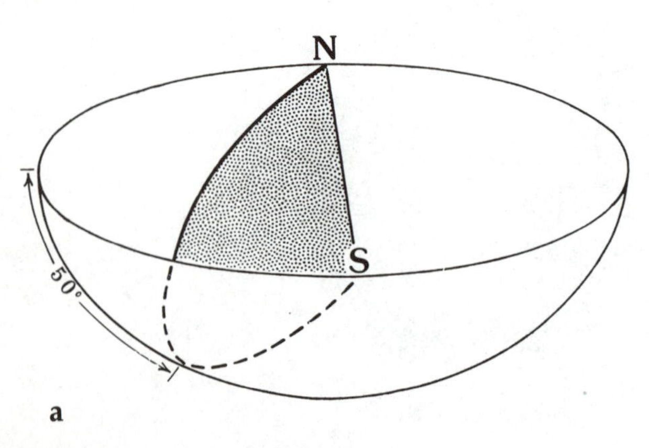

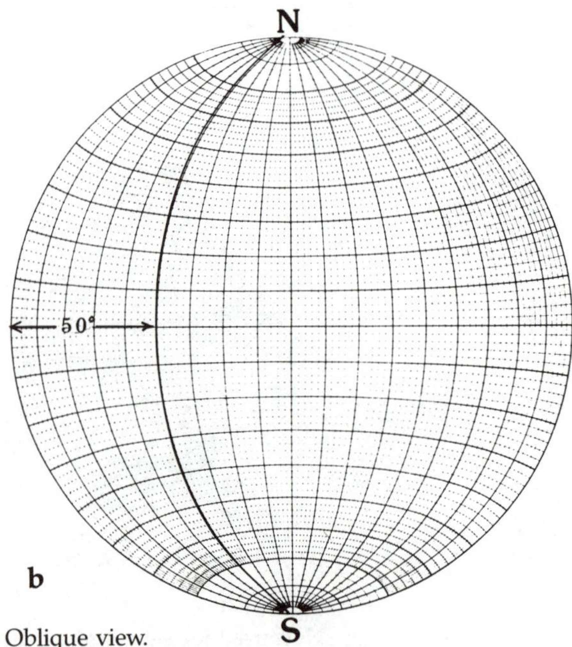

Figure 5-3
Projection of a plane striking north-south and dipping 50° west. (a) Oblique view. (b) Stereonet projection.

A Plane

Suppose a plane has an attitude of N45W, 60SW. It is plotted on the equal-area net as follows:

1. Stick a thumbtack through the center of the net from the back, and place a piece of tracing paper over the net such that the tracing paper will rotate on the thumbtack. A small piece of clear tape in the center of the paper will prevent the hole from getting larger with use.
2. Trace the primitive circle on the tracing paper (this step may be eliminated later), and mark the north and south poles on the paper.
3. Find N45W on the primitive circle, mark it with a small tick mark, and label it on the tracing paper (Fig. 5-4a).
4. Rotate the tracing paper so that the N45W mark is at the north pole of the net (Fig. 5-4b).
5. Southwest is now on the left side of the tracing paper, so count 60° down from the primitive circle along the east-west line of the net and put a mark on that point.
6. Without rotating the tracing paper, draw the great circle that passes through that point (Fig. 5-4b).
7. Finally, rotate the paper back to the original position (Fig. 5-4c).

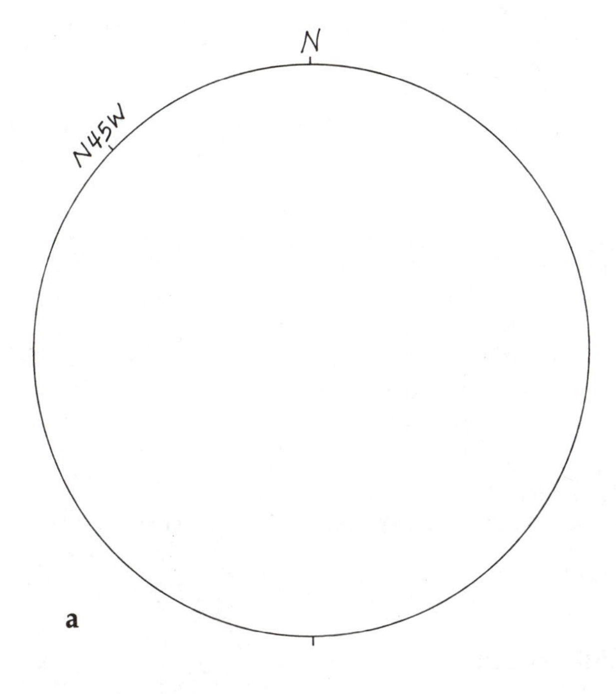

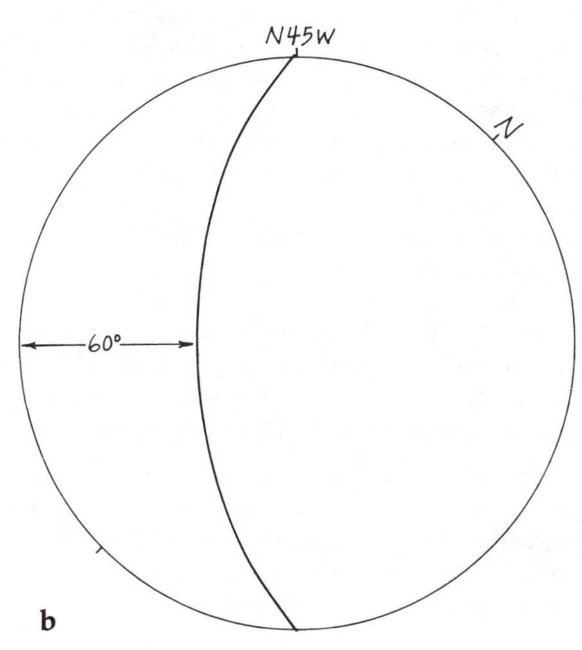

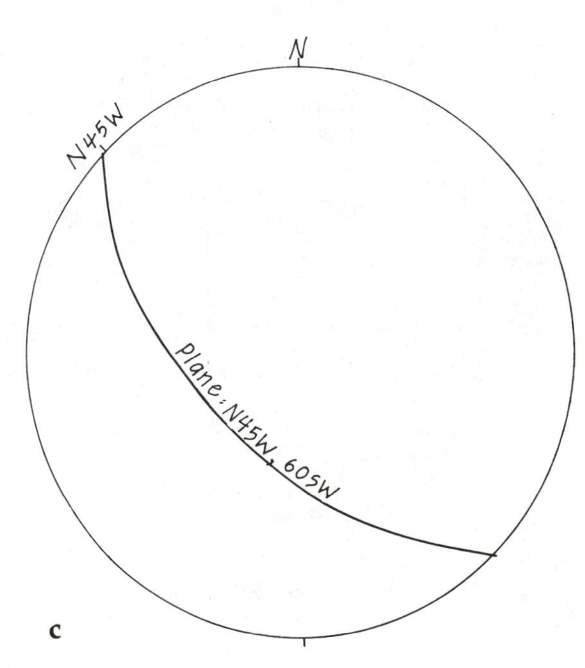

Figure 5-4
Projection of a plane striking N45W (315) and dipping 60SW. (a) Plotting of strike. (b) Projection of plane with tracing paper rotated so that strike is at top of net. (c) Tracing paper rotated back to original position.

A Line

While a plane intersects the hemisphere as a line, a line intersects the hemisphere as a single point. Figure 5-5a is an oblique view of a line that trends due west and plunges 30°. Figure 5-5b is an equal-area net projection of the same line. Lines can be imagined to pass through the center of the sphere and then pierce through the lower hemisphere.

Consider a line of attitude 32, S20E.

1. Mark the north pole and S20E on the tracing paper.
2. Rotate the S20E point to the bottom ("south") point on the net. The top, left, or right point works just as well. It is only from one of these four points on the net that the plunge of a line may be measured.

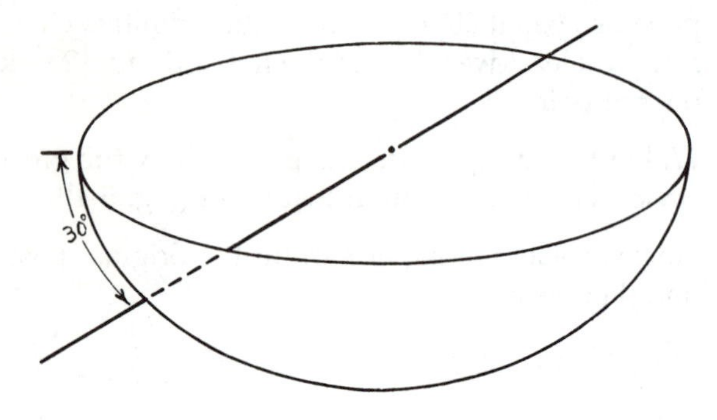

a

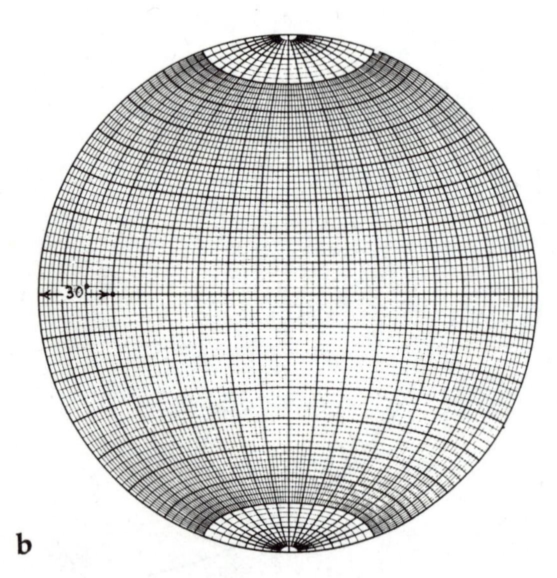

b

Figure 5-5
Projection of a line that plunges 30° due west. (a) Oblique view. (b) Stereonet projection.

3. Count 32° from the primitive circle inward, and mark that point (Fig. 5-6).
4. Rotate the paper to its original orientation.

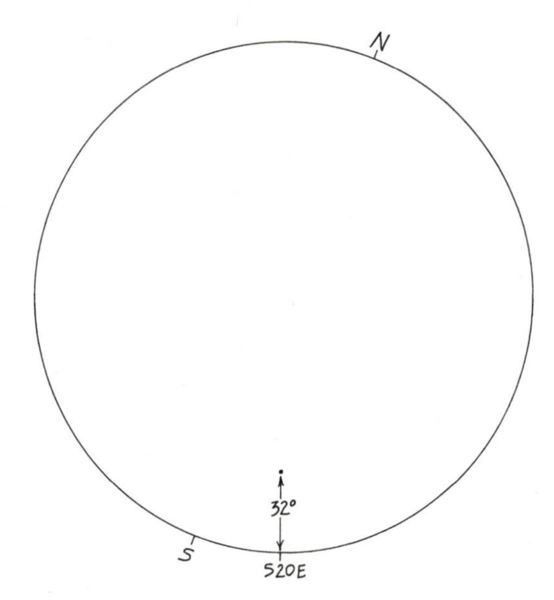

Figure 5-6
Projection of a line that plunges 32, S20E (160).

Pole of a Plane

It is possible to describe the orientation of a plane with a single point on the net. This is done by plotting the pole to the plane rather than the plane itself. The pole to a plane is the straight line perpendicular to the plane. As shown in Figure 5-7, when a plane strikes north-south and dips 40° west, its pole plunges 50° due east.

Suppose a plane has an attitude of N74E, 80N. Its pole is plotted as follows:

1. Rotate N74E on the tracing paper to the north pole of the net, as if you were going to plot the plane itself. The great circle representing this plane is shown by the dashed lines in Figure 5-8.
2. Find the point on the east-west line of the net where the great circle for this plane passes, and count 90° in a straight line across the net. This point is the pole to the plane. As shown in Figure 5-8, the pole to this plane plunges 10, S16E.

In this way the orientation of numerous planes may be displayed on one diagram without cluttering up the tracing paper with a lot of lines.

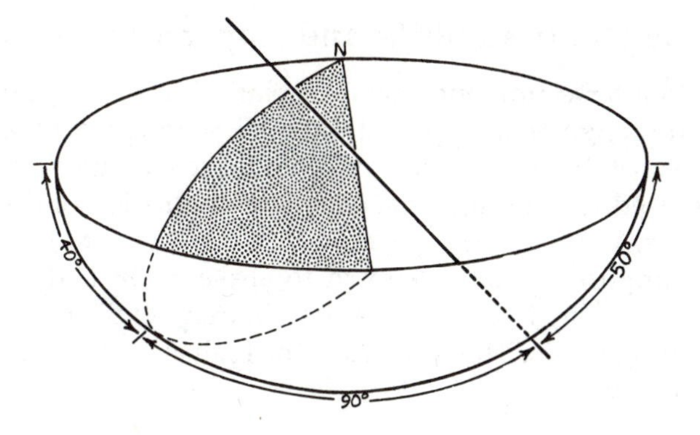

a

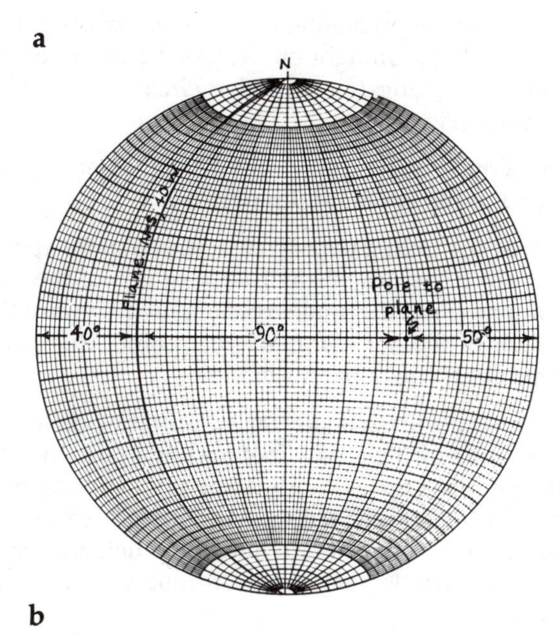

b

Figure 5-7
Projection of a plane (N-S, 40W) and the pole to the plane. (a) Oblique view. (b) Stereonet projection.

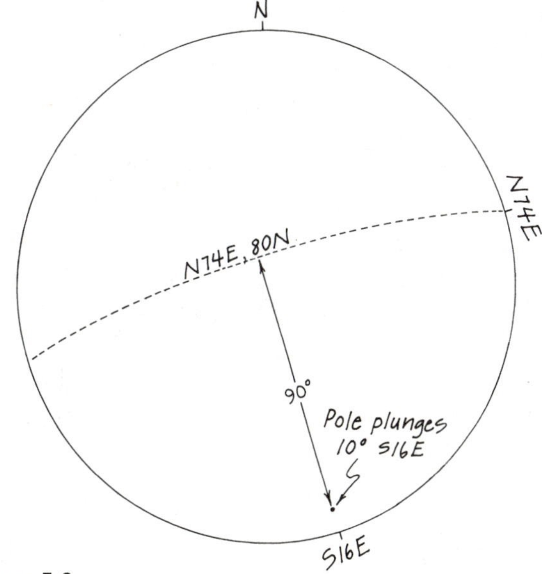

Figure 5-8
Projection of pole to a plane.

Line of Intersection of Two Planes

Many structural problems involve finding the orientation of a line common to two intersecting planes. Suppose we wish to find the line of intersection of a plane N38W, 65SW with another plane N60E, 78NW.

1. Draw the great circle for each plane (Fig. 5-9).
2. Rotate the tracing paper so that the point of intersection lies on the east-west line of the net. Mark the primitive circle at the closest end of the east-west line.
3. Before rotating the tracing paper back, count the number of degrees on the east-west line from the primitive circle to the point of intersection. This is the plunge of the line of intersection.
4. Rotate the tracing paper back to its original orientation. Find the bearing of the mark made on the primitive circle in step 2. This is the trend of the line of intersection. The line of intersection for this example plunges 61, S84W, as seen in Figure 5-9.

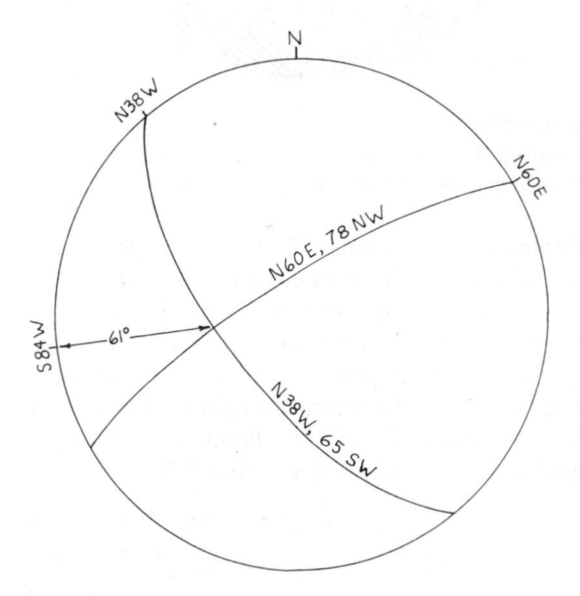

Figure 5-9
Projection of the line of intersection of two planes. Attitude of the line is indicated.

Angles within Planes

Angles within a plane are measured along the great circle of the plane. In Figure 5-10, for example, each of the two points represents a line in a plane that strikes north-south and dips 50E. The angle between these two lines is 50°, measured directly along the plane's great circle.

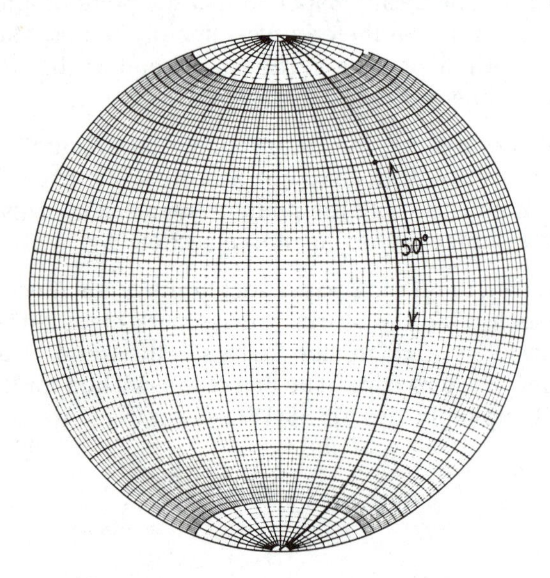

Figure 5-10
Measuring the angle between two lines in a plane using its great circle.

The more common need is to plot the pitch or rake of a line within a plane. Plotting pitches may be useful when working with rocks containing lineations. The lineations must be measured in whatever outcrop plane they may occur. Suppose, for example, that an outcrop surface of N52W, 20NE contains a lineation with a pitch of 43° to the east (Fig. 5-11a). Figure 5-11b shows the lineation plotted on the equal-area net.

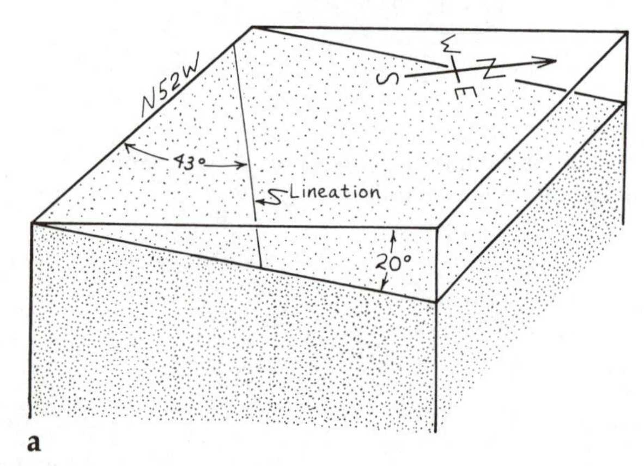

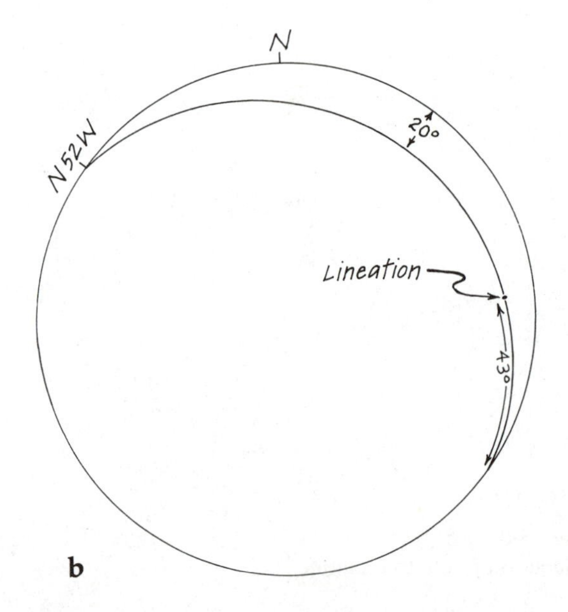

Figure 5-11
Pitch of a line in a plane. (a) Block diagram. (b) Stereonet projection.

True Dip from Strike and Apparent Dip

Two intersecting lines define a plane, so if the trend and plunge of an apparent dip are known, and if the strike of the plane is known, then these two lines can be used to determine the complete orientation of the plane.

Suppose a fault is known to strike N10E and an apparent dip is measured to have a trend of S26E and a plunge of 35. The true dip of the fault is determined as follows:

1. Draw a line representing the strike line of the plane. This will be a straight line across the center of the net intersecting the primitive circle at the strike bearing (Fig. 5-12a).
2. Put a mark on the primitive circle representing the trend of the apparent dip (Fig. 5-12a).
3. Rotate the tracing paper so that the point for apparent dip trend lies on the east-west line of the net. Count the number of degrees of plunge toward the center of the net and mark that point. This point represents the apparent dip line.
4. We now have two points on the primitive circle (the two ends of the strike line) and one point not on the primitive circle (the apparent dip point), all three of which lie on the fault plane. Turn the strike line to lie on the north-south line of the net, and draw the great circle that passes through these three points.
5. Before rotating the tracing paper back, measure the true dip along the east-west line of the net. As shown in Figure 5-12b, the true dip is 50°.

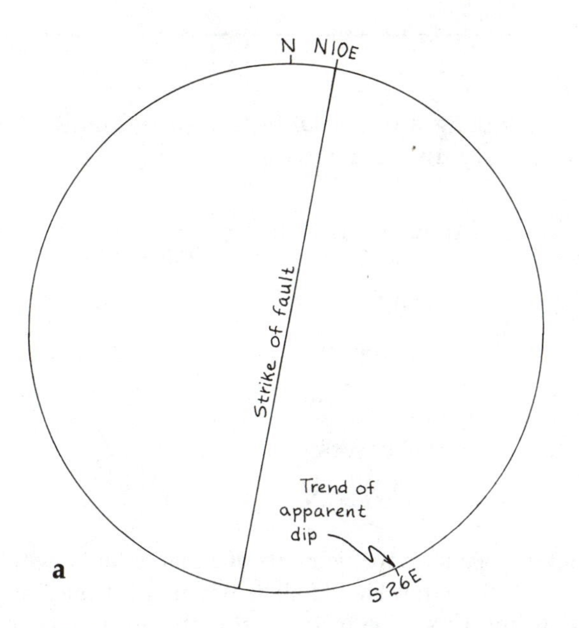

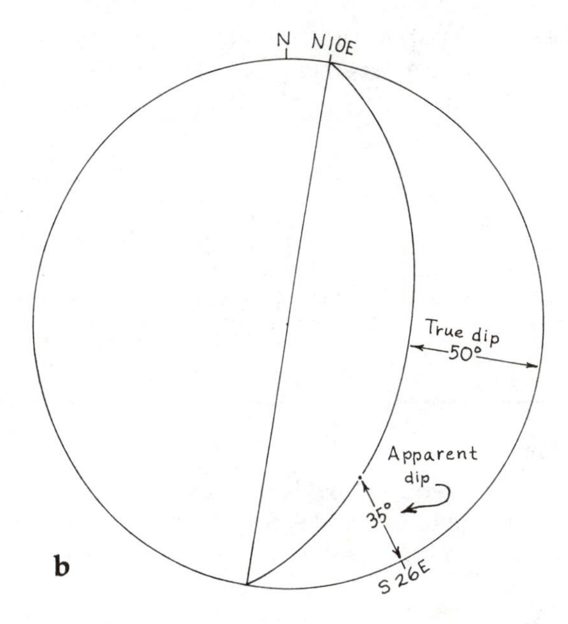

Figure 5-12
Determining true dip from strike and apparent dip. (a) Draw strike line and trend of apparent dip. (b) Completed diagram showing direction and amount of true dip.

Strike and Dip from Two Apparent Dips

Even if the strike of a plane is not known, two apparent dips are sufficient to find the complete attitude. Suppose two apparent dips of a bed are 13, S18E and 19, S52W.

1. Plot points representing the two apparent dip lines (Fig. 5-13a).
2. Rotate the tracing paper until both points lie on the same great circle. This great circle represents the plane of the bed, and the strike and dip are thus revealed. As shown in Figure 5-13b, the attitude of the plane in this problem is N57W, 20SW.

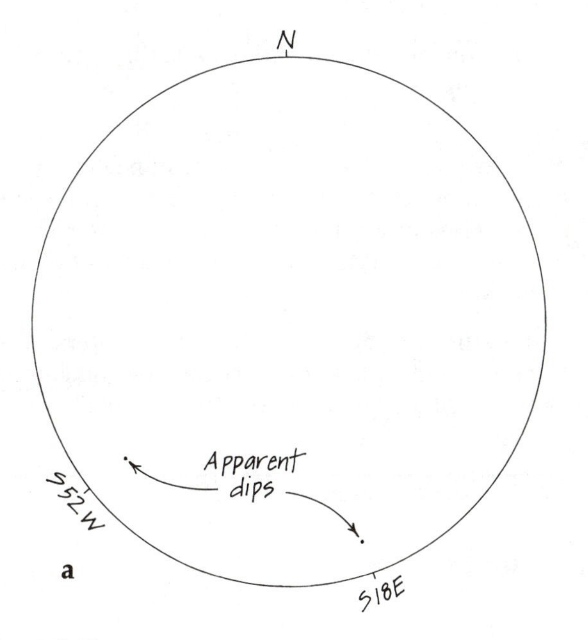

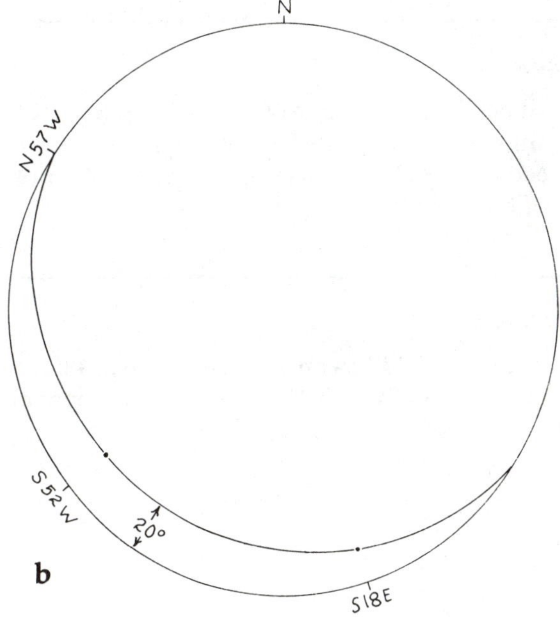

Figure 5-13
Determining strike and dip from two apparent dips. (a) Apparent dips plotted as points on net. (b) Completed diagram showing strike and true dip.

Use stereographic projection to solve the following problems. Use a separate piece of tracing paper for each problem.

Problem 5-1

Along a vertical railroad cut a bed has an apparent dip of 20, N62W. The bed strikes N67E. What is the true dip?

Problem 5-2

In a mine a tabular dike has an apparent dip of 14, N90W in one tunnel and 9, S70W in another. What is the attitude of the dike?

Problem 5-3

A fault strikes due north and dips 70E. A limestone bed with an attitude of N35W, 25SW is cut by the fault. Hydrothermal alteration along the fault has resulted in an ore shoot at the intersection of the two planes.

1. What is the orientation of the ore shoot?
2. What is the pitch of the ore shoot in the plane of the fault?
3. What is the pitch of the ore shoot in the plane of the limestone bed?

Problem 5-4

A coal bed has an attitude of N68E, 40S. At what two bearings may mine adits be dug along the bottom of the bed such that the adits slope 10° and water drains out of the mine?

Problem 5-5

One limb of a fold has an attitude of N61E, 48SE and the other limb N28E, 55NW. What is the orientation of the fold axis?

Problem 5-6

The following are measurements of five lineations, taken at five different outcrops.

Locality	Attitude of outcrop surface	Pitch of lineation
1	N60W, 84NE	76°E
2	N10W, 30E	50°N
3	N40E, 70SE	63°SW
4	N23W, 30W	50°S
5	N88W, 45N	59°E

If these lineations are elements of a planar fabric within the rock, then they should all lie within the same plane. If you find this to be true, what is the attitude of this plane?

Problem 5-7

Two intersecting shear zones have the following attitudes: N80E, 75S and N60E, 52NW.

1. What is the orientation of the line of intersection?
2. What is the orientation of the plane perpendicular to the line of intersection?
3. What is the obtuse angle between the shear zones within this plane?
4. What is the orientation of the plane that bisects the obtuse angle?
5. A mining adit is to be dug to the line of intersection of the two shear zones. For maximum stability the adit is to bisect the obtuse angle between the two shear zones and intersect the line of intersection perpendicularly. What should the trend and plunge of the adit be?
6. If the adit is to approach the line of intersection from the south, will the full ore carts be going uphill or downhill as they come out of the mine?

Rotation of Lines

In subsequent examples it will be necessary to use stereographic projection to rotate lines and planes. Imagine three lines A, B, and C with plunges of 30°, 60°, and 90°, respectively, all lying within a vertical, north-south-striking plane. Figure 5-14a is an oblique view of these lines intersecting a lower hemisphere. Figure 5-14b shows the same three lines projected on the equal-area net.

As a plane rotates around the horizontal north-south axis of the net, the projection points of lines within the plane move along the small circles. Figure 5-14c shows points A, B, and C moving in unison 40° to points A′, B′, and C′ as the plane rotates 40°. As the plane rotates 90° to a horizontal position, the projected points A″, B″, and C″ lie on the primitive circle. Notice on Figure 5-14c that when the plane is horizontal each line is represented by two points 180° apart on the primitive circle, one in the northeast quadrant and one in the southwest quadrant.

As the plane continues to rotate, A and B leave the northern half of the hemisphere and appear in the lower half. Figure 5-14d shows the projection points of the three lines as the plane in which they lie rotates 180° from its original orientation in Figure 5-14a.

Although in this example the lines being rotated are coplanar, this need not be the case. Lines representing the poles of several variously oriented planes, for example, can be rotated together. The only requirement is that all points must move the same number of degrees along their respective small circles.

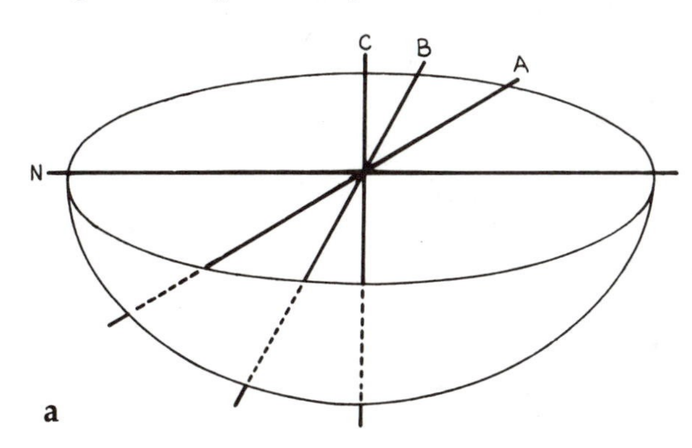

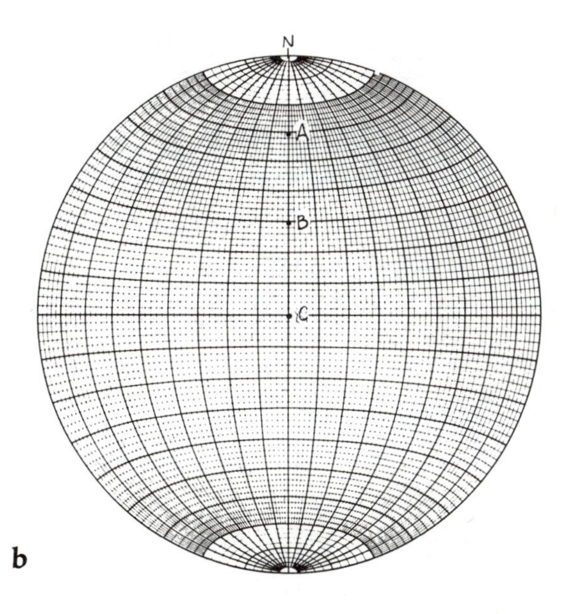

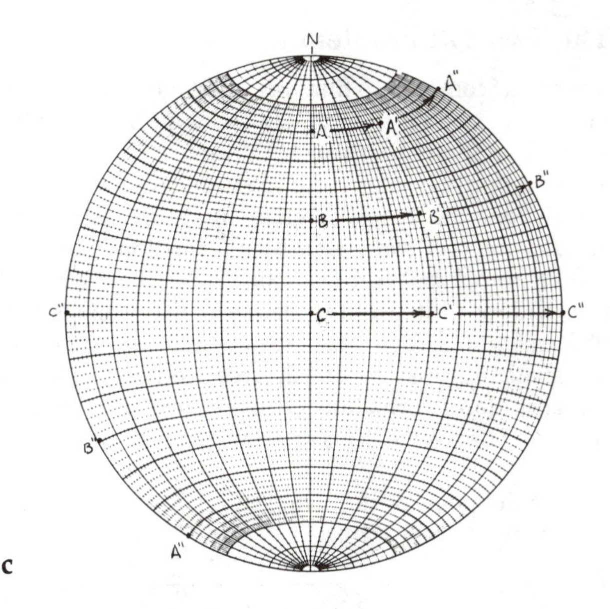

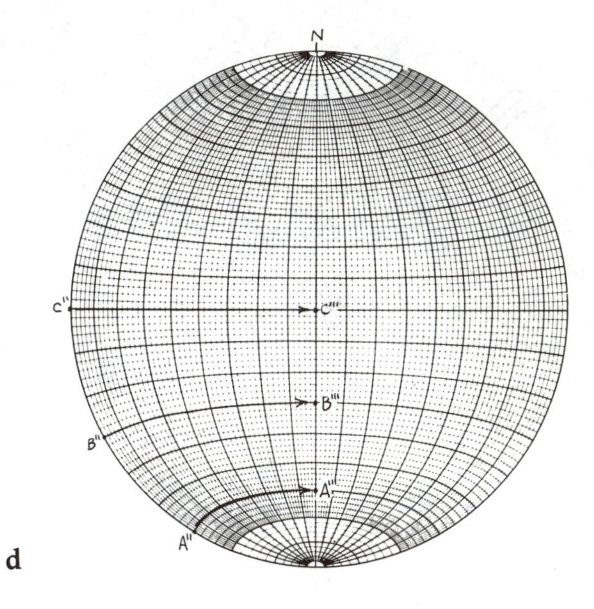

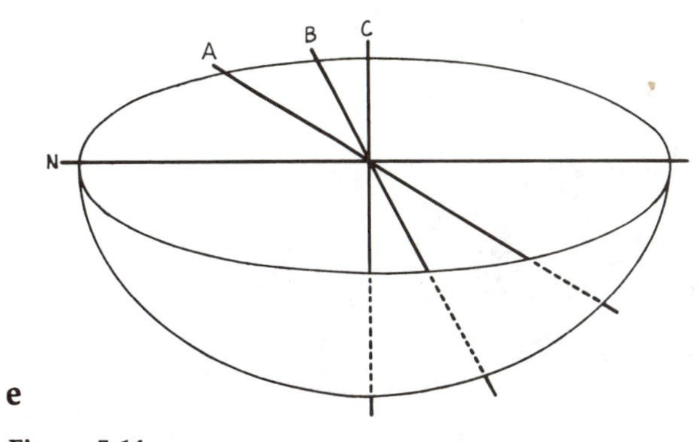

Figure 5-14
Rotation of three coplanar lines. (a) Oblique view. (b) Projection of lines on net. (c) Projection of lines rotated 40° (A′, B′, C′) and 90° (A″, B″, C″). (d) Projection of lines rotated 180° from their original positions. (e) Oblique view of final position of lines rotated 180°.

The Two-Tilt Problem

It is not uncommon to find rocks that have undergone more than one episode of deformation. In such situations it is sometimes useful to remove the effects of a later deformation in order to study an earlier one.

Consider the block diagram in Figure 5-15a. An angular unconformity separates formation Y (N60W, 35NE) from formation O (N50E, 70SE). Formation O was evidently tilted and eroded prior to the deposition of formation Y, then tilted again. In order to unravel the structural history of this area we need to know the attitude of formation O at the time formation Y was being deposited. This problem is solved as follows:

1. Plot the poles of the two formations on the equal-area net (Fig. 5-15b).
2. We want to return formation Y to horizontal and measure the attitude of formation O. The pole of a horizontal bed is vertical, so if we move the point Y to the center of the net, formation Y will be horizontal. Rotate the tracing paper so that Y lies on the east-west line of the net.
3. Y can now be moved along the east-west line to the center of the net (Fig. 5-15c). This involves 35° of movement. O, therefore, must also be moved 35° along the small circle on which it lies, to O′ (Fig. 5-15c).
4. O′ is the pole of formation O prior to the last episode of tilting. As shown in Figure 5-15d, the attitude of formation O at that time was N58E, 86SE.

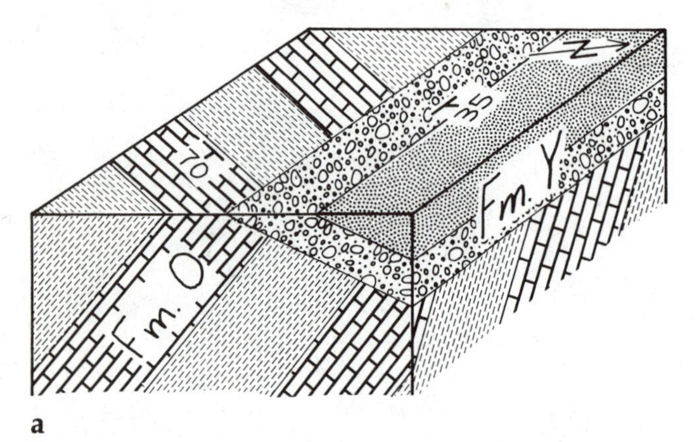

a

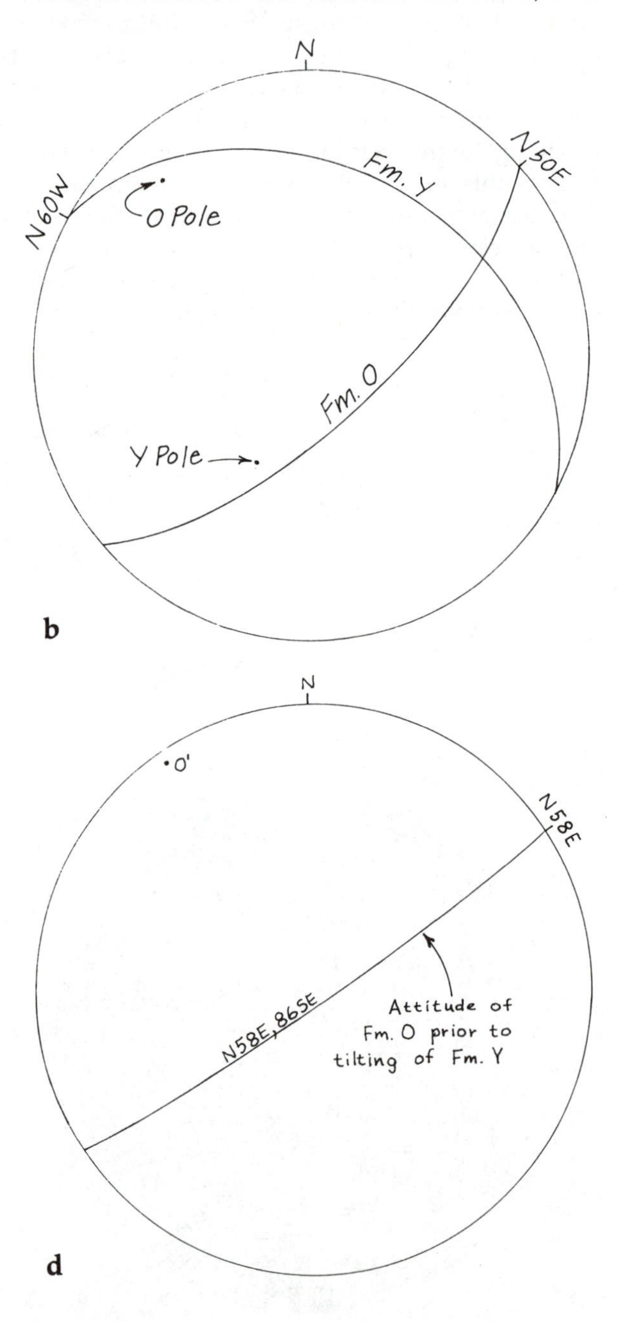

b

d

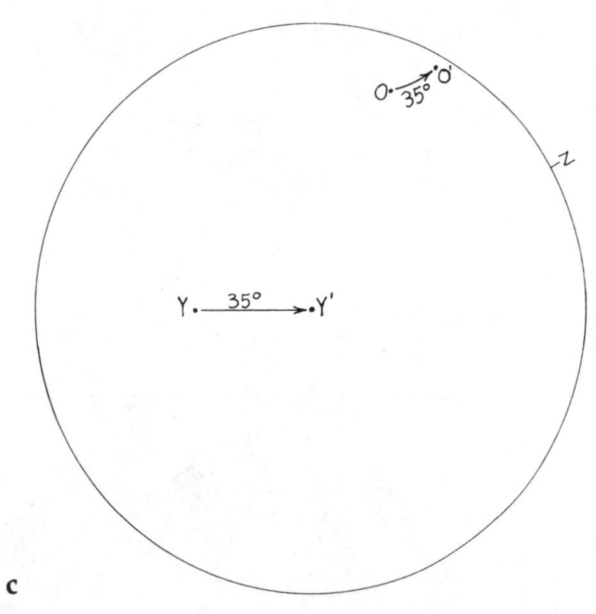

c

Figure 5-15
Two-tilt problem. (a) Block diagram. (b) Plotting attitudes and poles of formations O and Y. (c) Removing dip of formation Y from that of formation O. (d) Plotting attitude of formation O before tilting of formation Y.

Cones—The Drill-Hole Problem

In certain problems it is necessary to project a cone onto the net. Suppose, for a simple example, that a hole has been drilled horizontally due north into a cliff. The drill core is shown in Figure 5-16a. The rock is layered, and the angle between the core axis and the bedding plane is 30°. The angle between the pole to the bedding plane and the core axis is 60°, the complement of 30°.

Because the core rotated as it was extracted from the hole, the exact orientation of the bedding plane cannot be determined. A locus of possible orientations can be defined, however. Figure 5-16b is an oblique view of the situation, showing a cone with its axis horizontal and trending north-south. The cone represents all possible lines 30° from the axis. Lines perpendicular to the sides of the cone, representing poles to the bedding plane, pass through the center of the sphere and intersect the lower hemisphere as two half-circles. As shown in Figure 5-16c, the equal-area plot of the possible poles to bedding is two small circles, each 60° from a pole. If a second hole is drilled, oriented differently from the first, two more small circles can be drawn. The second set of circles will have two, three, or four points in common with the first pair of small circles. A third hole results in a unique solution, establishing the pole to bedding.

Consider the following example:
Data from three drill holes are:

Hole no.	Orientation of hole	Angle between axis of core and bedding	Angle between axis and pole to bedding
1	74, N80W	17°	73°
2	70, S30E	18°	72°
3	62, N67E	51°	39°

What is the attitude of the beds? The solution involves consideration of the holes in pairs. In part A we will consider holes 1 and 2 together, and in part B holes 1 and 3 together. The two parts should be done on separate pieces of tracing paper.

A-1. Plot each hole (Fig 5-17a).

A-2. Rotate the tracing paper so that both points lie on the same great circle. In this position the two holes lie in a common plane that in this example dips 82° SW.

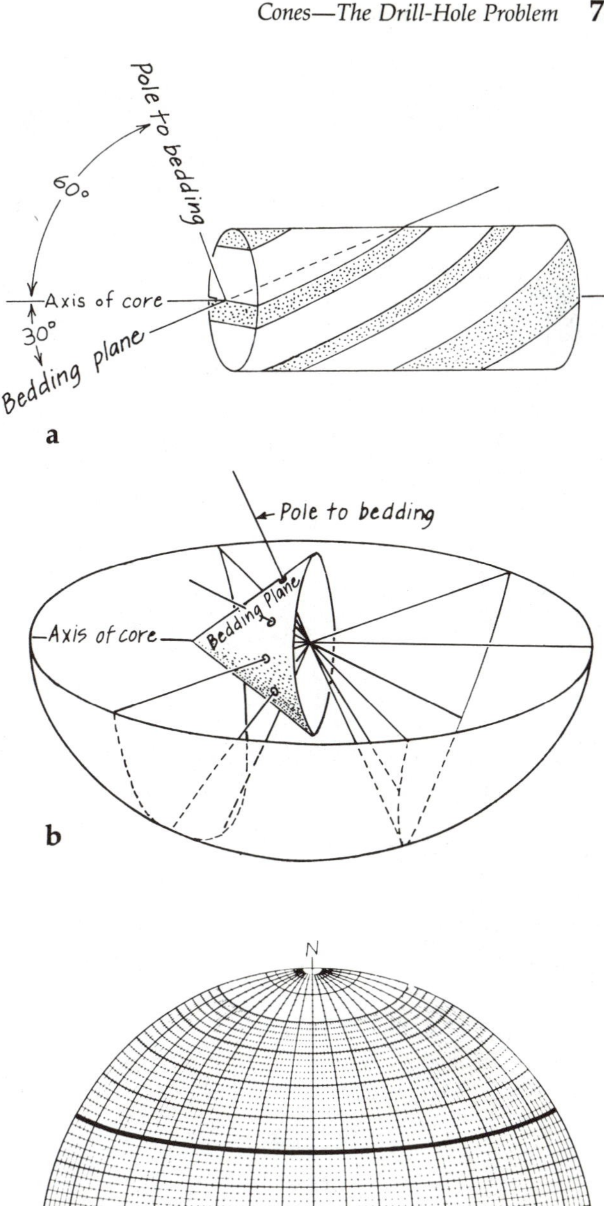

Figure 5-16
Drill-hole problem. (a) Drill core. (b) Oblique view of projection of all possible poles to bedding of drill core. (c) Stereonet projection of all possible poles to bedding of drill core.

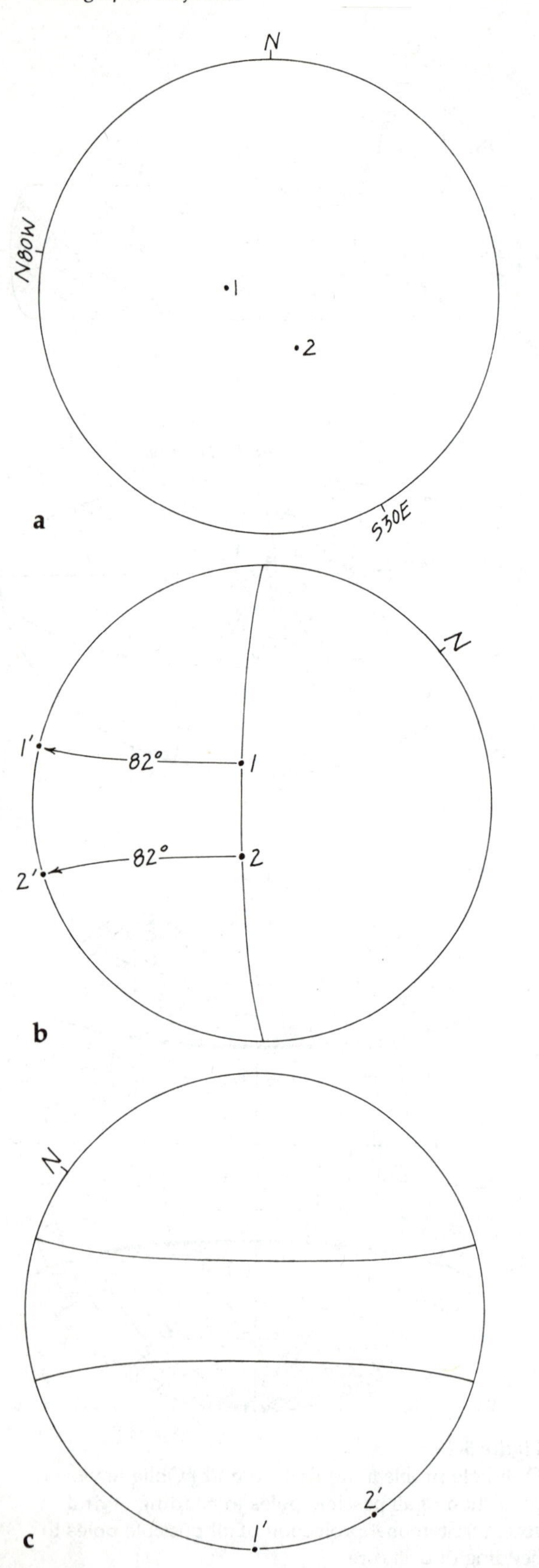

A-3. Rotate the common plane to horizontal. This step moves points 1 and 2 along their respective small circles 82° to points 1′ and 2′ on the primitive circle (Fig. 5-17b).

A-4. Rotate the tracing paper so that point 1′ lies at a pole of the net. In this position hole 1 has effectively been oriented horizontal and north-south, similar to the core shown in Figure 5-16a. The angle between the axis of hole 1 and the pole to bedding is 73°, so two small circles, each 73° from a pole, describe all of the possible pole orientations. Figure 5-17c shows these two small circles.

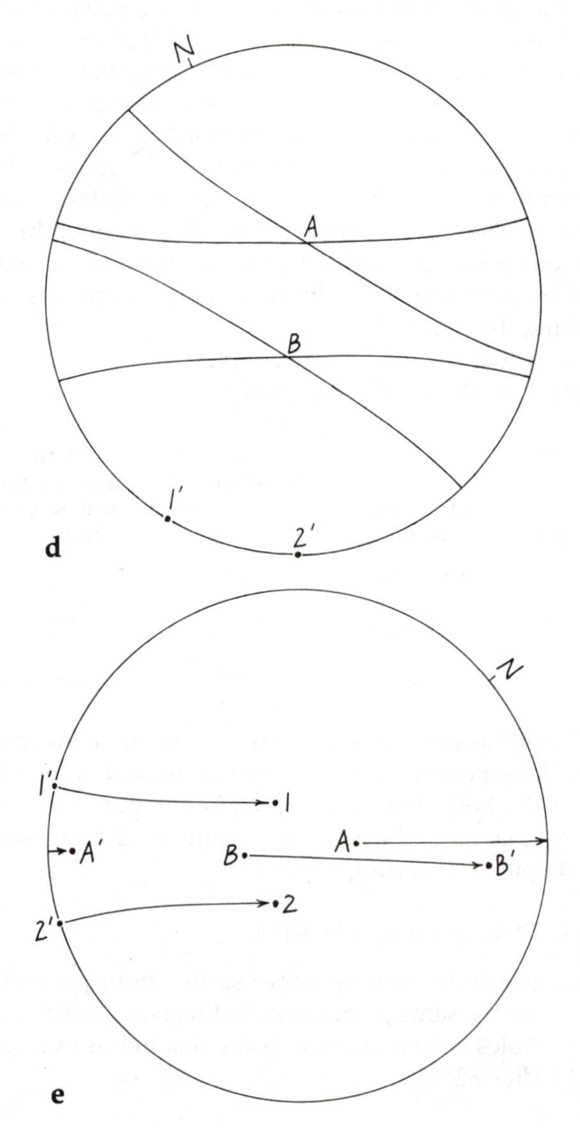

Figure 5-17
Part A of drill-hole solution. (a) Plot orientation of holes 1 and 2. (b) Locate common plane and rotate to horizontal. (c) Plot cone of possible bedding-plane poles relative to hole 1. (d) Do the same for hole 2. (e) Determine orientation of common poles A and B. Another hole must be analyzed to choose the correct pole.

A-5. Now rotate the tracing paper so that point 2′ is at a pole of the net. The angle between the axis of hole 2 and the pole to bedding is 72°, so two small circles, each 72° from a pole, describe the possible orientations of the pole to bedding. As shown in Figure 5-17d, the two sets of small circles cross at points A and B. Depending on the orientations of the two cores, there may be from one to four points of intersection.

A-6. The points of intersection have been determined with the cores in a horizontal plane. The cores must be returned to their proper orientation for the orientations of the points of intersection to be determined. To do this, rotate the tracing paper so that points 1 and 2 again lie on a common great circle, as in step A-2. In this position points 1′ and 2′ are imagined to move 82° back to 1 and 2, and points A and B move 82° along small circles in the same direction to points A′ and B′, as shown in Figure 5-17e. Notice that point A, after moving 72°, encounters the primitive circle. In order to complete its 82° excursion it reappears 180° around the primitive circle and travels 10° more. Points A′ and B′ thus located are both possible poles to bedding in this problem.

B-1. Holes 1 and 3 will now be considered together on a separate piece of tracing paper. Figure 5-18a shows cores 1 and 3 plotted, rotated to a common great circle, and moved to the primitive circle as 1′ and 3′. Notice that 1′ here is in a different location than 1′ in step A-3.

B-2. Small circles 73° and 39° from points 1′ and 3′, respectively, are drawn, as shown in Figure 5-18b. The intersection of the two pairs of small circles are points C and D.

B-3. As points 1′ and 3′ travel 83° back to their original positions as points 1 and 3, points C and D travel on small circles 83° also (Fig. 5-18c).

One of the points C′ or D′ should be in the same position on the net as A′ or B′. In this example points B′ and D′ are the same point with an orientation of 26, N41E. This is the pole to the bedding plane, thus establishing a bedding attitude of N50W, 64SW. A third solution could be done, with points 2 and 3 considered together, for further confirmation. If no two points are coincident, then either a mistake has been made or the attitude is not consistent from one hole to the next.

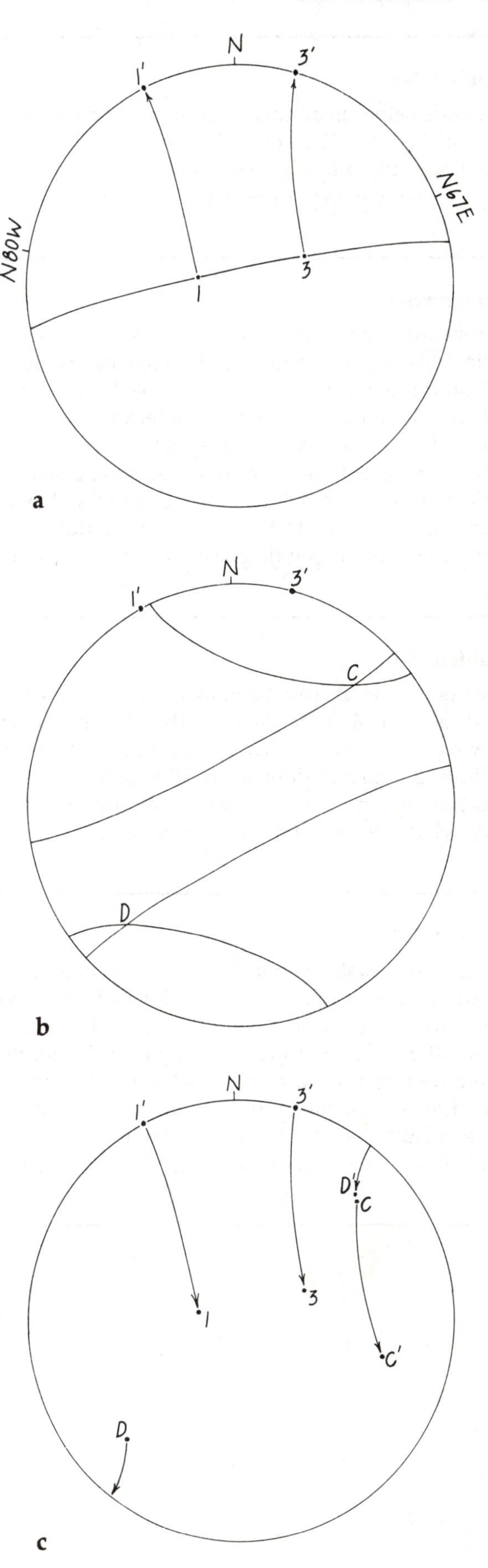

Figure 5-18
Part B of drill-hole solution. (a) Plot orientation of holes 1 and 3 and rotate common plane to horizontal. (b) Plot cones of possible bedding-plane poles relative to holes 1 and 3. (c) Determine orientation of common poles C and D; pole D′ matches pole B′ and is the solution.

Problem 5-8

The beds below an angular unconformity have an attitude of N26W, 74W, and those above N30E, 54NW. What was the attitude of the older beds while the younger were being deposited?

Problem 5-9

Iron-bearing minerals in volcanic rocks contain magnetic fields that were acquired when the magma flowed out onto the earth's surface and cooled. These magnetic fields can be measured to determine the orientation of the lines of force of the earth's magnetic field at the time of cooling. If the north-seeking paleomagnetic attitude in a basalt is 32, N67E, and the flow has been tilted to N12W, 40W, what was the attitude (trend and plunge) of the pre-tilt paleomagnetic orientation?

Problem 5-10

The orientation of cross-bedding in sandstones can be used to determine the direction that the current was flowing when the sand was deposited. If crossbeds indicating a current direction of S38E occur within an overturned sandstone bed whose attitude is N16W, 77W, what direction did the current flow?

Problem 5-11

In the northeastern fault block of the Bree Creek Quadrangle is a hill capped by Helm's Deep Sandstone overlying Rohan Tuff. In Problem 3-1 you determined the attitude of these two units at this locality. Using stereographic projection determine the amount and direction of tilting that occurred in the northeastern fault block after deposition of the Rohan Tuff but before deposition of the Helm's Deep Sandstone.

Problem 5-12

Using the data from three drill holes shown below, determine the attitude of bedding.

Hole no.	Orientation of hole	Angle between axis of core and bedding	Angle between axis and pole to bedding
1	70, N20W	40°	50°
2	76, N80E	65°	25°
3	68, S30W	54°	36°

Further Reading

Dennison, J. M. 1968. *Analysis of Geologic Structures.* New York: Norton. Chapter 11 provides an introduction to stereographic projection, with numerous problems and an extensive bibliography.

Hobbs, B. E., Means, W. D., and Williams, P. F. 1976. *An Outline of Structural Geology.* New York: Wiley. Appendix A contains a useful comparison of equiangular and equal-area projection.

Phillips, F. C. 1960. *The Use of Stereographic Projection in Structural Geology.* London: Edward Arnold Ltd., 2nd ed. Contains transparent overlays of sample problems.

6

Fold Geometry

OBJECTIVES

Describe the orientation and geometry of folds

Classify folds on the basis of dip isogons

Chapters 1 through 5 have been devoted to various techniques of structural analysis; now it is time to use the techniques on folded and faulted rocks. Chapter 4 included techniques for drawing vertical structure sections of folded and faulted rocks, but the emphasis was on the mechanics of drawing the structure sections rather than on the structures themselves. Chapters 6 through 10 are devoted to an analysis of folds and faults.

Rocks may undergo large amounts of permanent deformation either brittlely, by faulting, or ductilely, by folding. In general, faulting occurs under conditions of relatively low temperature, low pressure, and fast deformation, while folding occurs under the opposite conditions. In this chapter we examine the geometry and classification of folds.

An understanding of the formation of geologic structures begins with a precise description of the structures themselves. Listed below are the principal terms used to describe folds:

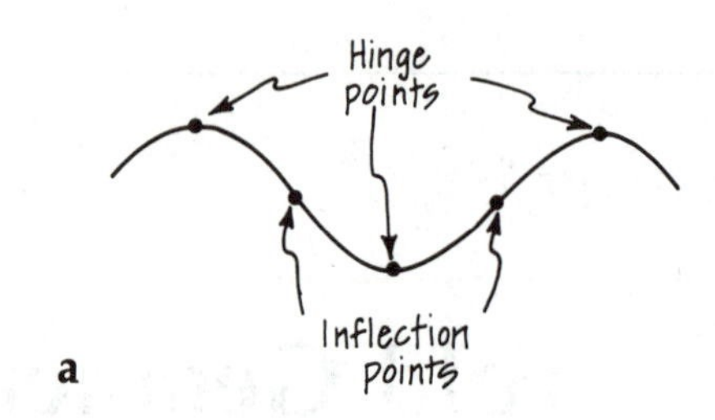

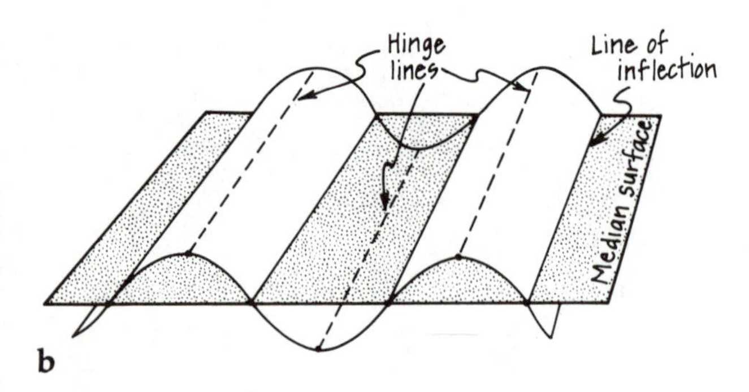

Figure 6-1
Some terms for describing the geometry of folds. (a) Profile view. (b) Block diagram.

Hinge point The point of minimum radius of curvature on a fold (Fig. 6-1a).

Hinge line The locus of hinge points on a folded surface (Fig. 6-1b).

Inflection point The point on a fold where the rate of change of slope is zero, usually chosen as the midpoint of each straight section (Fig. 6-1a).

Line of inflection The locus of inflection points of a folded surface (Fig. 6-1b).

Median surface The surface that joins the successive lines of inflection of a folded surface (Fig. 6-1b).

Crest and trough The high and low points, respectively, of a fold, usually in reference to folds with gently plunging hinge lines (Fig. 6-2a).

Crestal surface and trough surface The surfaces joining the crests and troughs, respectively, of nested folds (Fig. 6-2b).

Crestal trace and trough trace The lines representing the intersections of the crestal and trough surfaces, respectively, with another surface, usually the surface of the earth (Fig. 6-2b).

Figure 6-2
More terms for describing the geometry of folds. (a) Profile view. (b) Block diagram.

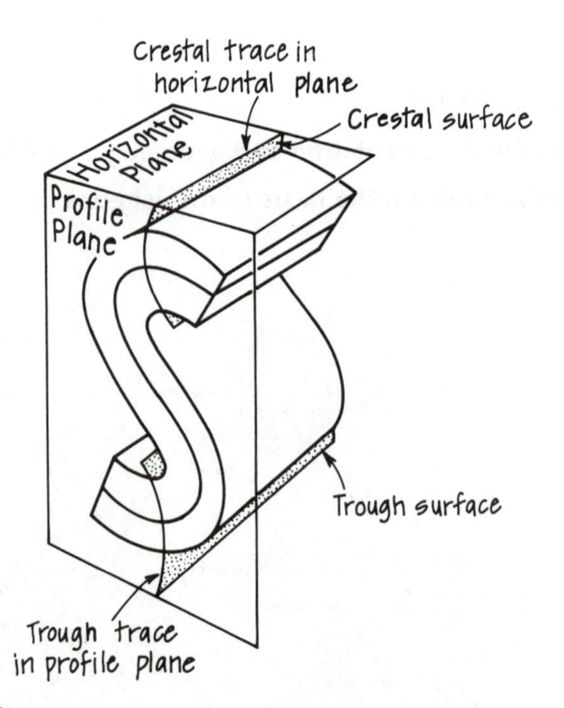

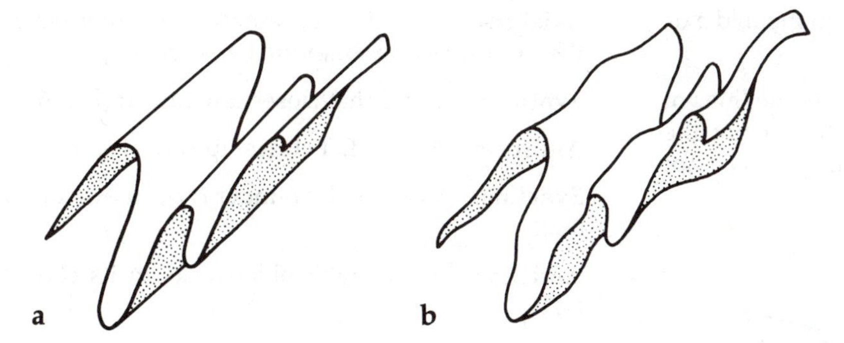

Figure 6-3
Cylindrical folds (a) and noncylindrical (b) folds.

Cylindrical fold A fold generated by a straight line moving parallel to itself in space (Fig. 6-3a).

Noncylindrical fold A fold that cannot be generated by a straight line moving parallel to itself in space (Fig. 6-3b).

Cylindroidal folds Folds that are approximately cylindrical; although no real fold is perfectly cylindrical, many folds approximate the cylindrical condition quite closely at least for part of their length. For them it is common to use the term "fold axis," even though, strictly speaking, only cylindrical folds can have a fold axis.

Fold axis The straight line that generates a cylindrical fold. Unlike the hinge line, the fold axis is not a specific line but rather a hypothetical line defined by its attitude. Only cylindrical folds, or cylindrical segments of folds, have fold axes.

Symmetric folds Folds that meet the following criteria: (1) the median surface is planar, (2) the axial plane is perpendicular to the median surface, and (3) the folds are bilaterally symmetrical about their axial planes (Fig. 6-4a).

Asymmetric folds Folds that are not symmetric (Fig. 6-4b).

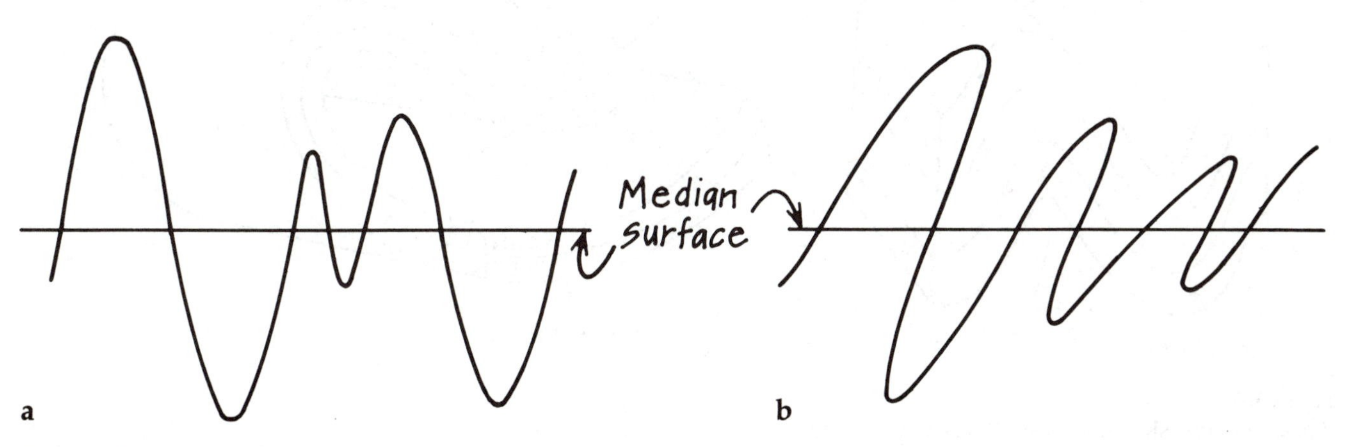

Figure 6-4
Symmetric (a) and asymmetric (b) folds with varying amplitudes.

Profile plane A plane perpendicular to the fold axis (Fig. 6-5).

Axial surface The surface joining the hinge lines of a set of nested folds (Fig. 6-5). Whether or not the folds are cylindrical, the axial surface may or may not be planar.

Axial plane A planar axial surface.

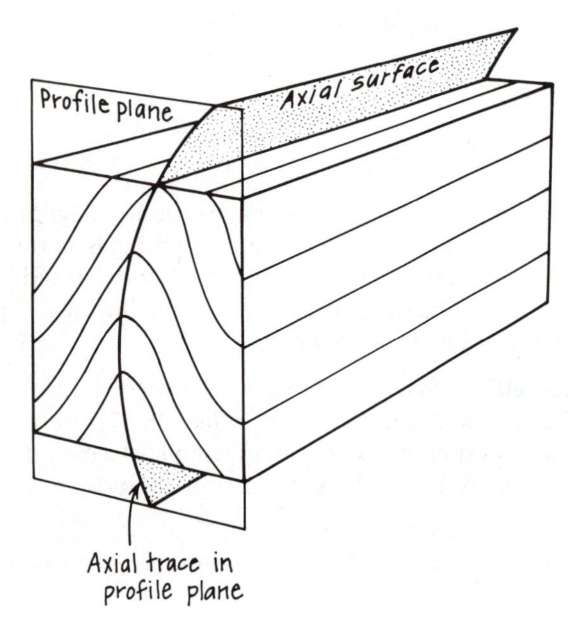

Figure 6-5
Profile plane and axial surface of folds.

Axial trace The line representing the intersection of the axial surface and another surface (Fig. 6-5).

Synform A fold that closes downward (Fig. 6-6).

Antiform A fold that closes upward (Fig. 6-6).

Syncline A fold with younger rocks in its core (Fig. 6-7).

Anticline A fold with older rocks in its core (Fig. 6-7).

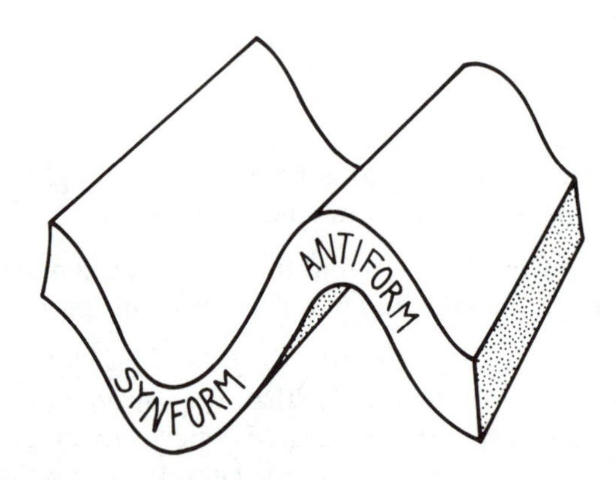

Figure 6-6
Block diagram showing synform and antiform.

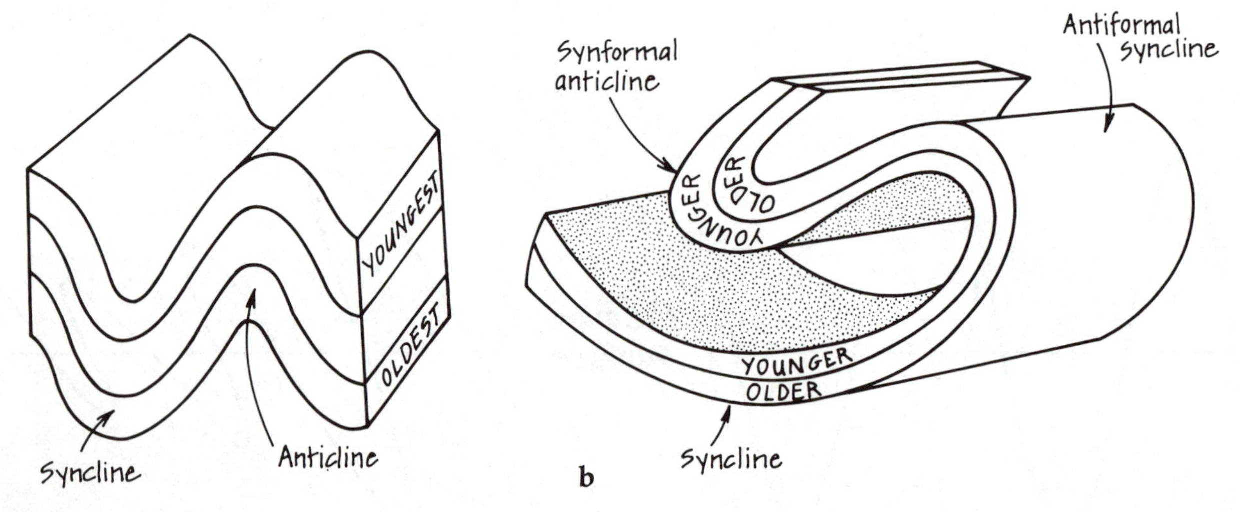

Figure 6-7
Block diagrams showing (a) synclines and anticlines and (b) how they can differ from synforms and antiforms.

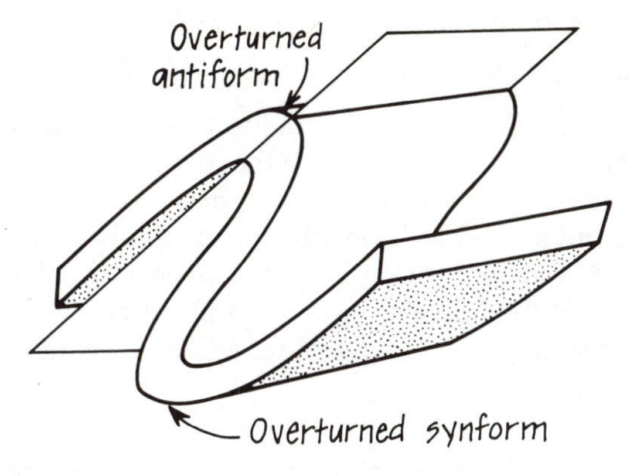

Figure 6-8
Block diagram showing overturned folds.

Overturned fold A fold in which one limb, and only one limb, has been tilted more than 90°, resulting in both limbs dipping the same direction (Fig. 6-8).

Vertical fold A fold whose hinge line is vertical or nearly so (Fig. 6-9a).

Reclined fold A fold whose axial surface is gently dipping (Fig. 6-9b).

Recumbent fold A fold whose axial surface is horizontal or nearly so (Fig. 6-9c).

Interlimb angle The angle between adjacent fold limbs. Fig. 6-10 shows terms used to describe this angle.

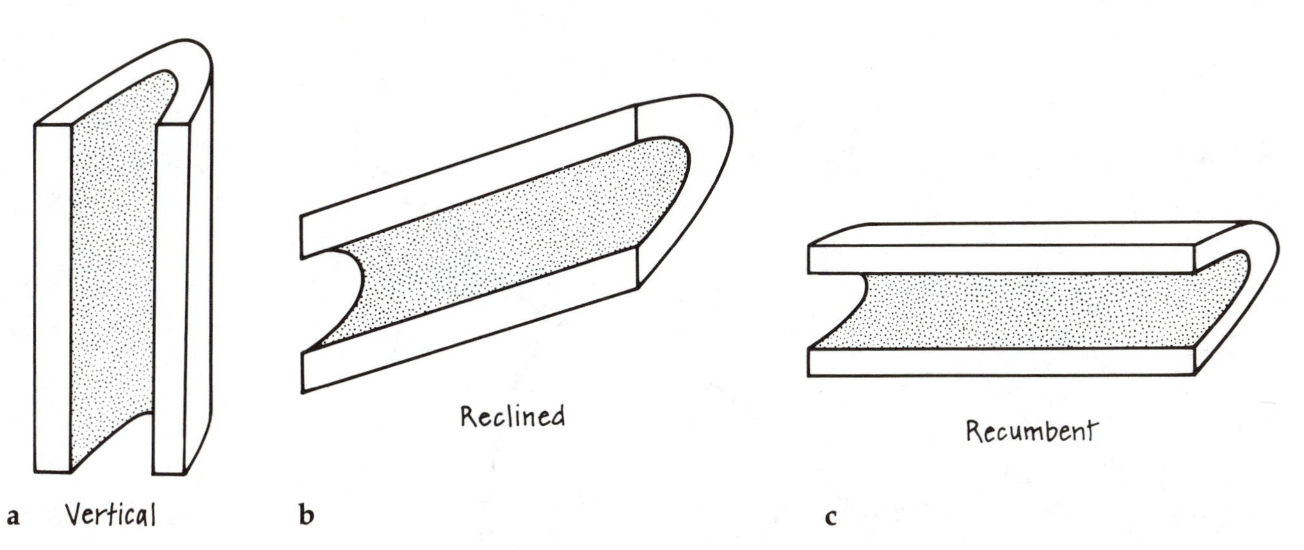

Figure 6-9
Vertical (a), reclined (b), and recumbent (c) folds.

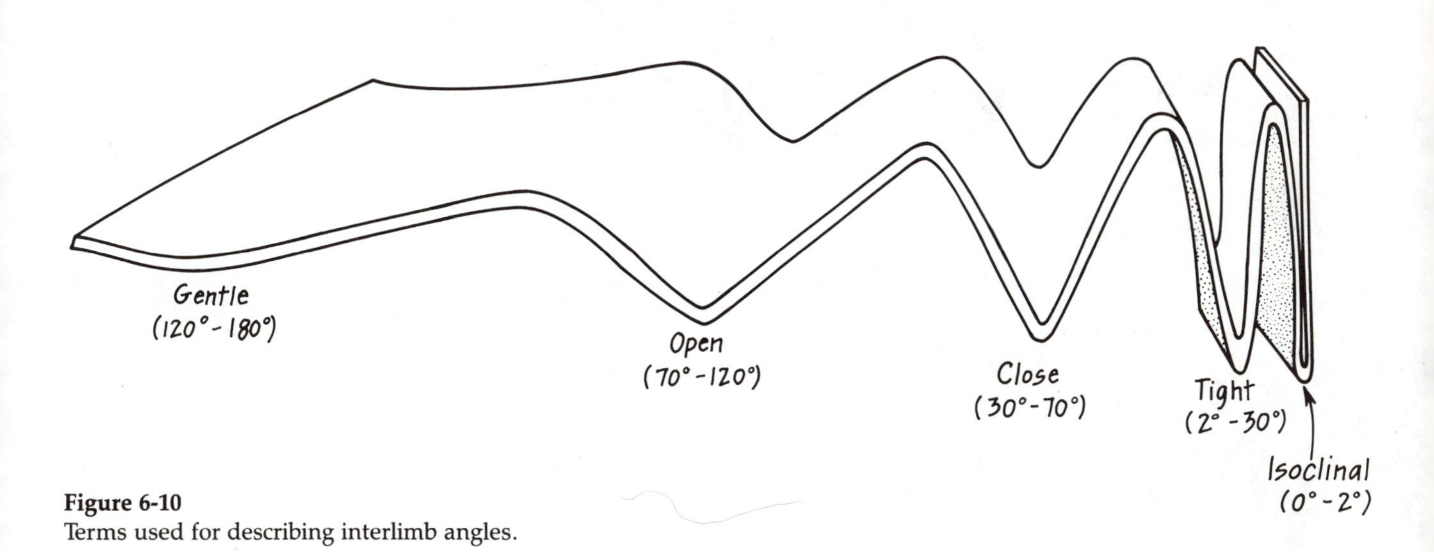

Figure 6-10
Terms used for describing interlimb angles.

Geometric axes Three orthogonal axes of a fold, used to refer to various lineations and foliations that accompany folding (Fig. 6-11). The *ab*-plane is the axial plane of the fold, and the *c*-axis is perpendicular to the axial plane. Joints that are parallel to the *ac*-plane, for example, may be described as *ac*-joints, and lineations that parallel the *b*-axis may be referred to as *b*-lineations.

Problem 6-1

In Figure 6-12 is a block model to be cut out or photocopied and folded. Describe the folds in this model. Your description should be one concise complete sentence and should include whether the folds are cylindrical or noncylindrical, symmetric or asymmetric, the attitude of the fold axis and the axial surface, and the interlimb angle. This same block will be used again in Problem 6-4.

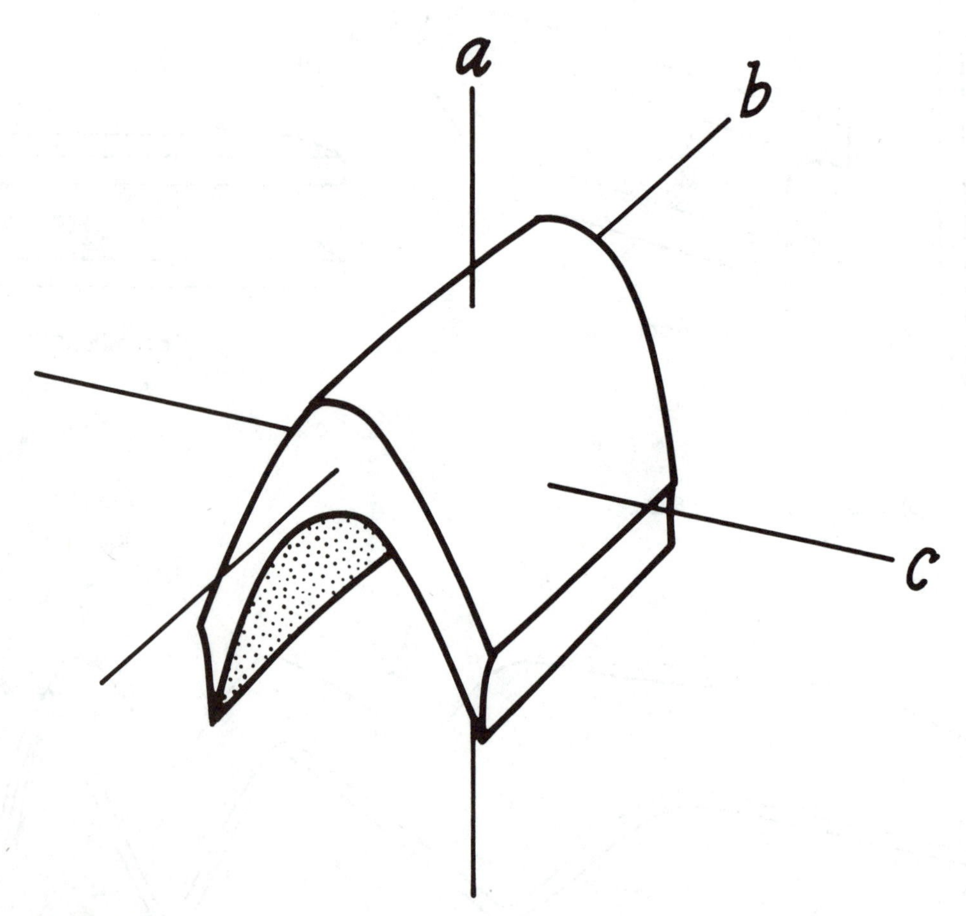

Figure 6-11
Geometrical axes of a fold.

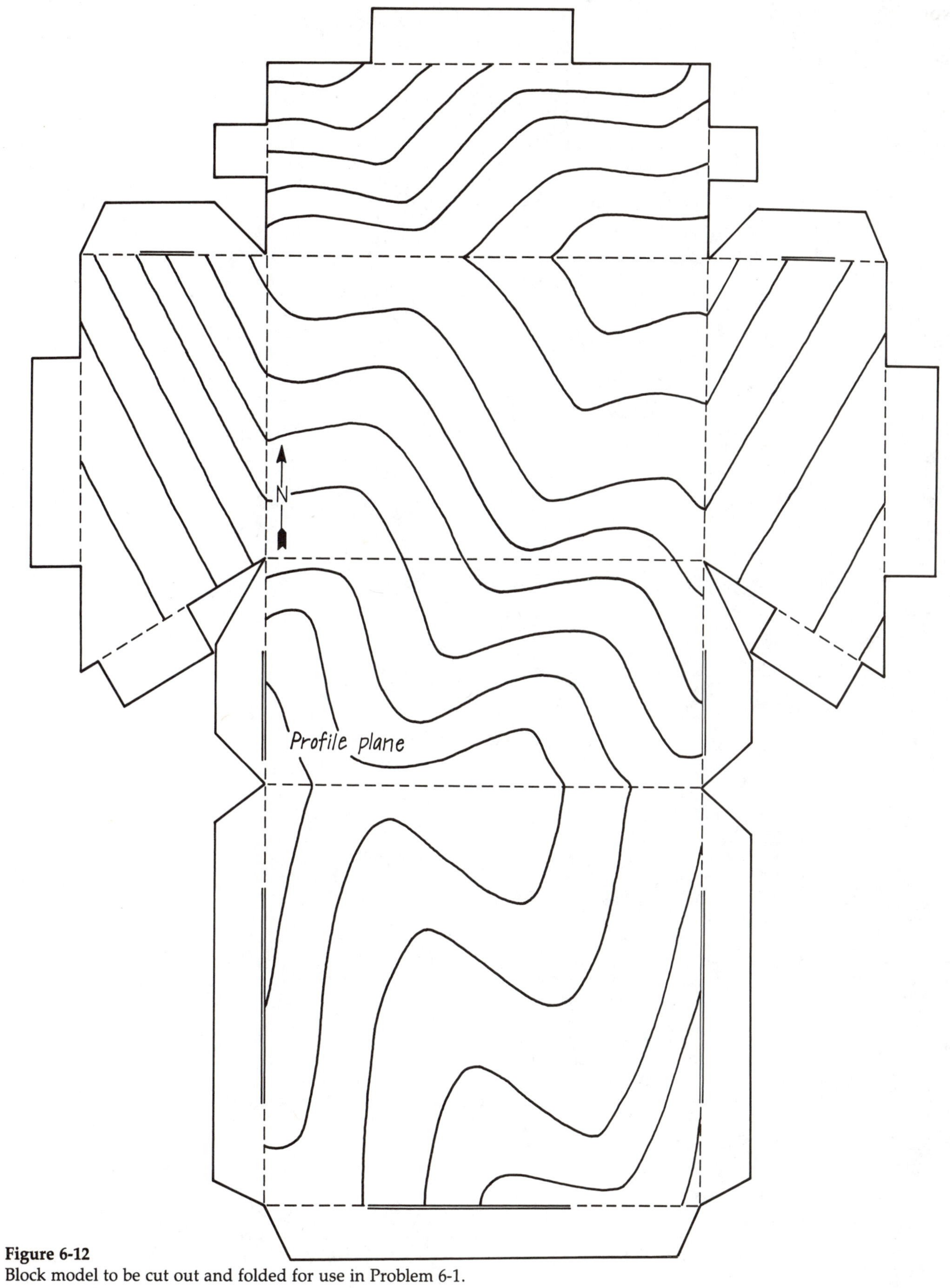

Figure 6-12
Block model to be cut out and folded for use in Problem 6-1.

Notes

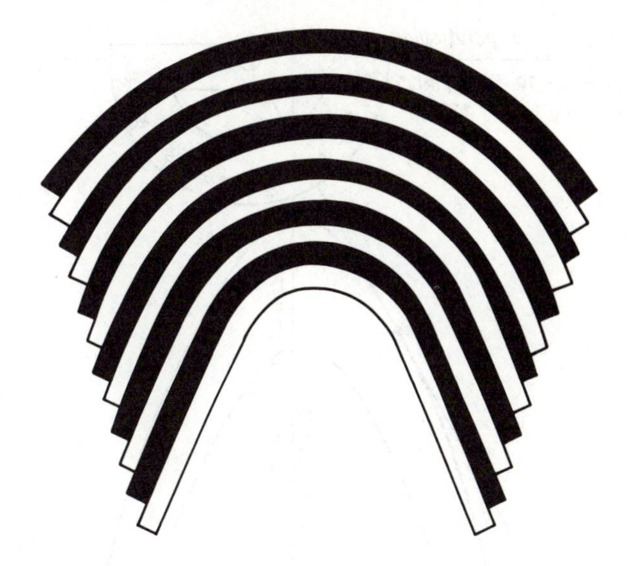

Figure 6-13
Development of concentric folds by flexure-slip folding.

Fold Classification Based on Dip Isogons

It was once thought that there are two basic processes for the development of folds in rocks and two types of folds produced. The first process, **flexure-slip folding,** involves slip between adjacent layers during buckling and bending (Fig. 6-13). The second process, **passive folding,** involves slip on surfaces that are at an angle to the layering in the rock (Fig. 6-14). Flexure-slip, it was thought, produces **concentric folds** (also called parallel folds) such as the ones you drew using the arc method in Chapter 4 (Problem 4-2). Folds produced by passive folding were called **similar folds.**

These two categories of folds were first described in 1896 and, though they may be intellectually satisfying, they have not proved to be useful for most naturally occurring folds. Nor has this system of classification been very helpful in our attempts at understanding the process of folding in rocks. A more promising fold classification based on **dip isogons** was proposed by J. G. Ramsay (1967) and has become widely used.

Dip isogons are lines connecting points of equal dip on adjacent folded surfaces. They are constructed as shown in Figure 6-15. The axial plane trace is drawn

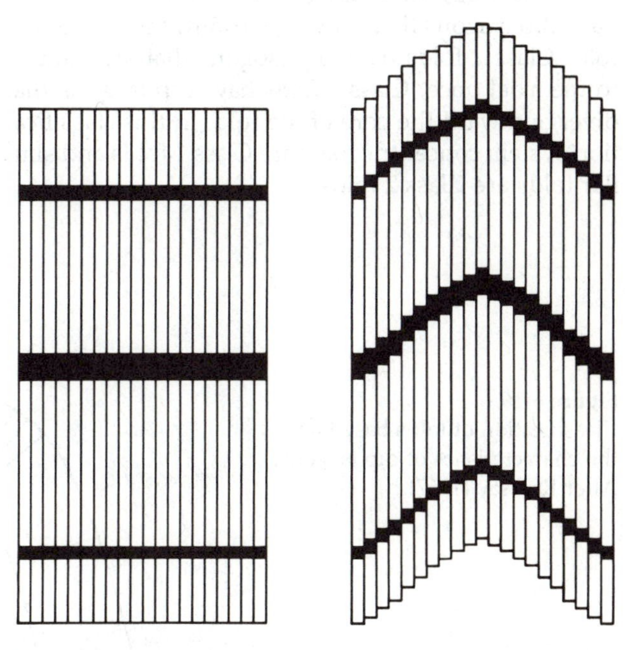

Figure 6-14
Development of similar folds by passive folding.

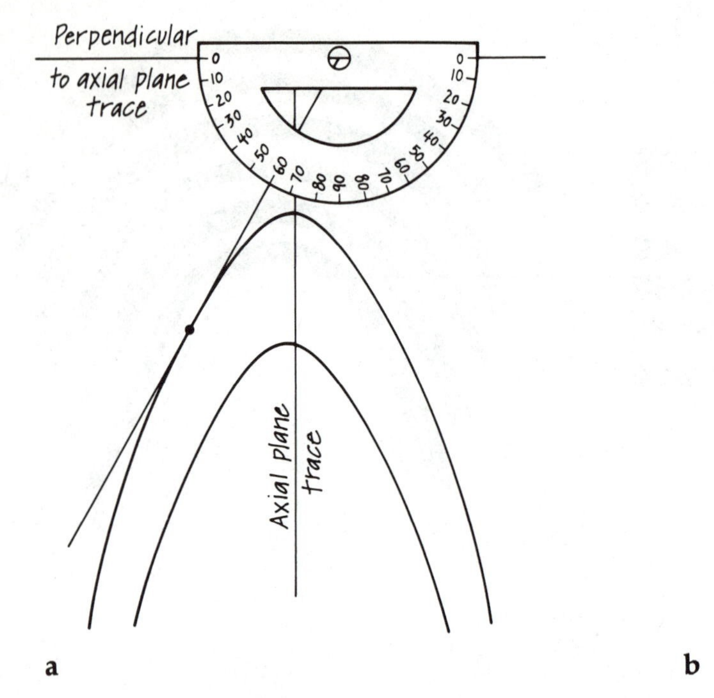

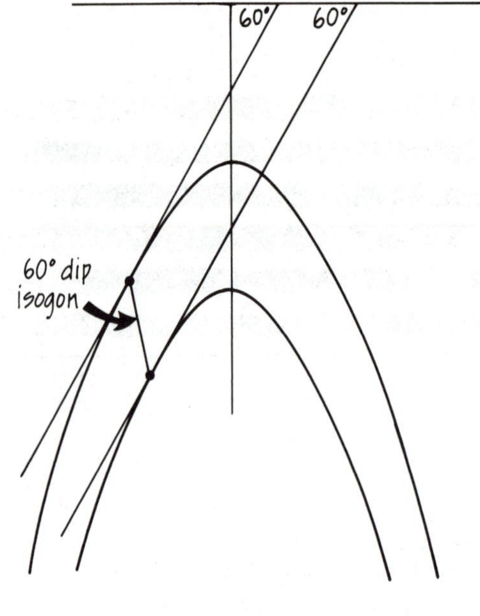

Figure 6-15
Construction of dip isogons. (a) Drawing tangents at a particular angle. (b) Drawing the isogon.

on a profile view of the fold, and another line is drawn perpendicular to the axial plane trace. With a protractor, points along the folded surface are located whose tangents intersect the horizontal line at specific angles (Fig. 6-15a). A set of points on adjacent folded surfaces permits the drawing of dip isogons (Fig. 6-15b).

From the characteristics of the dip isogons three classes of folds are defined (Fig. 6-16). Class 1 folds have dip isogons that converge toward the core of the fold. Class 2 folds have dip isogons that are parallel to the axial trace. Class 3 folds have dip isogons that diverge toward the core of the fold. In this classification system concentric folds are Class 1 folds and similar folds are Class 2 folds.

Problem 6-2

Figure 6-17 is a sketch of the profile view of a set of folds exposed in the face of a cliff. Draw dip isogons at 10° intervals for each of the three layers. Indicate the class of each layer.

Problem 6-3

Figure 6-18 contains photographs of four rock slabs. Without actually drawing dip isogons try to determine what types of folds are represented in each slab. Then tape a piece of tracing paper over the photographs and draw a few dip isogons on each fold. Indicate the class to which each fold belongs.

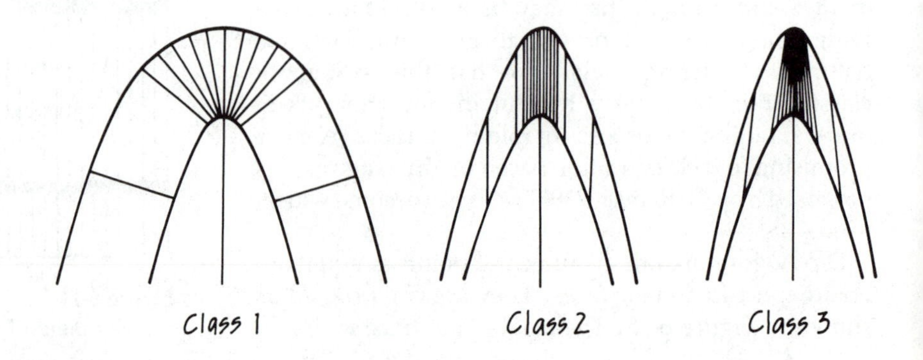

Figure 6-16
Classification of folds based on the characteristics of dip isogons. After Ramsay (1967).

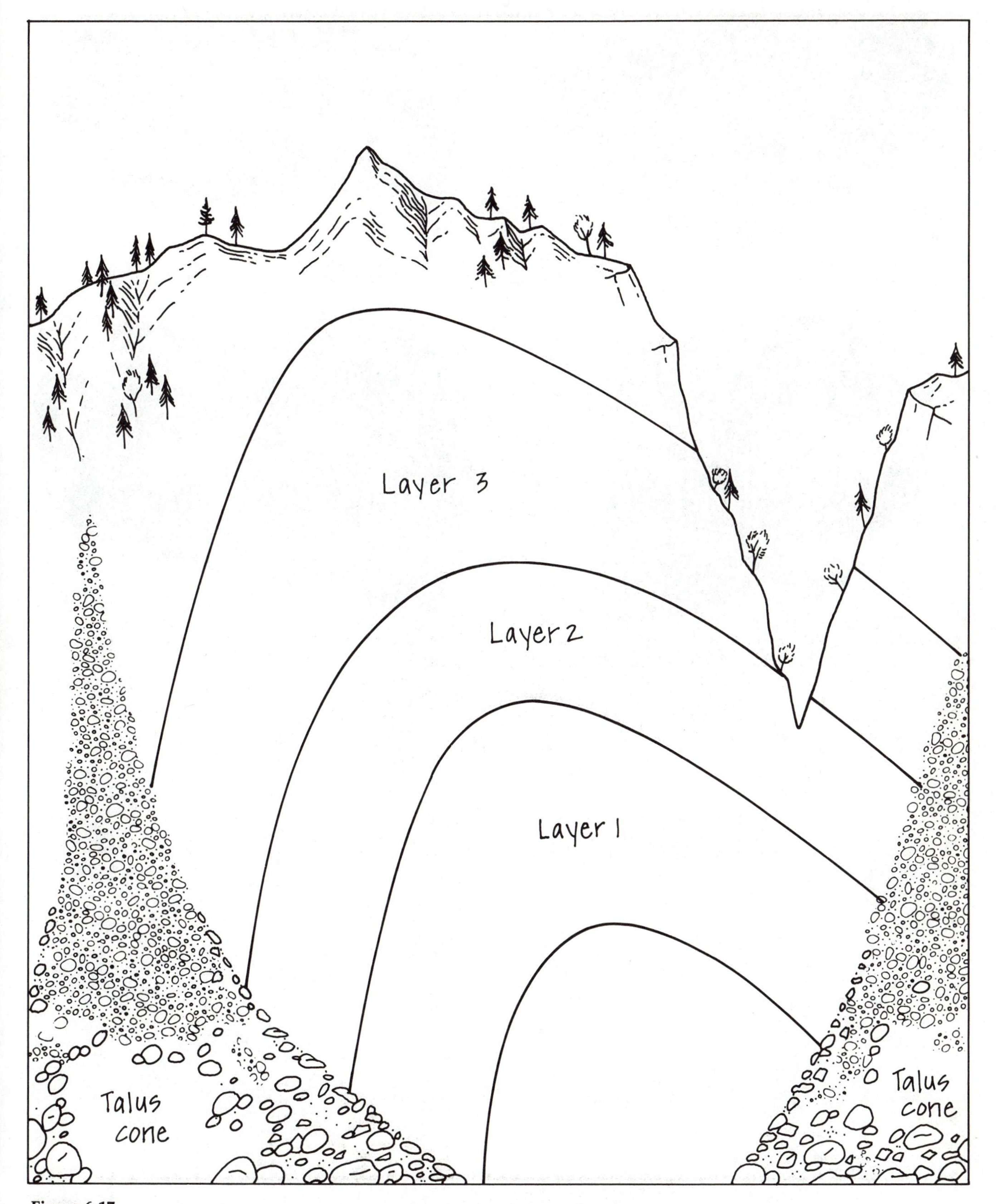

Figure 6-17
Drawing of folds to be used in Problem 6-2. Adapted from *Internal Processes,* The Open University (1972), Fig. 99.

Figure 6-18
Slabs of folds for use in Problem 6-3. From the collection of O. T. Tobisch.

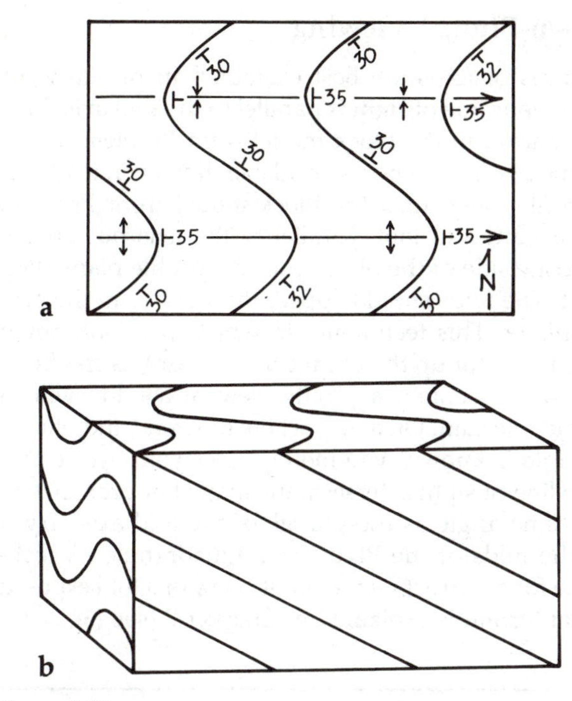

Figure 6-19
Eastward-plunging folds. (a) Map view. (b) Block diagram.

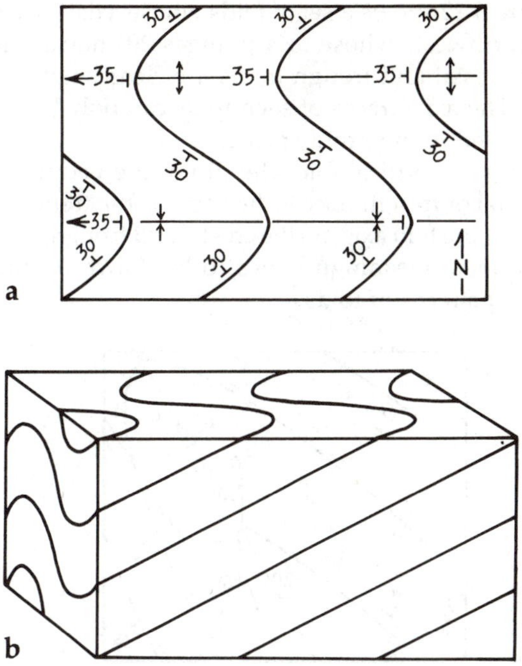

Figure 6-20
Westward-plunging folds. (a) Map view. (b) Block diagram.

Outcrop Patterns of Folds

Figures 6-19 and 6-20 show sets of folds in which the outcrop patterns are identical, but the folds are plunging in opposite directions. Clearly, outcrop pattern alone is not sufficient to determine the orientation of a fold. The direction and amount of dip must also be determined. Symmetric folds with vertical axial surfaces, such as those in Figures 6-19 and 6-20, are symmetrical on opposite sides of the crestal and trough traces. The crestal and trough traces of such folds are also axial traces, and the dip at the axial trace is equal to the plunge of the fold axis. The folds shown in Figure 6-19 plunge 35° due east, and those in Figure 6-20 plunge 35° due west.

Figure 6-21 shows an outcrop pattern identical to that in Figures 6-19 and 6-20, but here the folds are overturned and plunging 45° south. Fold a piece of paper and tilt it to simulate one of the folded layers in Figure 6-21. Notice that these folds, and all overturned folds, contain vertical beds—the strike of the vertical beds in overturned folds is parallel to the trend of the fold axis.

Because the folds in Figure 6-21 have no unique high and low points, they have no crestal and trough traces. In most cases of plunging folds with tilted axial surfaces, the crestal and trough traces are not axial traces. Sometimes the crestal and trough traces are not even parallel to the axial traces.

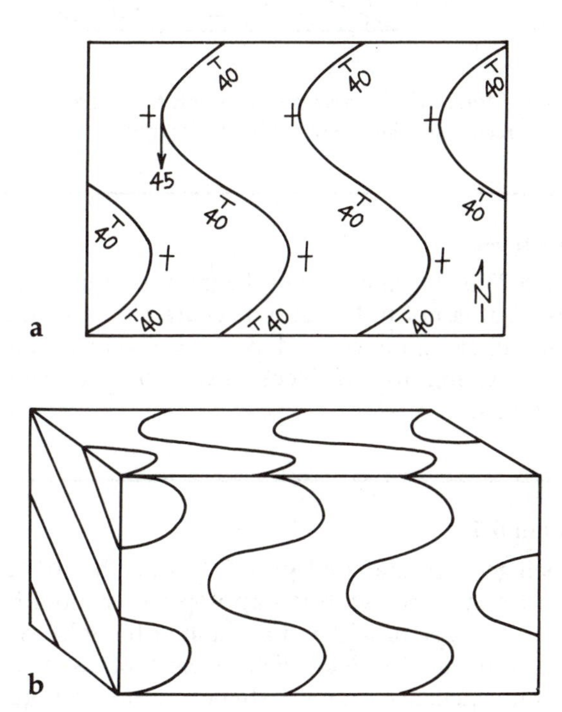

Figure 6-21
Southward-plunging folds. (a) Map view. (b) Block diagram.

Figure 6-22 shows a set of folds whose axial surface dips northwest, whose axis plunges 20° north, and whose crestal and trough traces are clearly not axial traces. The axial traces of such folds can only be reliably located in the profile plane.

For any cylindrical fold, the dip of the bedding at the crestal or trough trace is the same as the trend and plunge of the fold axis. So the crestal and trough traces are the most meaningful lines to be drawn on map outcrop patterns of folds.

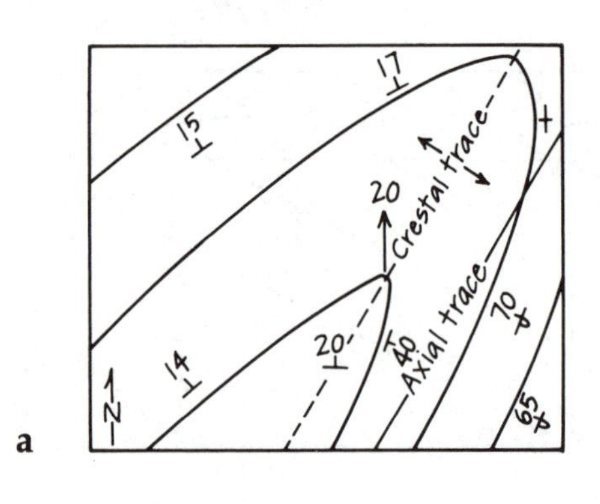

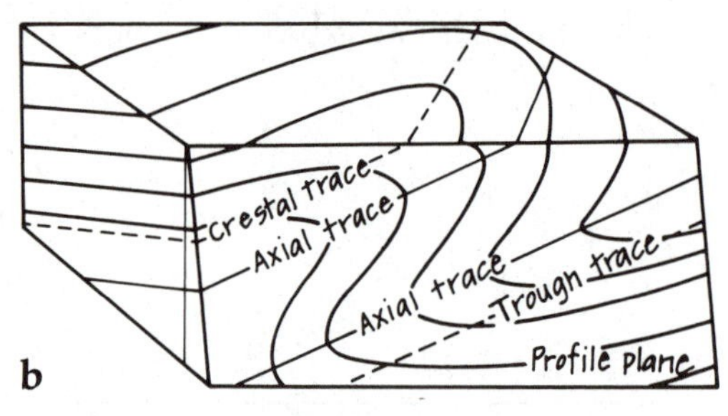

Figure 6-22
Example of folds whose crestal and trough traces are not axial traces. (a) Map view. (b) Block diagram.

Problem 6-4

Figure 6-23 is the same layout diagram used in Problem 6-1 but is not to be cut out. Draw and label the crestal and trough traces and axial traces on this diagram, referring to the block that you've already constructed.

Problem 6-5

On each geologic map on Figure 6-24 draw the crestal and/or trough trace with the appropriate symbol to indicate the type of fold and attitude of the fold axis (symbols are in Fig. 3-13). Fold a piece of paper to help you visualize each fold's shape and orientation.

Determine the exact trend and plunge of the fold axis on each map and write it in the space provided.

Down-Plunge Viewing

Features of folds are best examined in profile view, when your line of sight is parallel to the fold axis. This is apparent in the block model from Problem 6-1. A profile plane need not be available, however, to obtain a profile view. Turn the block model from Problem 6-1 around and look parallel to the axis but on the opposite side of the block from the profile plane (Fig. 6-25). The folds should appear the same as in the profile plane. This technique, in which you look down the plunge (or up the plunge in this case), is an effortless way to obtain a profile view of a fold even in irregular terrain. On a map where the trend and plunge of a fold is known, you merely place your eye so that your line of sight intersects the map at approximately the same angle as the plunge of the fold axis. Try it on the folds of the Bree Creek Quadrangle. A technique for constructing the profile view of a fold exposed in flat terrain is explained in Chapter 7 (see Fig. 7-6).

Problem 6-6

Figure 6-26 shows the sides and top of a block model of folded layered rocks. Photocopy it or cut it out, and fold it into a block. Look at the block from different angles until you see a set of cylindrical folds. At this point you are looking down-plunge. If you have trouble, try coloring one or more units.

1. Make a drawing of the block as it appears in the down-plunge view, showing the folds.
2. What is the approximate attitude of the axial surfaces?
3. What is the approximate attitude of the fold axis?
4. By comparing the folds with those in Figure 6-16, determine the classes of folds in the block model and label each fold on your drawing accordingly.

Further Reading

Marjoribanks, R. W. 1974. "An instrument for measuring dip isogons and fold layer shape parameters." *Journal of Geological Education,* v. 22, 62–64. Design of an isogon plotter and discussion of procedure for measuring fold parameters.

Ramsay, J. G. 1967. *Folding and Fracturing of Rocks.* New York: McGraw-Hill. A standard reference on the geometry and mechanisms of folding.

Suppe, J. 1985. *Principles of Structural Geology.* Englewood Cliffs, NJ: Prentice-Hall. Contains a good discussion of fold mechanics and mechanisms and many photographs and drawings of folds.

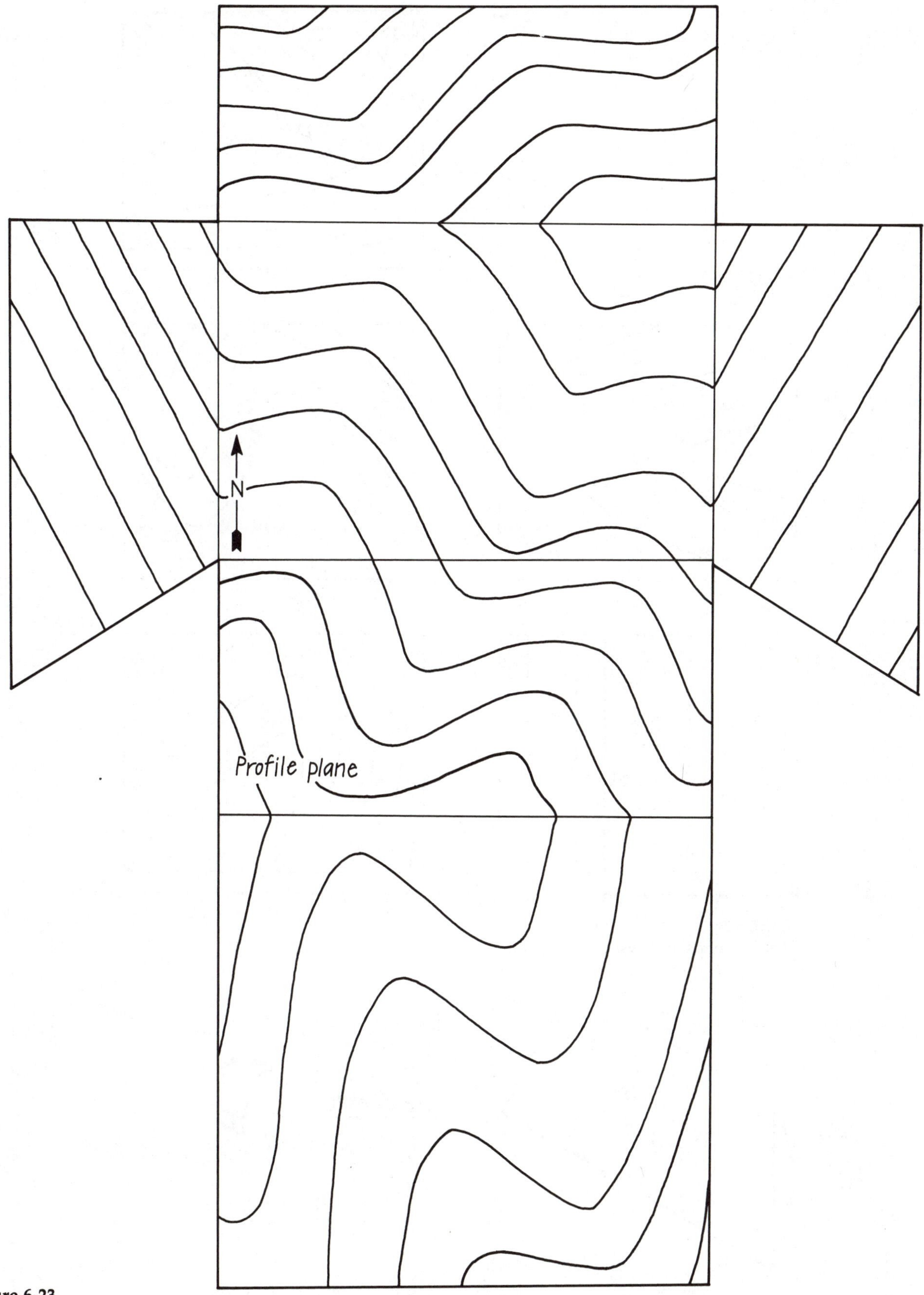

Figure 6-23
Layout of block diagram to be used in Problem 6-4. Use this diagram in conjunction with the block constructed in Problem 6-1.

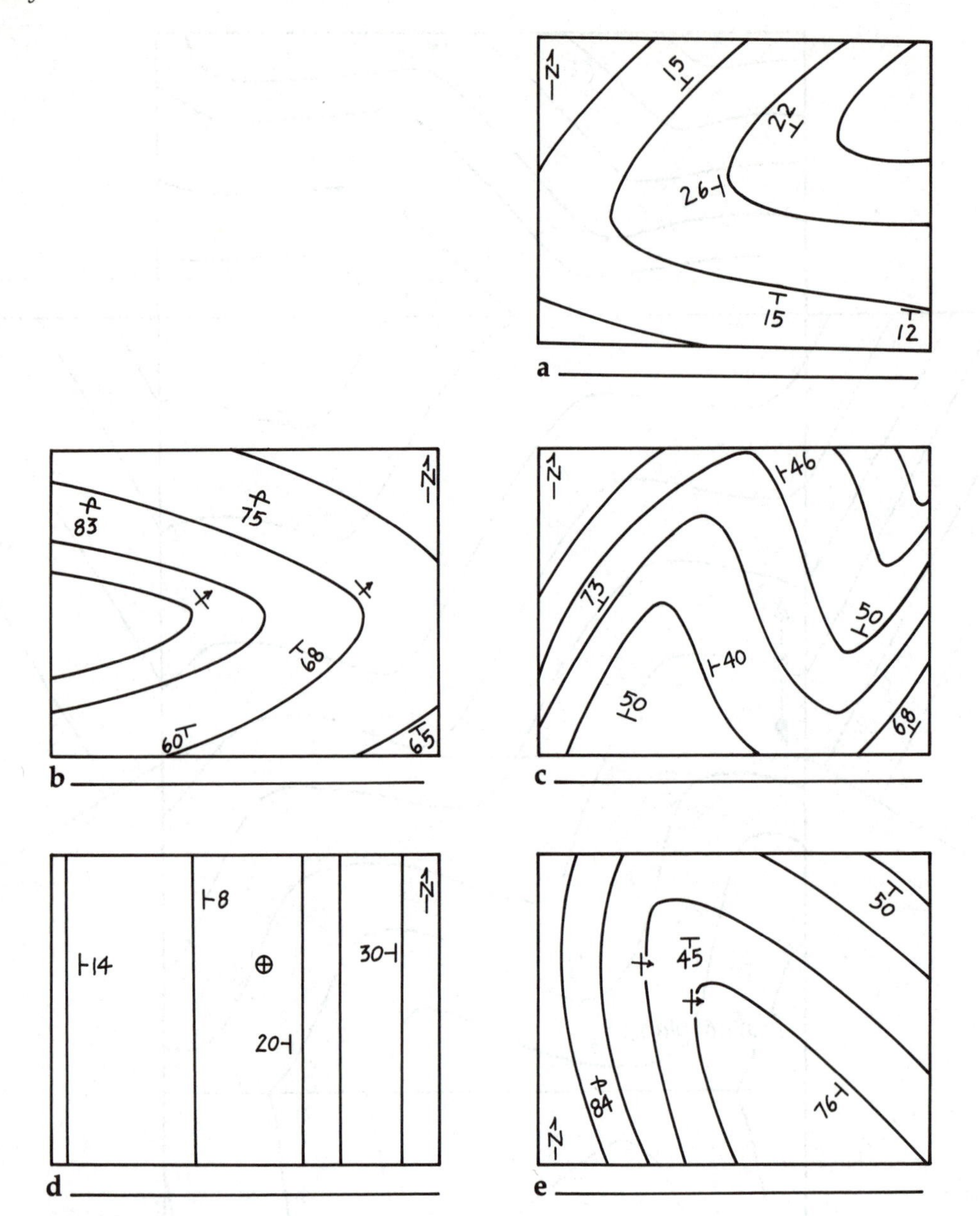

Figure 6-24
Geologic maps to be used in Problem 6-5.

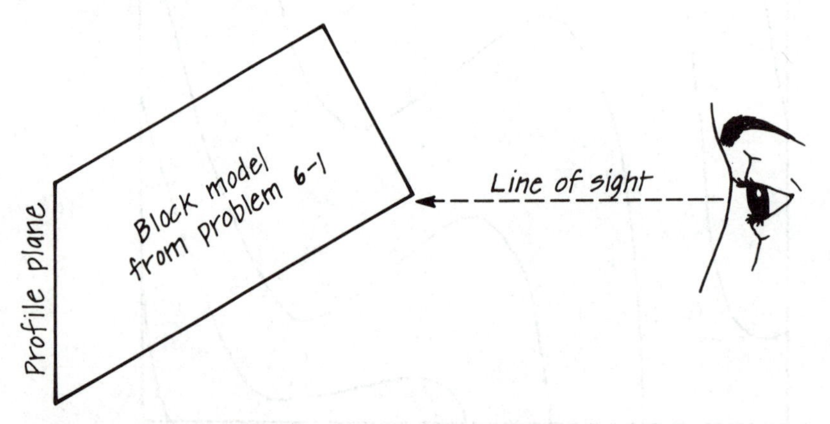

Figure 6-25
Illustration of down-plunge viewing technique for obtaining a profile view of a fold.

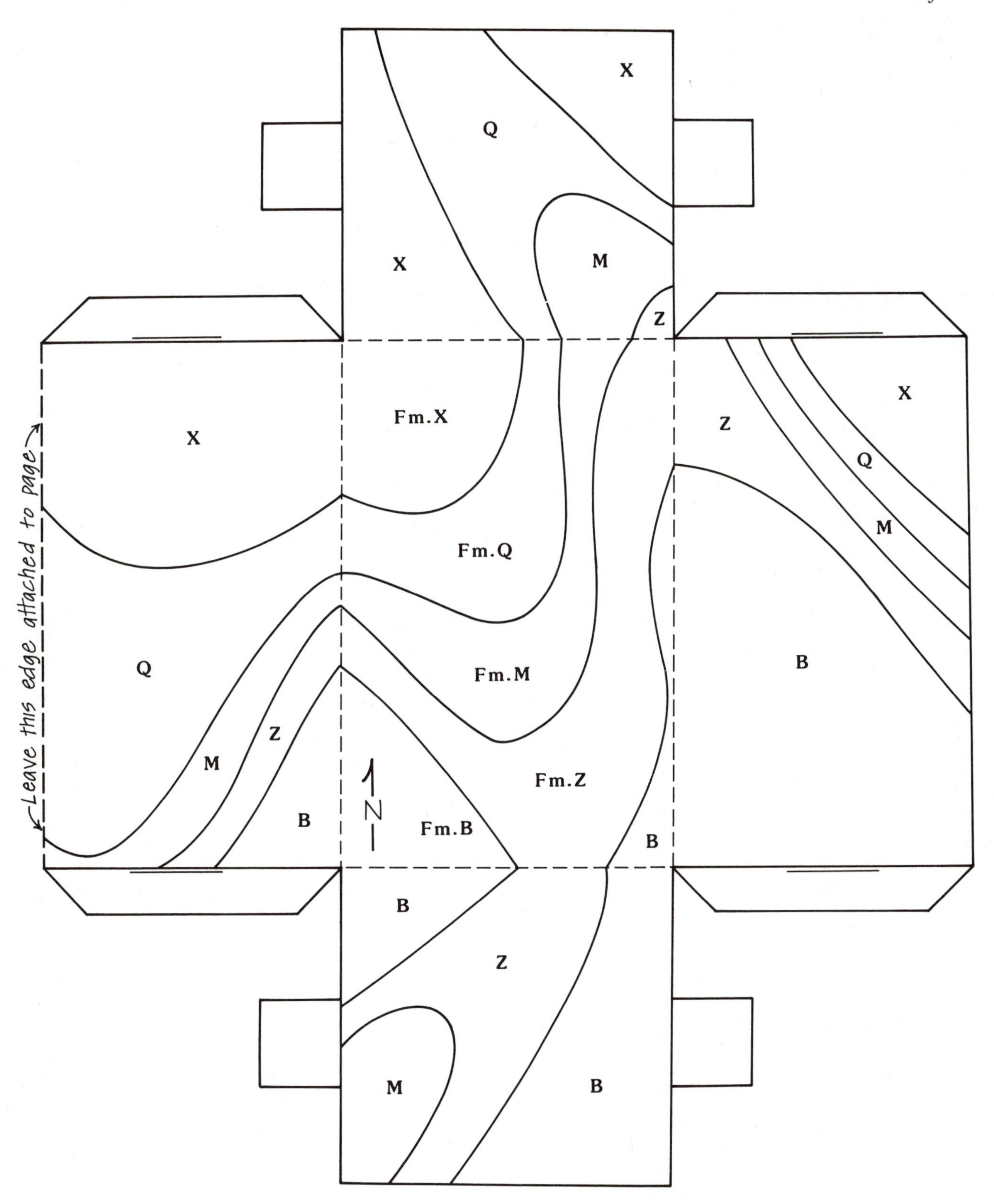

Figure 6-26
Layout of block diagram to be constructed in Problem 6-6. After Dahlstrom (1954) in Whitten (1966).

7

Stereographic Analysis of Folded Rocks

OBJECTIVES

Construct beta diagrams and pi diagrams of folds

Construct and interpret a contoured equal-area diagram of structural data

In situations where the rocks are not well exposed, it may be impossible to analyze folds using the techniques discussed in Chapter 6. By stereographic projection, data from isolated outcrops may be combined to characterize the fold geometry. Stereographic analysis of the folded Paleogene rocks of the Bree Creek Quadrangle will serve as the major exercise in this chapter.

While the techniques described in this chapter will be applied specifically to the analysis of folds, these same techniques are also commonly used for analyzing such features as joints and cleavage.

Beta Diagrams

A simple method for determining the orientation of the axis of a cylindrical (or cylindroidal) fold is to construct a beta diagram (or β-diagram). Any two planes tangent to a folded surface intersect in a line that is parallel to the fold axis (Fig. 7-1). Such a line is called a β-axis. A β-axis is found by plotting attitudes of foliations (such as bedding or cleavage) on an equal-area net; the β-axis is the intersection point of the planes.

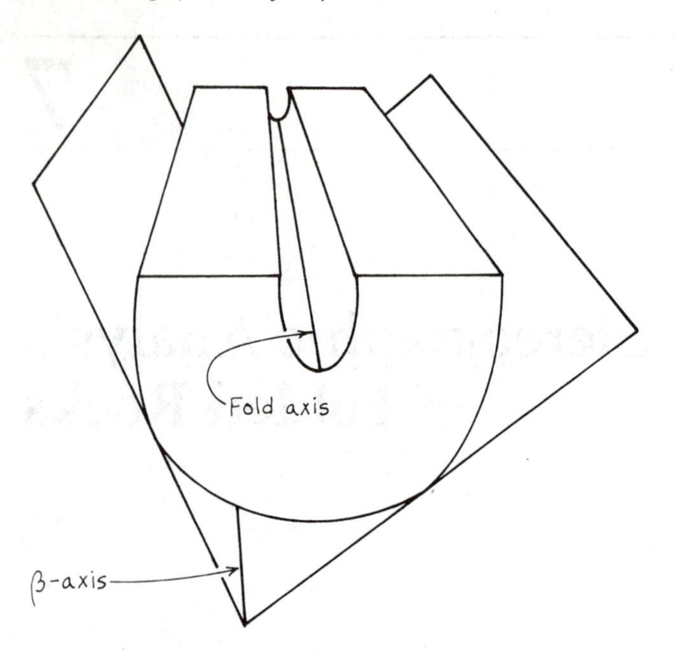

Figure 7-1
Beta axis as the intersection of planes tangent to a folded surface.

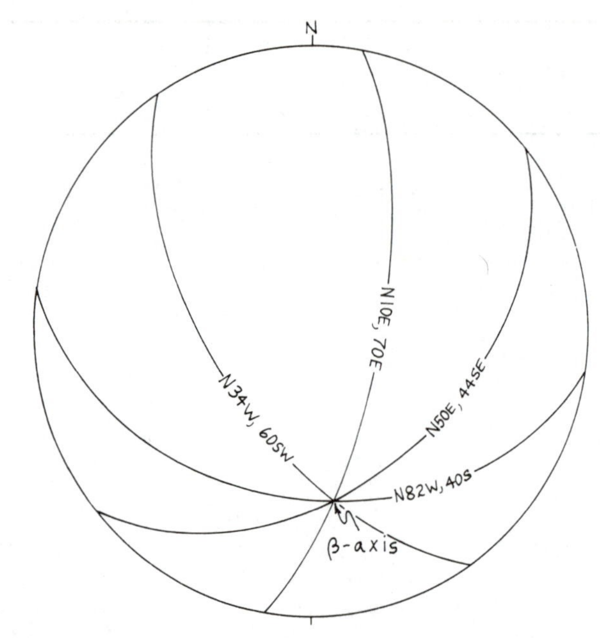

Figure 7-2
Great circles representing four foliation attitudes of a cylindrical fold intersect at the β-axis.

Suppose, for example, that the following four foliation attitudes are measured: N82W, 40S; N10E, 70E; N34W, 60SW; N50E, 44SE. These attitudes are shown plotted on an equal-area net in Figure 7-2. The β-axis, and therefore the fold axis, plunges 39, S7E.

Few folds are perfectly cylindrical, and usually the great circles will not intersect perfectly. If data from areas with different folding histories are plotted together, different β-axes will appear.

Pi Diagrams

A less tedious method of plotting large numbers of attitudes is to plot the poles (called s-poles) of the foliations (s-surfaces) on an equal-area net. In a cylindrical fold these s-poles will lie on one great circle, called the pi circle (π-circle). The pole to the π-circle is the π-axis, which like the β-axis is parallel to the fold axis. Figure 7-3 shows the same four attitudes that were plotted on the β-diagram in Figure 7-2, and the corresponding π-circle and π-axis. In most cases π-diagrams are more revealing, as well as more quickly constructed, than β-diagrams.

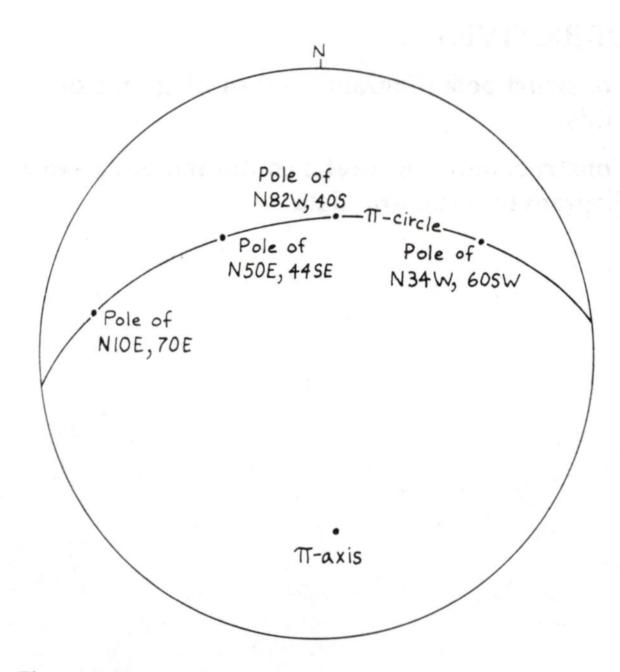

Figure 7-3
Pi circle as the great circle common to four poles of foliations of a cylindrical fold.

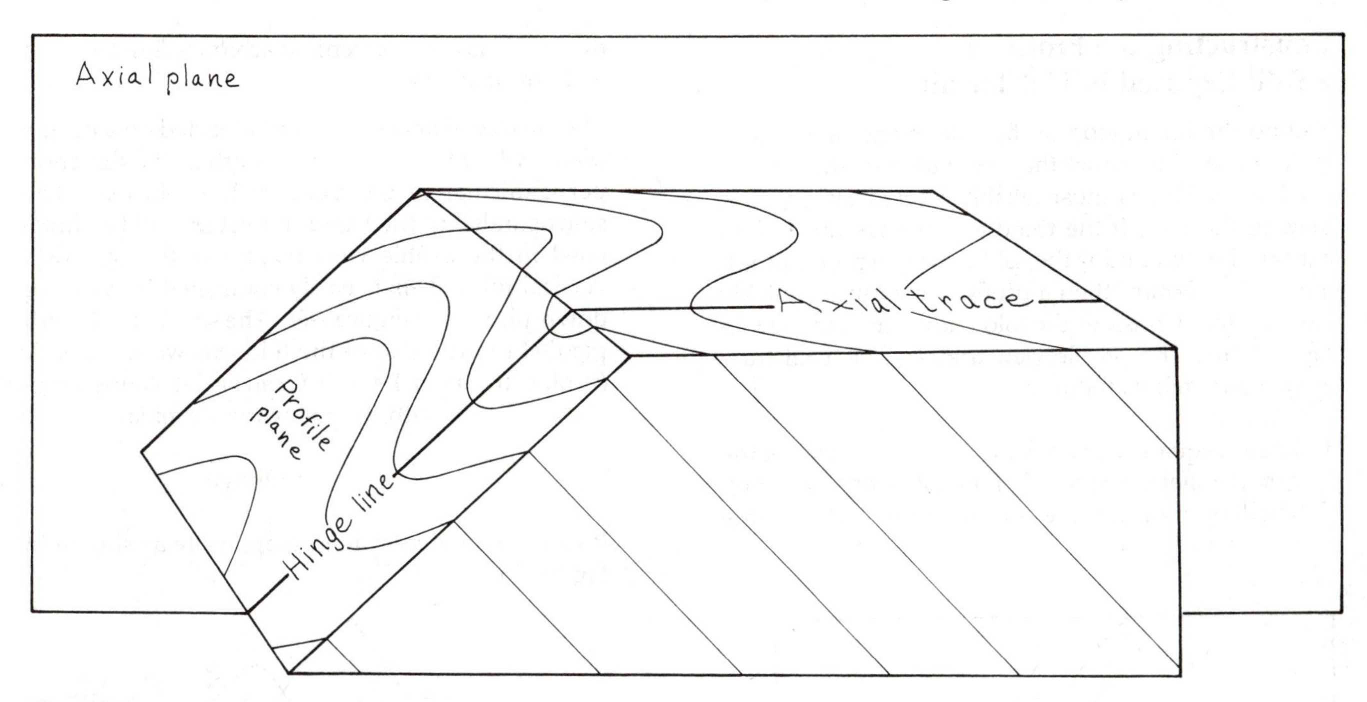

Figure 7-4
Block diagram of folds showing profile plane and axial plane.

Determining the Orientation of the Axial Plane

The orientation of a fold is defined not only by the trend and plunge of the fold axis, but also by the attitude of the axial plane. The axial plane can be thought of as a set of coplanar lines, one of which is the hinge line and another the surface axial trace (Fig. 7-4). If the axial trace can be located on a geologic map, and if the trend and plunge of the hinge line can be determined, then the orientation of the axial plane can easily be determined stereographically: it is the great circle that passes through the π-axis (or β-axis) and the two points on the primitive circle that represent the surface axial trace (Fig. 7-5).

Often the surface axial trace cannot be reliably located on a geologic map. If the folds are well exposed, then a profile view can be constructed (as explained in the next section), and the axial trace can be located in the profile plane and transferred to the geologic map.

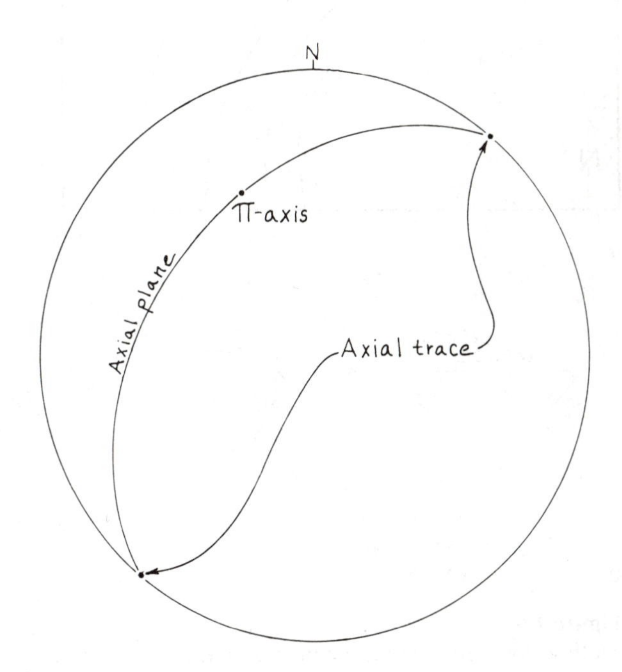

Figure 7-5
Equal-area net plot. Axial plane is the great circle common to the π-axis and the two points on the primitive circle that represent the axial trace.

Constructing the Profile of a Fold Exposed in Flat Terrain

To find the orientation of the axial plane of a fold, it is very useful to know the orientation of the surface axial trace. This is most reliably located on a profile view of the fold. If the trend and plunge of the fold axis are known and if the fold is well exposed in relatively flat terrain, then a profile view can be quickly constructed. Consider the fold shown in plan view in Figure 7-6a. The profile view and surface axial trace are constructed as follows:

1. Draw a square grid on the map with one axis of the grid parallel to the trend of the fold axis (Fig. 7-6b). The length of the sides of each square, d_s (surface distance), is any convenient arbitrary length, such as 1 cm or 10 cm.
2. The surface distance d_s when projected on a profile plane will remain the same length in the direction perpendicular to the trend of the fold axis. The sides parallel to the trend, however, will be shortened in the profile view (except in the case of a vertical fold). This is easily confirmed by viewing down-plunge on Figure 7-6b. The shortened length parallel to the trend of the fold axis we will call d_p (profile distance). Length d_p can be determined trigonometrically with the following formula:

$$d_p = d_s \sin \text{plunge}$$

It can also be determined graphically as shown in Figure 7-6c.

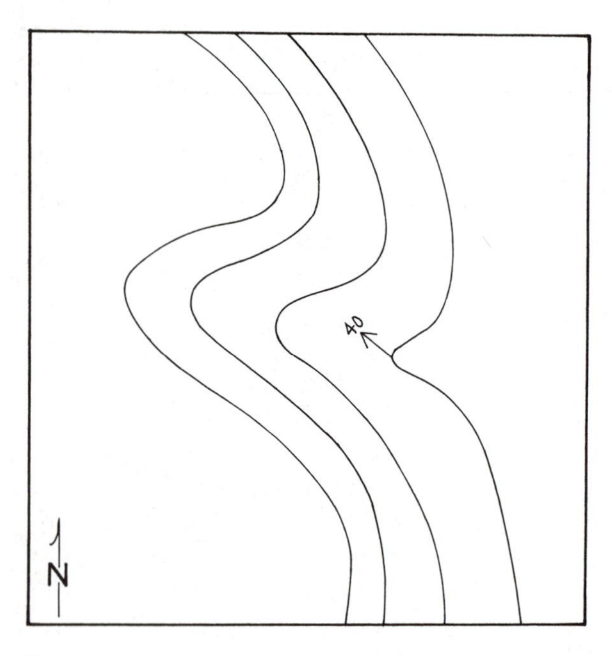

a

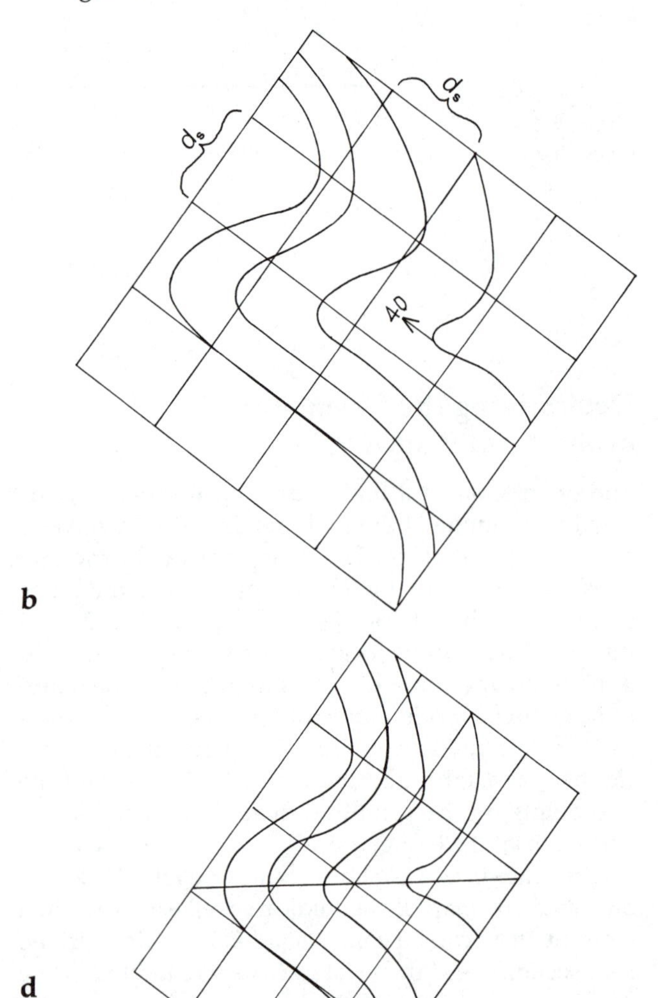

b

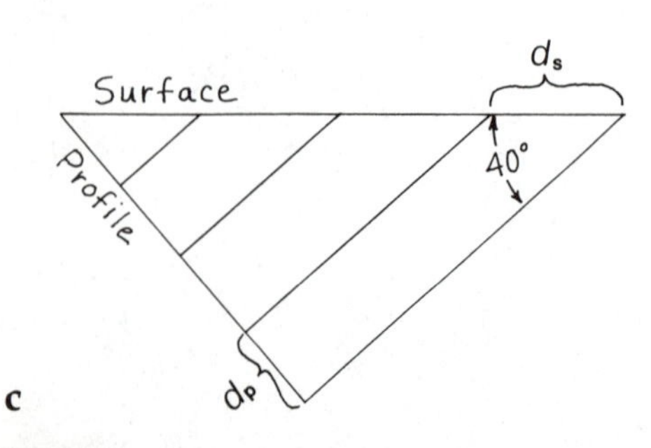

c

d

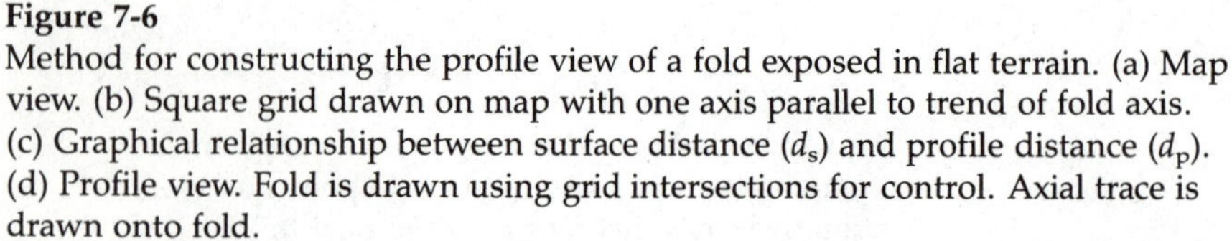

Figure 7-6
Method for constructing the profile view of a fold exposed in flat terrain. (a) Map view. (b) Square grid drawn on map with one axis parallel to trend of fold axis. (c) Graphical relationship between surface distance (d_s) and profile distance (d_p). (d) Profile view. Fold is drawn using grid intersections for control. Axial trace is drawn onto fold.

3. With d_p now determined, a rectangular grid is drawn which represents the square grid projected onto the profile plane. Points on the square grid in the map view are then transferred to corresponding points on the rectangular grid. The profile view of the fold is then sketched freehand, using the transferred points for control (Fig. 7-6d).
4. The axial trace can now be drawn on the profile plane (Fig. 7-6d), and then transferred back to corresponding points on the square grid.

Simple Equal-Area Diagrams of Fold Orientation

The orientation of a fold can be simply and clearly characterized by an equal-area diagram showing the axial plane and fold axis. Examples of various folds are shown in Figure 7-7. Picture in your mind's eye what each of these folds would look like. What characteristics of a fold are *not* displayed in such a diagram?

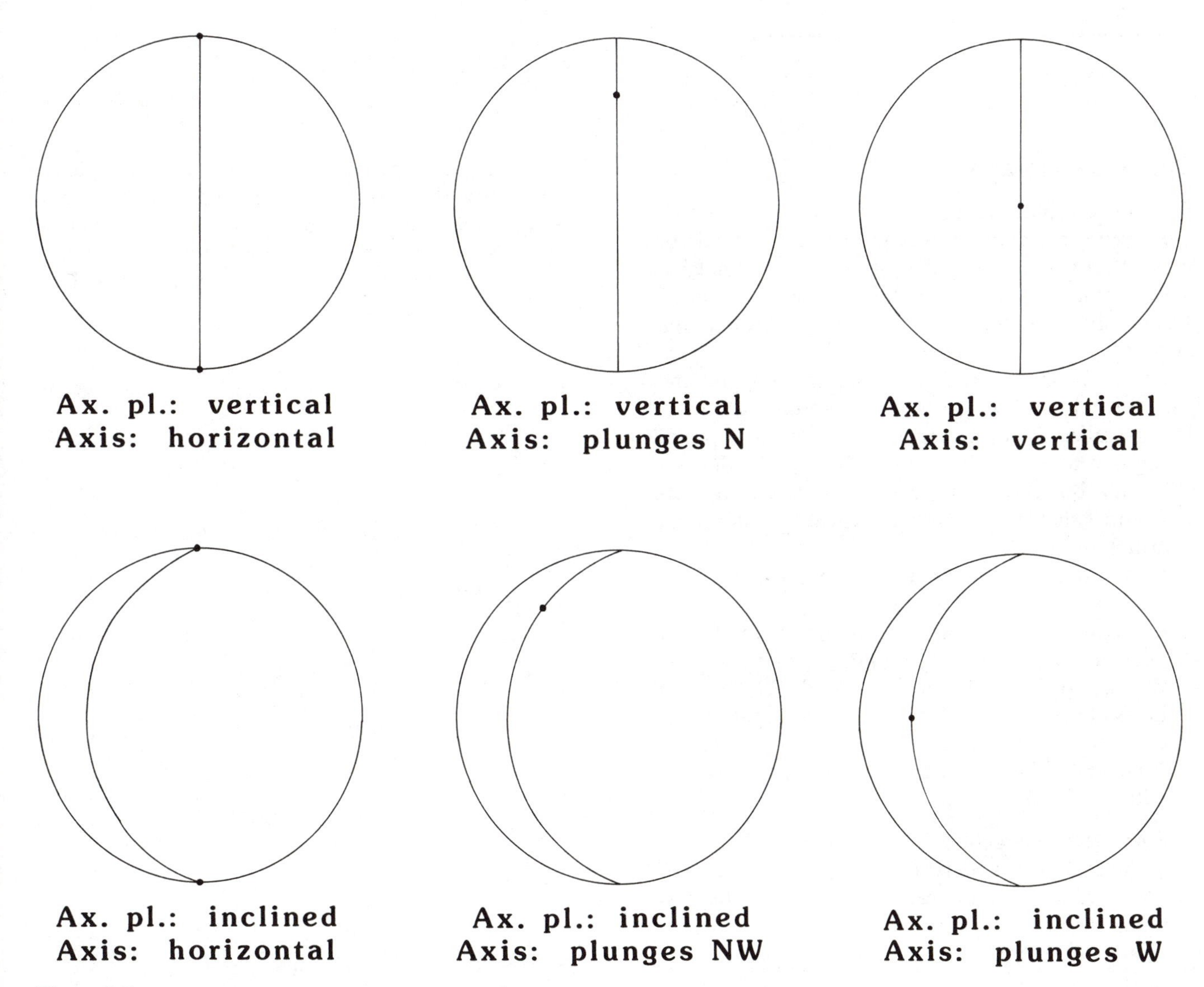

Figure 7-7
Simple equal-area diagrams showing orientation of folds. AP is axial plane.

Problem 7-1

1. On separate pieces of tracing paper construct a β-diagram and π-diagram for the folds in Figure 7-8. Determine the trend and plunge of the fold axis.
2. Construct a profile view as shown in Figure 7-6, and draw surface axial traces on the plan view. Determine the strike of the axial plane.
3. Draw a simple equal-area diagram such as those in Figure 7-7 showing the orientation of these folds.
4. Concisely but completely describe these folds. Include the attitude of the fold axis, attitude of the axial plane, interlimb angle, symmetry, and fold class.

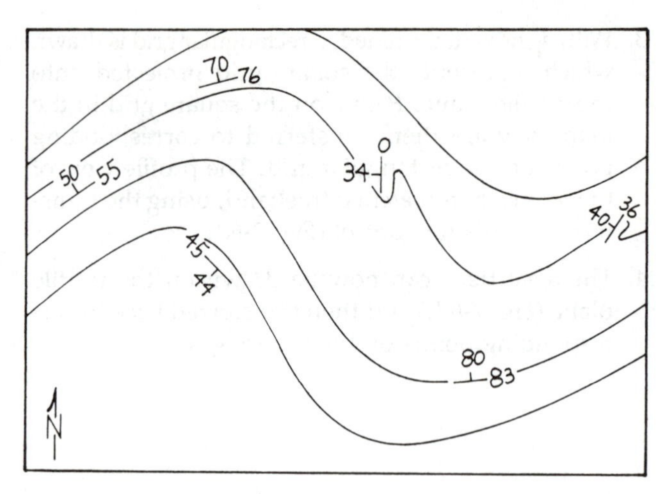

Figure 7-8
Geologic map for use in Problem 7-1.

Contour Diagrams

Because real folds are not exactly cylindrical, when a β-diagram or π-diagram is made no single β-axis or π-axis emerges. If a large number of data are available the orientation of the hinge line can be determined statistically through the use of a contoured equal-area diagram.

Figure 7-9 is an equal-area diagram showing the piercing points of the poles of 50 bedding attitudes. This is called a point diagram. You could approximately locate a π-circle through the highest density of points, but contouring makes the results repeatable and reliable, as well as providing additional information.

A point diagram is contoured as follows:

1. Cut out or photocopy the center counter and peripheral counter in Figure 7-15 at the end of this chapter. With a razor blade, carefully cut out the holes in the counters and cut a slit in the peripheral counter as indicated. The holes in the counters are 1 percent of the area of the equal-area net provided with this book.
2. Remove or photocopy the grid in Figure 7-16, also at the end of this chapter, and tape it to a piece of thin cardboard to increase its longevity. The distance between grid intersections is equal to the radius of the holes in the counters.
3. Tape the tracing paper containing the point diagram onto the grid such that the center of the point diagram lies on a grid intersection. Tape a second, clean piece of tracing paper over the point diagram. The two pieces of tracing paper do not move while you are counting points.

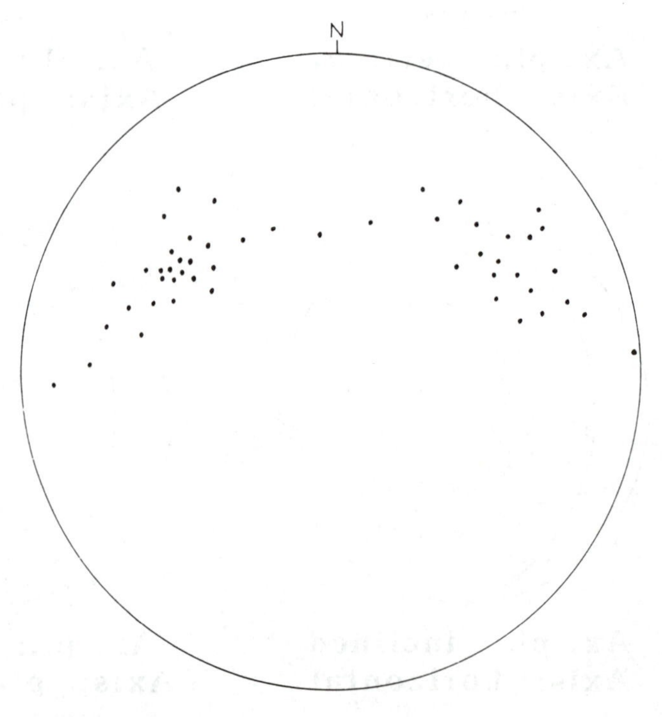

Figure 7-9
Point diagram with 50 attitudes plotted.

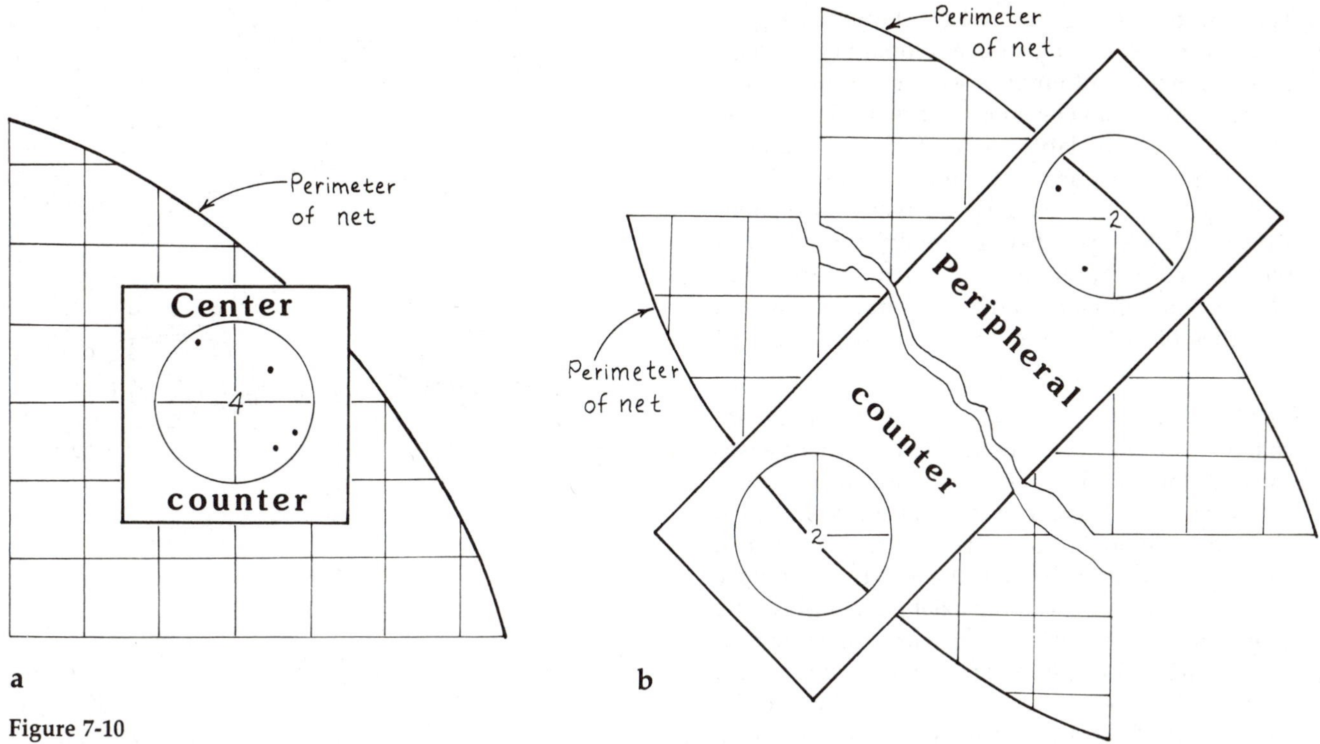

Figure 7-10
Technique for counting points for the purpose of contouring. (a) Use of center counter. (b) Use of peripheral counter. Total number of points in both circles is written at the center of both circles.

4. You are now ready to start counting points. This is done by placing the center counter on the point diagram such that the hole is centered on a grid intersection (Fig. 7-10a). Count the number of points within the circle, and write that number in the center of the circle on the clean sheet of tracing paper. Systematically move the counter from one grid intersection to the next, recording the number of points within the 1 percent circle at each intersection. Each point will be counted several times.

5. On the periphery of the point diagram, where part of the circle of the center counter lies outside the net, the peripheral counter is used (Fig. 7-10b). When the peripheral counter is used the points in both circles are counted, added together, and that number is written at the center of both circles. Figure 7-11 shows the results of counting the points in the point diagram in Figure 7-9. Each number is a sample of 1 percent of the area of the point diagram.

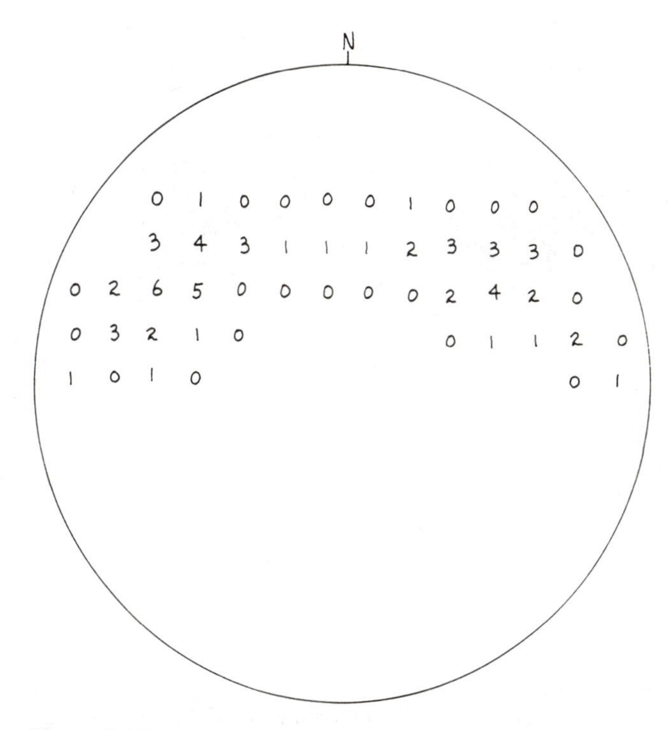

Figure 7-11
Results of counting point diagram in Figure 7-9.

6. The numbers are contoured as shown in Figure 7-12a. The result is much like a topographic map, except that the contour lines separate point-density ranges rather than elevation ranges. In the interest of simplicity and clarity some of the contours can be eliminated. Figure 7-12b shows contours 3 and 5 eliminated and a stippled pattern of varying density added. Fifty points were involved in this sample, so each point represents 2 percent of the total. The contours in Figure 7-12b, therefore, represent densities of 2, 4, 8, and 12 percent per 1-percent area.

7. The highest density regions on such a diagram are called the π maxima. The great circle that passes through them is the π-circle, and its pole is the π-axis (Fig. 7-12c). For cylindroidal folds the great circle representing the axial plane must pass between the π maxima and through the π-axis. For mildly folded, symmetric folds the axial plane bisects the acute angle between the π maxima, as shown in Figure 7-12c. The attitude of the axial plane is most reliably determined from a combination of the π-axis attitude and the strike of the surface axial trace, the latter being determined by the technique shown in Figure 7-6.

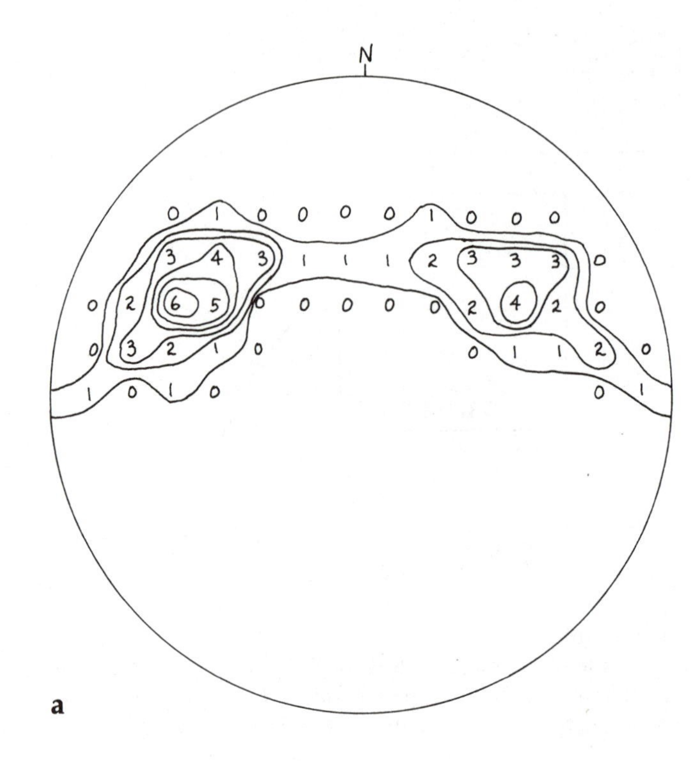

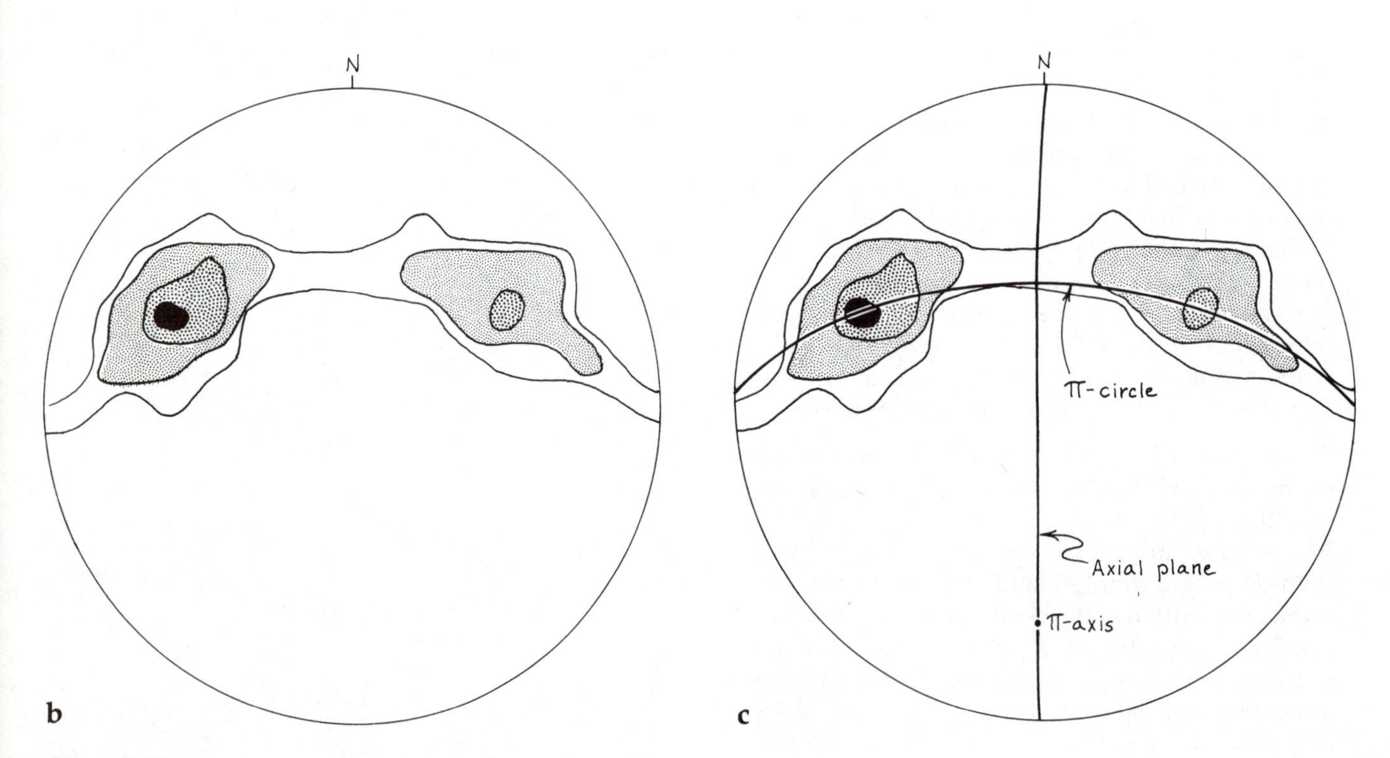

Figure 7-12
Deriving π-circle. (a) Contours drawn on point grid. (b) Selected and shaded contours. (c) Pi axis and π-circle determined from contour diagram. Axial plane cannot be located with certainty without additional information.

Folding Style and Interlimb Angle from Contoured Pi Diagrams

In addition to providing a statistical π-axis, contoured π-diagrams indicate the style of folding and the interlimb angle. The band of contours across the diagram is referred to as the girdle, and the shape of the girdle reflects the shape of the folds. Figure 7-13a shows a profile and π-diagram of an extreme case of long limbs and narrow hinge zones. Figure 7-13b shows asymmetric folds whose eastward-dipping limbs are longer than the westward-dipping limbs. Figure 7-13c, d, and e are other examples of different styles of folding and corresponding contoured π-diagrams.

The interlimb angle of a fold can be measured between the two maxima along the π-circle directly off the contoured π-diagram. It is the supplement of the angle through which the axial plane passes.

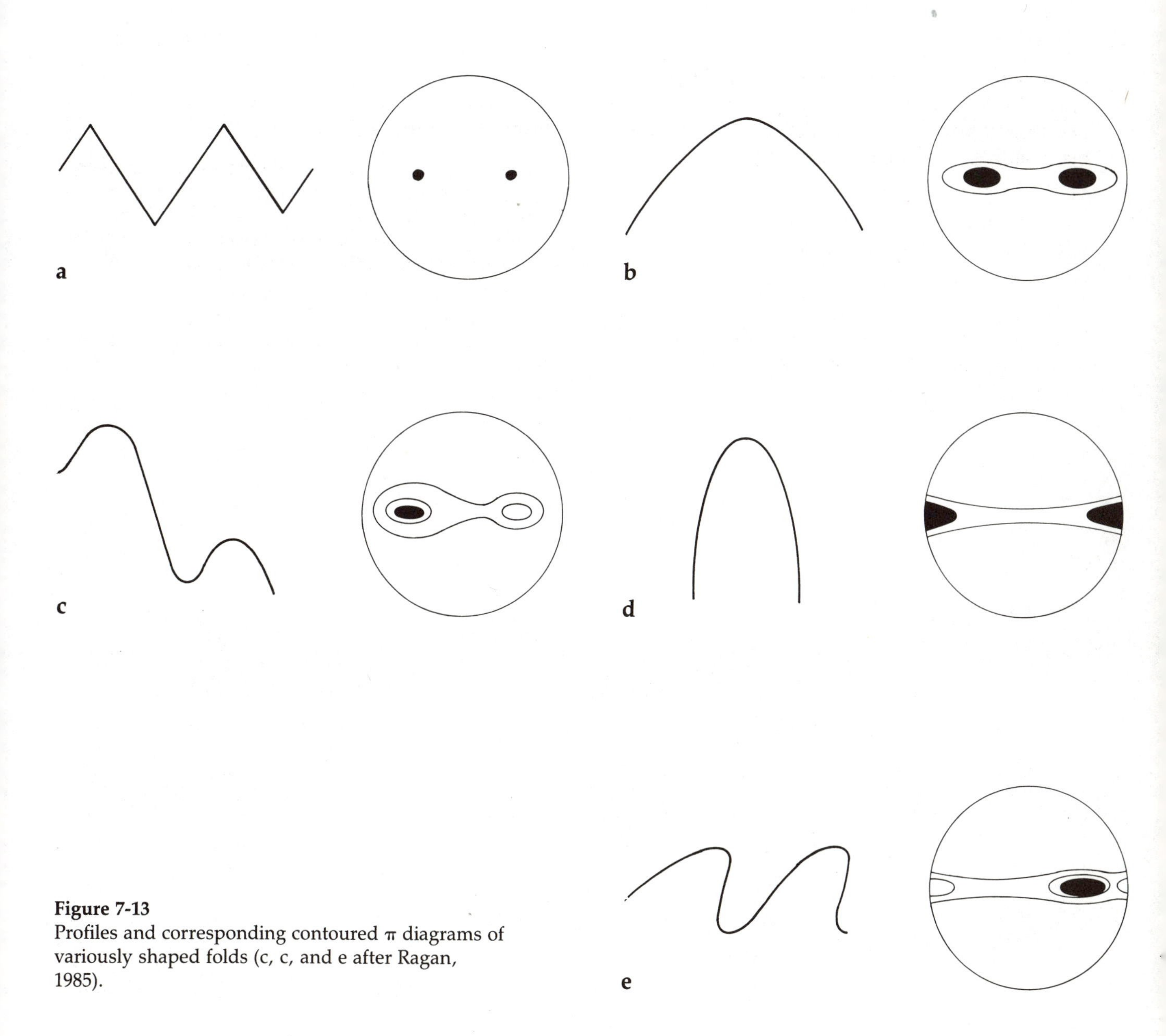

Figure 7-13
Profiles and corresponding contoured π diagrams of variously shaped folds (c, c, and e after Ragan, 1985).

Problem 7-2

Three of the four fault blocks on the Bree Creek Quadrangle contain folded Paleogene rocks for which bedding attitudes are shown on the map. It may be useful to work in groups of three students, with each student doing all of the following for only one of these three fault blocks. Make copies of your results so that each student in the group has data for all three of the fault blocks. These diagrams will be used in Chapter 11 for a structural synthesis of the Bree Creek Quadrangle.

1. Construct a contoured π-diagram.
2. Construct a profile view of the folds as shown in Figure 7-6. In the northeastern block, do not include the beds exposed on the Gollum Ridge fault scarp, because this technique can only be used in areas of low relief.
3. Draw dip isogons on your profile view and determine the class of each folded layer.
4. Describe the folds as succinctly and completely as possible. Your description should include the trend and plunge of the π-axis, the attitude of the axial surface, interlimb angle, symmetry, class of folds, and age of folding.
5. Figure 7-14 is a reference map of the Bree Creek Quadrangle with a circle on each of the three fault blocks involved in this problem. Sketch the contour diagram for each of the three fault blocks in the corresponding circle similar to those in Figure 7-12c. Draw the π-axis and axial plane on each circle. Such a reference map is an effective way of summarizing the orientation and geometry of folds in separate areas.

Further Reading

Amenta, R. V. 1975. "Multiple deformation and metamorphism from structural analysis in the eastern Pennsylvania piedmont." *Geological Society of America Bulletin*, v. 85, 1647–1660. Good example of stereographic analysis of folds used to reconstruct the deformation history of an area.

Ragan, D. M. 1985. *Structural Geology: An Introduction to Geometrical Techniques* (3rd ed.). New York: Wiley. Good discussion of contoured equal-area diagrams and a description of an alternative method of counting points with the use of a Kalsbeek net.

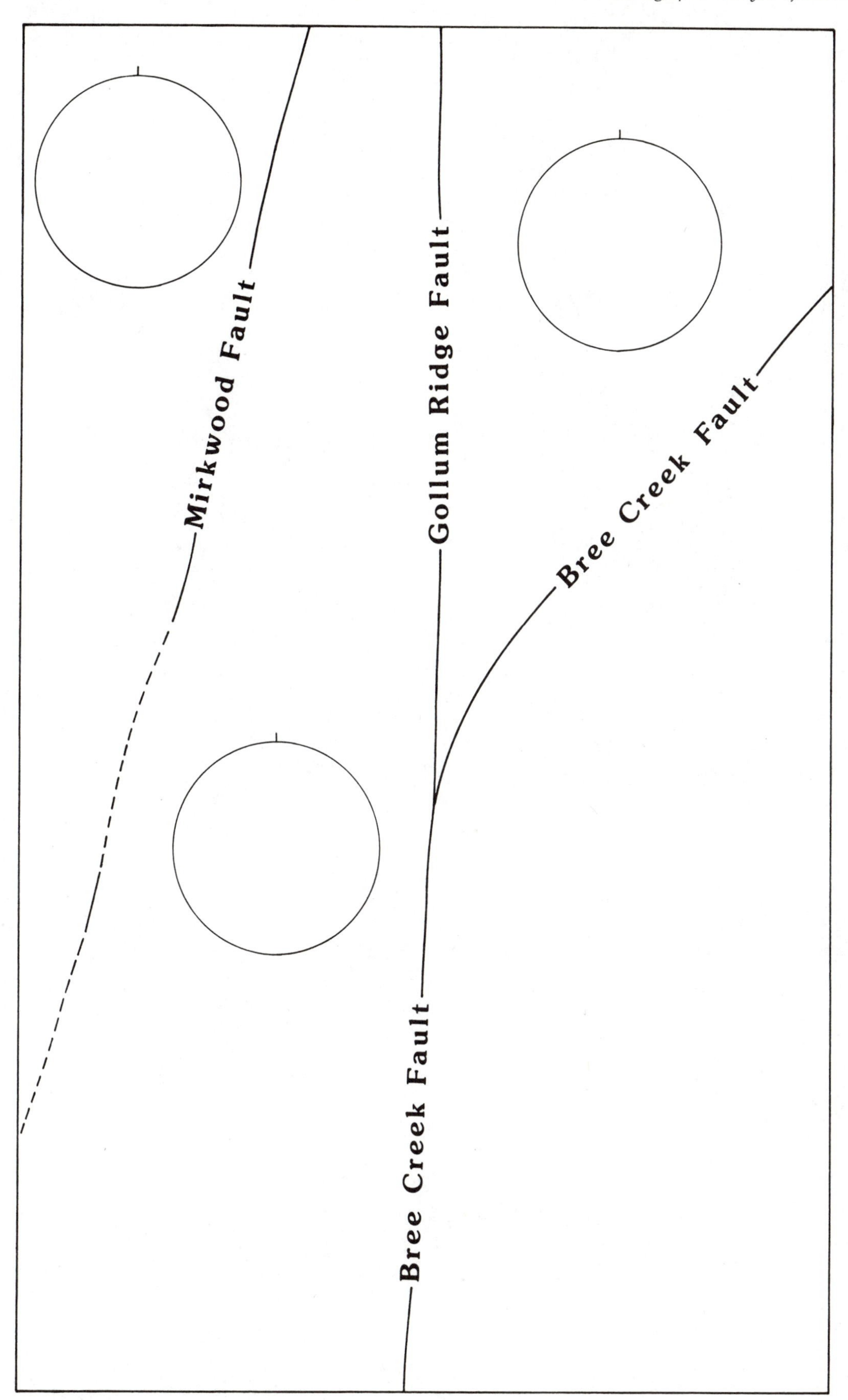

Figure 7-14
Map of separate fault blocks of the Bree Creek Quadrangle for use in Problem 7-2.

Notes

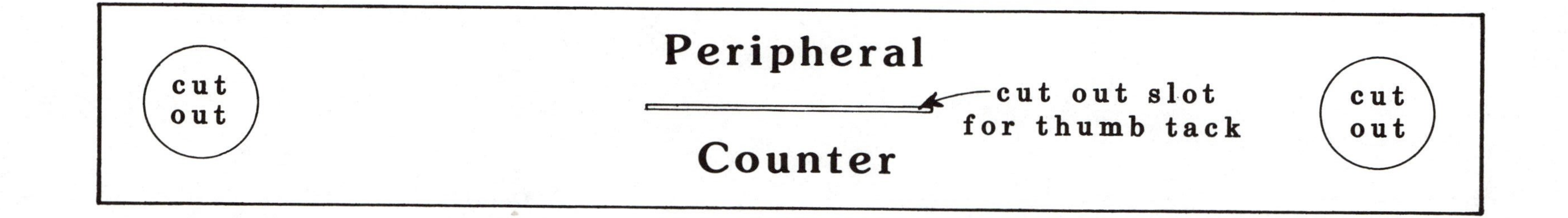

Figure 7-15
Center counter and peripheral counter to be cut out and used for constructing contour diagrams.

Notes

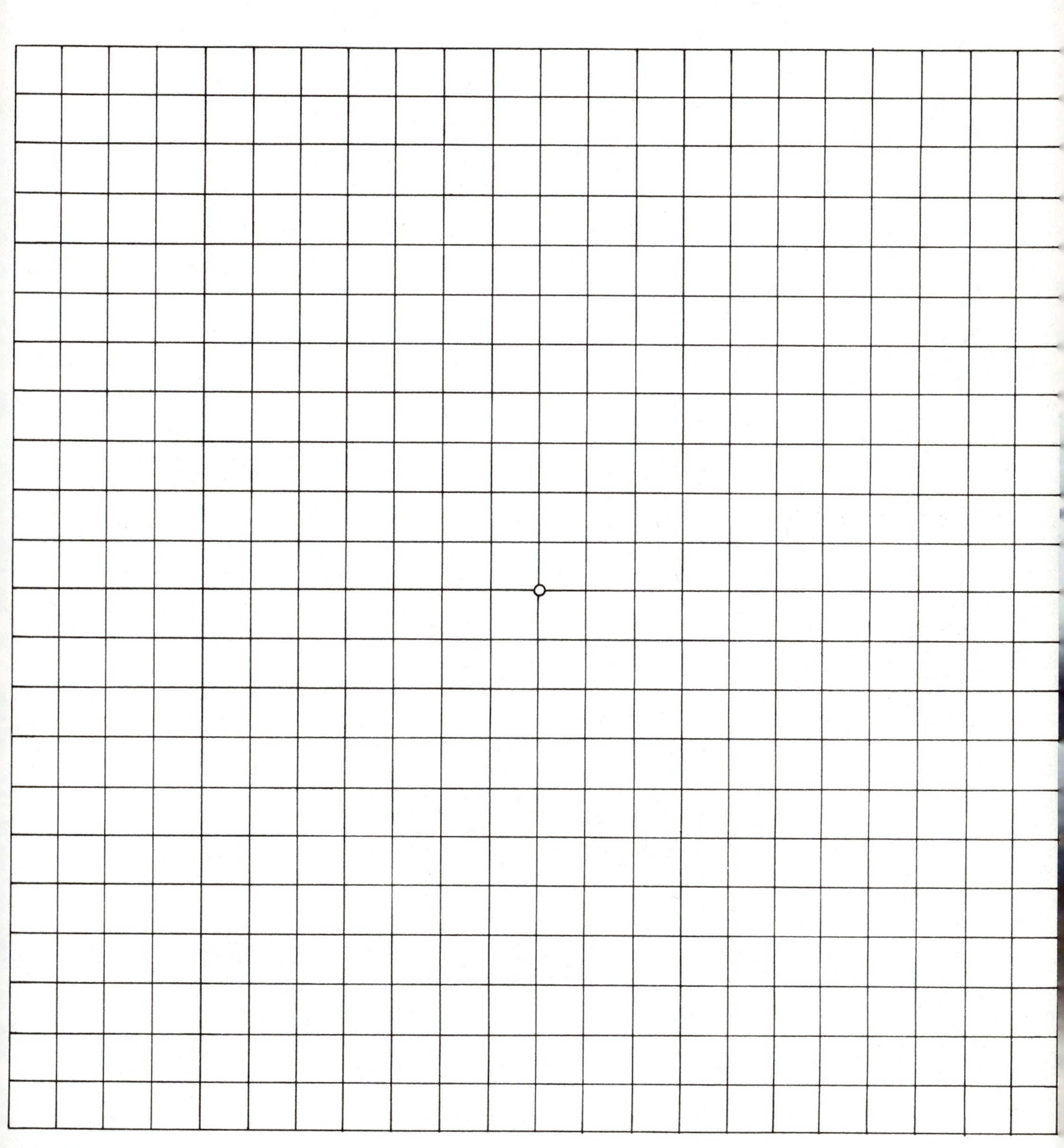

Figure 7-16
Grid for use in constructing contour diagrams.

Notes

8

Parasitic and Superposed Folds

OBJECTIVES

Use parasitic folds to locate axial traces of major folds

Reconstruct the structural history of an area that underwent two generations of folding

In this chapter we investigate folds within folds and refolded folds. The southeastern block of the Bree Creek Quadrangle, which up to now has been ignored, will be analyzed as the major exercise in this chapter.

Parasitic Folds

On the limbs and in the hinge zones of large folds one can often find small folds whose axes are parallel to the major fold axes. Such small folds are called parasitic folds. Parasitic folds that occur in the hinge zone of a larger fold are usually symmetric and are sometimes referred to as M folds because of their shape. Those that occur on the limbs of large folds are usually asymmetric and may be referred to as Z or S folds, depending on their shapes (Fig. 8-1). Of course, the same fold that appears as a Z on a south-facing exposure will be an S on a north-facing exposure.

Another way to describe parasitic folds is in terms of their sense of rotation. Z folds display a clockwise sense of rotation of the short limb and S folds a counterclockwise sense (Fig. 8-1). **S or counterclockwise parasitic folds consistently occur on the left limbs of synclines viewed in profile, and on the right limbs of anticlines.** Z or clockwise parasitic folds are found on the opposite limbs from S folds (Fig. 8-1).

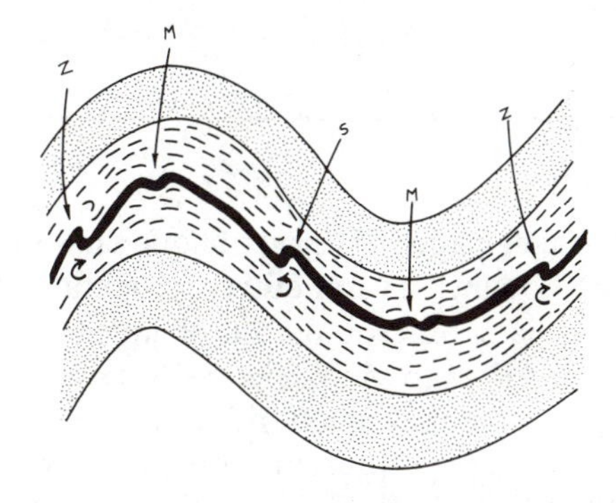

Figure 8-1
Types of parasitic folds. Arrows show sense of rotation.

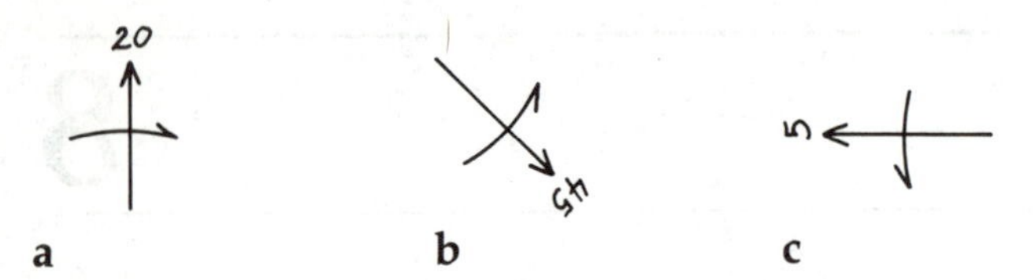

Figure 8-2
Examples of symbols representing variously oriented parasitic folds. (a) Axis plunges 20N; sense of rotation is clockwise. (b) Axis plunges 45SE; sense of rotation is counterclockwise. (c) Axis plunges 5W; sense of rotation is counterclockwise.

The map symbol commonly used to show the attitude of the axis and the sense of rotation of a parasitic fold is shown in Figure 8-2. The straight arrow represents the trend of the fold axis, and the curved arrow shows the sense of rotation. When viewed down-plunge on the map or in the field, the curved arrow shows either clockwise or counterclockwise rotation "at the top of the clock."

The chief importance of parasitic folds is that they may often be used to locate major fold axial traces in areas that are poorly exposed. Consider the map in Figure 8-3a. There are four bedding attitudes, all of which show a westward dip. Without considering the parasitic folds, the structure appears to be a westward-dipping homocline. The sense of rotation of the parasitic folds allows an anticline and syncline to be recognized and the axial traces to be approximately located (Fig. 8-3b). The sense-of-rotation arrows point **toward** anticlinal axes and **away from** synclinal axes. Figure 8-3c is a structure section showing the two folds.

Problem 8-1

Figure 8-7a (see page 115) is an oblique view of a small map area showing several outcrops, some with parasitic folds. Each outcrop on the oblique view is represented on the map view (Fig. 8-7b, page 115) by a strike-and-dip symbol or a trend-and-plunge symbol. The black layers represent thin limestone beds interbedded with shale. The limestone bed at one outcrop is not necessarily the same bed as at another outcrop.

1. Using a curved arrow, as in Figure 8-3, indicate on the map view (Fig. 8-7b) the sense of rotation of each parasitic fold.
2. Draw in the major fold axial traces on the map.
3. Sketch structure section A–A′ on Figure 8-7c, schematically showing parasitic folds.

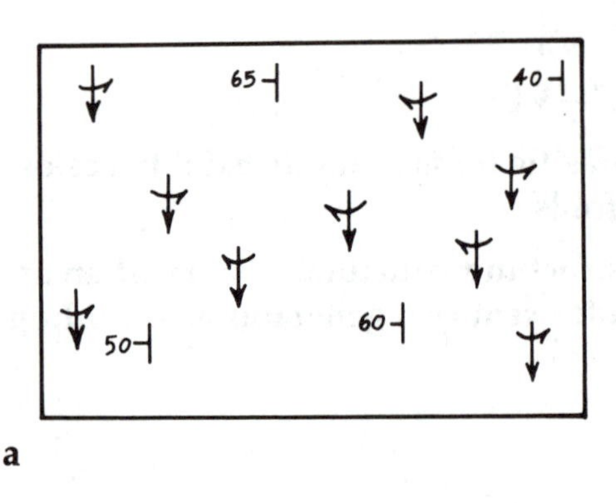

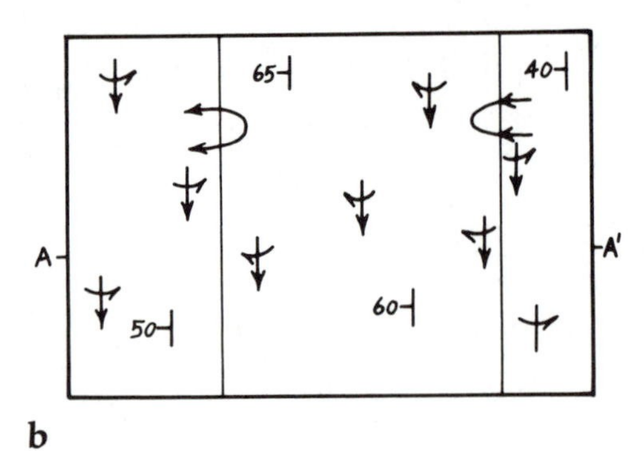

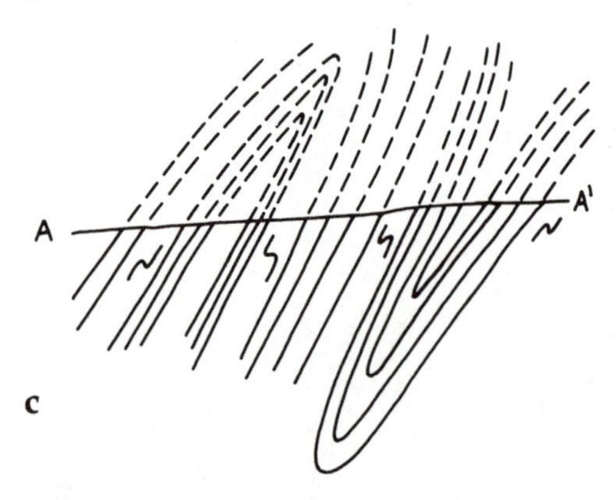

Figure 8-3
Attitudes of beds and parasitic folds. (a) Map view. (b) Same map with axial traces of overturned anticline and overturned syncline approximately located. (c) Structure section with parasitic folds shown schematically.

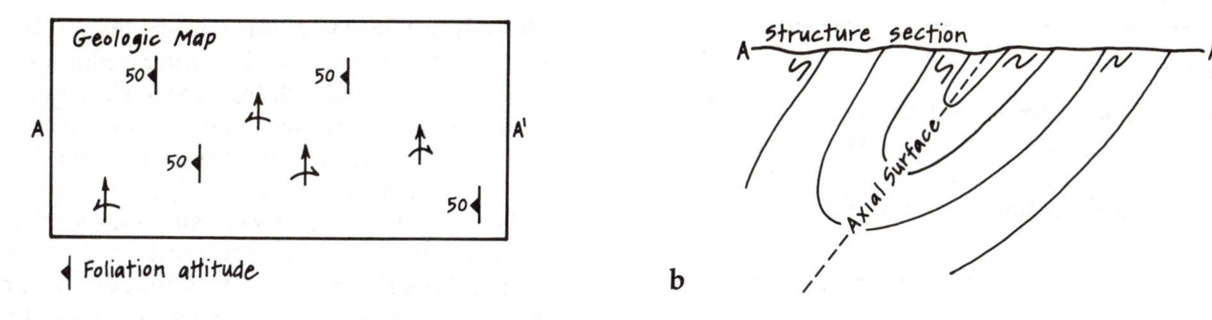

Figure 8-4
Axial planar foliations. (a) Map showing attitudes of foliations and parasitic folds. (b) Structure section A–A′.

Axial Planar Foliations

In tightly folded rocks, such as those shown in Figure 8-3c, foliations often occur that are parallel or subparallel to the axial surface. Such foliations may be relict bedding planes or they may be cleavage planes that developed during folding. In either case, if they can be determined to be parallel or subparallel to the axial surface of the fold, they are called axial planar foliations. Such foliations are very helpful in determining the orientation of the axial surface at outcrops where the folding itself is not obvious. Figure 8-4a shows a geologic map in which parasitic folds reveal the presence of a syncline, and the axial planar foliations reveal the attitude of the axial surface of the syncline.

Superposed Folds

It is common for a region to have undergone more than one episode of deformation. A variety of clues ranging from outcrop patterns to mineralogy are used to try to unravel such multiple deformed areas.

Consider the simple case shown in Figure 8-5. The first generation of folding, F_1, produced folds with east-west-trending axial surfaces (Fig. 8-5a). The second generation of folding, F_2, produced folds with north-south-trending axial surfaces (Fig. 8-5b). Such superimposed folds of different age and orientation are termed superposed folds. It is clear that the east-west axial surfaces, S_1, developed before the north-south axial surfaces, S_2, because the S_1 traces have been refolded and the S_2 traces are straight.

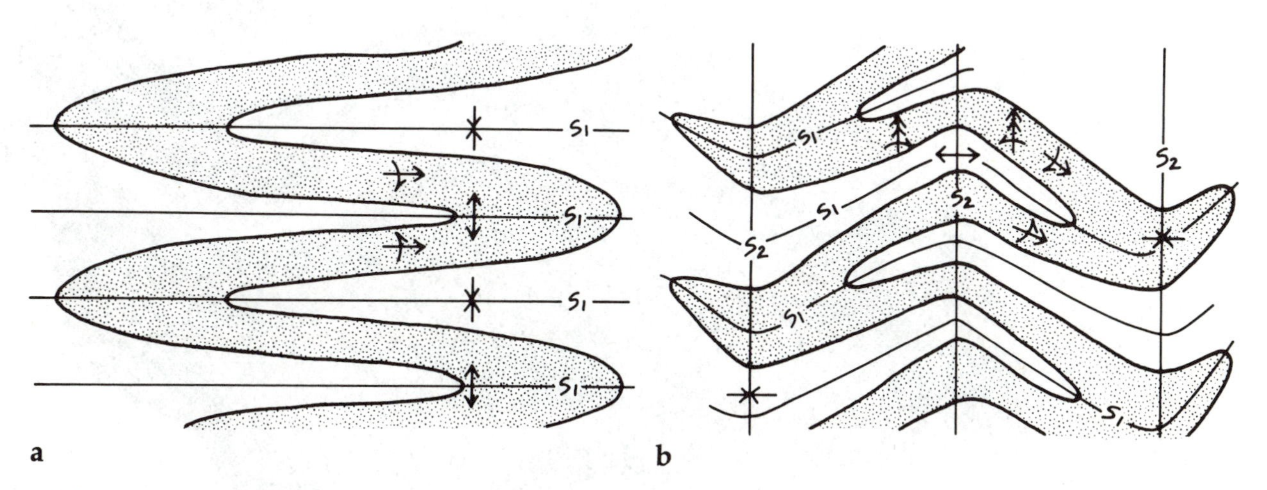

Figure 8-5
Superposition of folding. (a) First generation of folding (F_1). S_1 is axial surface trace of F_1 folds; symbols show parasitic folds. (b) Layers after two generations of folding (F_1 and F_2). S_2 is axial surface trace of F_2 folds, and double-headed arrows denote F_2 parasitic folds.

Problem 8-2

Figure 8-6 is a photograph of a slabbed rock that experienced two generations of folding. Draw and label the S_1 and S_2 axial surface traces on the photograph (or on a photocopy or transparent overlay).

Problem 8-3

In the southeastern fault block of the Bree Creek Quadrangle is a section of Paleozoic rocks that have experienced two episodes of folding, F_1 and F_2. The foliations shown on the map are axial planar. F_1 parasitic folds are indicated by the symbol with a single-headed arrow, and F_2 parasitic folds are indicated by the symbol with a double-headed arrow.

1. On your Bree Creek Quadrangle map draw the axial surface traces, S_1 and S_2, for both generations of folds, as was done in Figure 8-5, using the parasitic folds to help you locate them. Use the appropriate symbols to indicate synclines and anticlines.
2. Draw structure sections C–C′ and D–D′, using the axial planar foliations to help you determine the attitudes of the axial planes of the folds. It may be easiest to draw D–D′ first because it is simpler than C–C′. Because structure sections C–C′ and D–D′ are intersecting vertical planes, they have a common vertical line. Draw this vertical line on both structure sections and label it "Intersection of C–C′ and D–D′." The depth from the surface to each rock unit must be the same along this vertical line on both structure sections, as if you had cored down at this spot on your Bree Creek map and showed the data from the core on both structure sections. After you have drawn structure section D–D′, the depth and thickness of each rock unit on this line of intersection can be transferred to your C–C′ topographic profile, thereby providing stratigraphic control for your C–C′ structure section.
3. Completely and succinctly describe each generation of folding. Include attitudes of the fold axes, attitudes of the axial surfaces, interlimb angles, symmetry, class of folds, and age of folding.

Further Reading

Ramsay, J. G. 1967. *Folding and Fracturing of Rocks.* New York: McGraw-Hill. Chapter 10 is devoted to superposed folds.

Suppe, J. 1985. *Principles of Structural Geology.* Englewood Cliffs, NJ: Prentice-Hall. Contains an up-to-date review of superposed folds with several examples at different scales.

Figure 8-6
Slab of rock that experienced two generations of folding. For use in Problem 8-2.
From the collection of O. T. Tobisch.

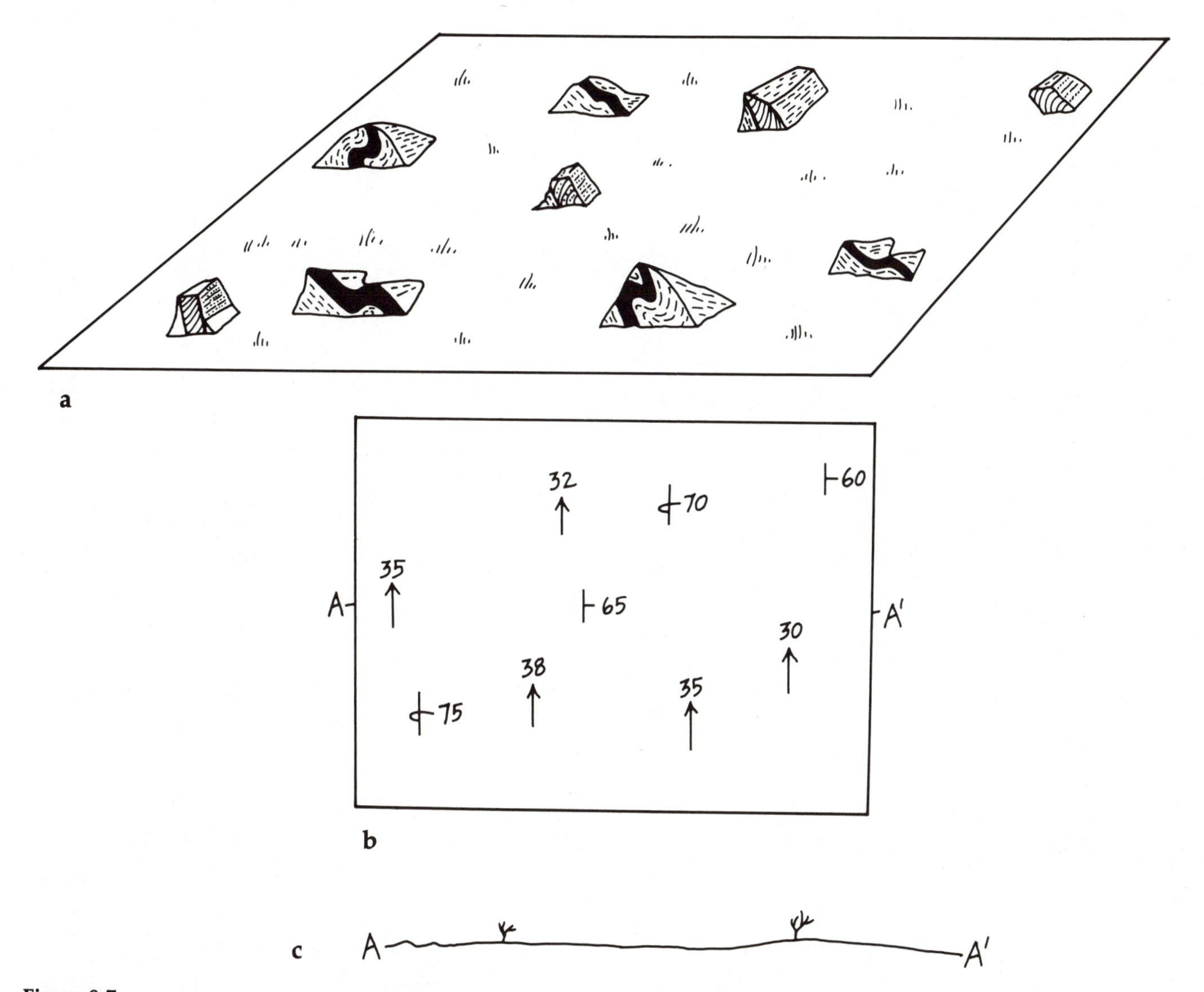

Figure 8-7
Diagrams for use in Problem 8-1. (a) Oblique view of map area showing nine outcrops. (b) Map view showing bedding attitude at each outcrop. (c) Topographic profile to be used for structure section A–A′.

Notes

9

Faults

OBJECTIVES

Measure net slip

Measure rotational slip

Describe geometry, sense of slip, and age of faults

A fault is a fracture along which movement has occurred. Sometimes there is a single discrete fault surface, or **fault plane,** but often movements take place on numerous subparallel surfaces resulting in a **fault zone** of fractured rock. The San Andreas fault in California, for example, in most places has a single recently active fault plane lying within a highly sheared fault zone that is tens or hundreds of meters wide.

Some faults are only a few centimeters long, while others are hundreds of kilometers long. On geologic maps it is usually impossible to show every fault. Only those faults that affect the outcrop pattern of two or more map units are usually shown. The scale of the map determines which faults can be shown.

Below are some terms used to describe faults and their movements:

Strike-slip fault A fault in which movement is parallel to the strike of the fault plane (Fig. 9-1). Strike-slip faults are sometimes called wrench faults, tear faults, or transcurrent faults. A right-lateral (dextral) strike-slip fault is one in which the rocks on one fault block appear to have moved to the right when viewed from the other fault block. A left-lateral (sinistral) strike-slip fault, shown in Figure 9-1, displays the opposite sense of displacement.

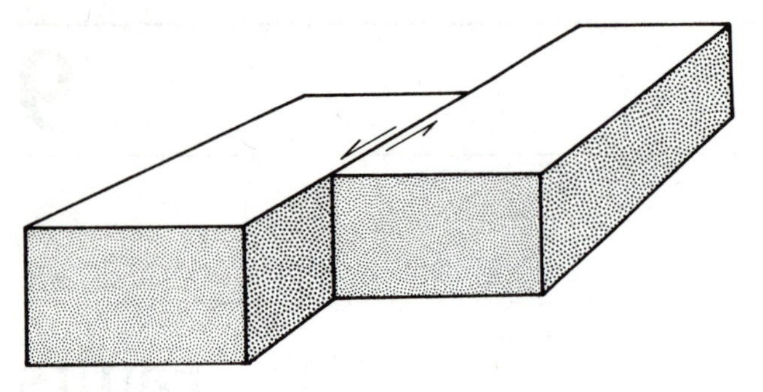

Figure 9-1
Block diagram showing left-lateral strike-slip fault.

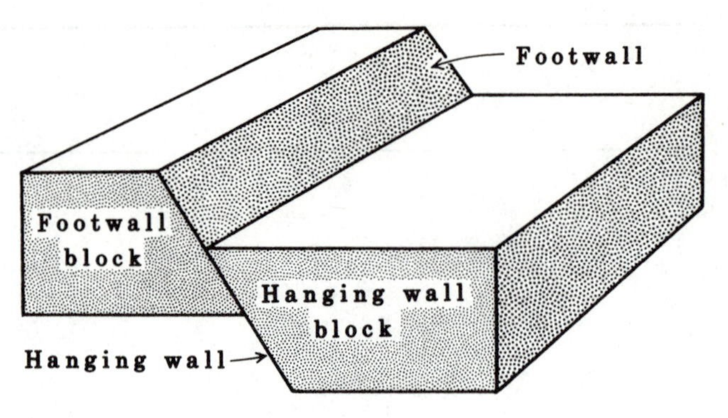

Figure 9-2
Block diagram showing dip-slip fault. This is a normal fault because hanging wall block has moved down relative to footwall block.

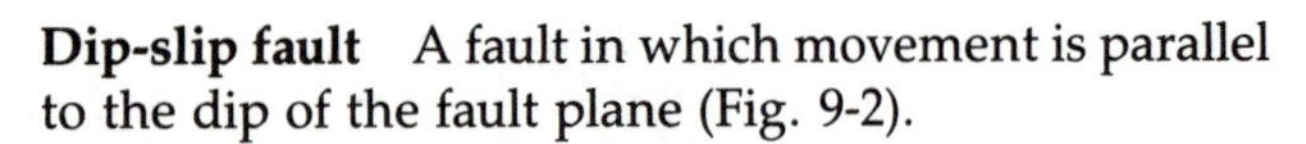

Dip-slip fault A fault in which movement is parallel to the dip of the fault plane (Fig. 9-2).

Oblique-slip fault A fault in which movement is parallel to neither the dip nor the strike of the fault plane.

Hanging wall block The fault block that overlies a high-angle, nonvertical fault (Fig. 9-2).

Footwall block The fault block that underlies a high-angle, nonvertical fault (Fig. 9-2).

Normal fault A dip-slip fault in which the hanging wall has moved down relative to the footwall (Fig. 9-2).

Reverse fault A dip-slip fault in which the hanging wall has moved up relative to the footwall.

Thrust fault A low-angle reverse fault.

Detachment fault A low-angle normal fault, also called a denudation fault.

Listric fault A fault shaped like a snow-shovel blade, steeply dipping in its upper portions, becoming progressively less steep with depth, and having a relatively straight surface trace (Fig. 9-3).

Translational fault One in which no rotation occurs during movement, so that originally parallel planes on opposite sides of the fault remain parallel (Figs. 9-1 and 9-2).

Rotational fault One in which one fault block rotates relative to the other (Fig. 9-4).

Scissor fault A rotational fault whose sense of displacement is reversed across a point of zero slip and whose amount of displacement increases away from this point (Fig. 9-4).

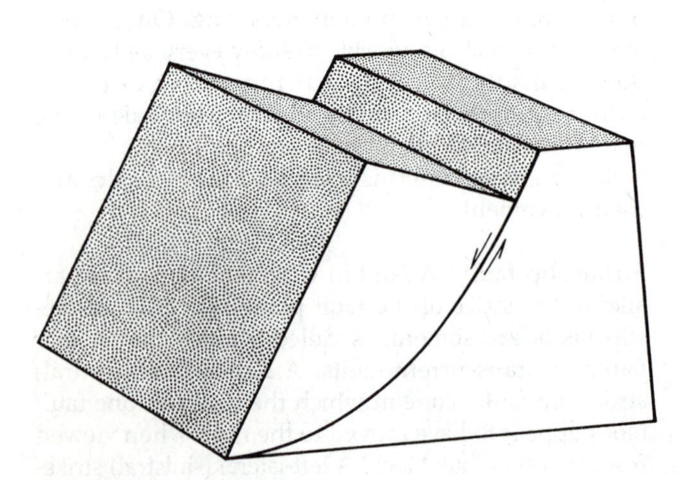

Figure 9-3
Block diagram showing listric fault.

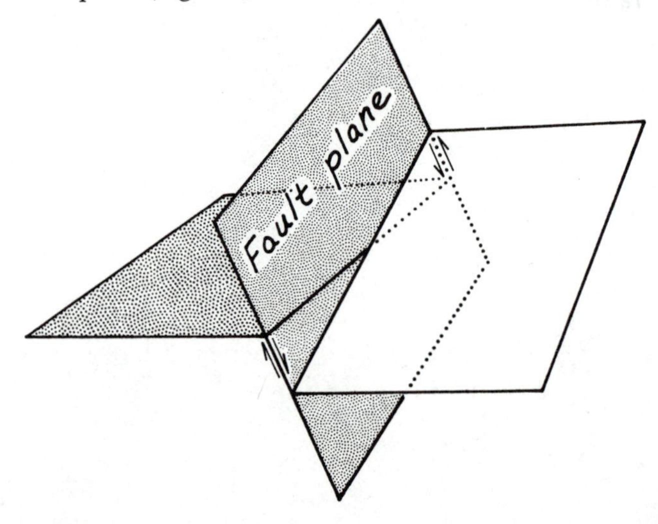

Figure 9-4
Block diagram showing rotational fault. This is a scissor fault because there is a reversed sense of displacement across a point of zero slip.

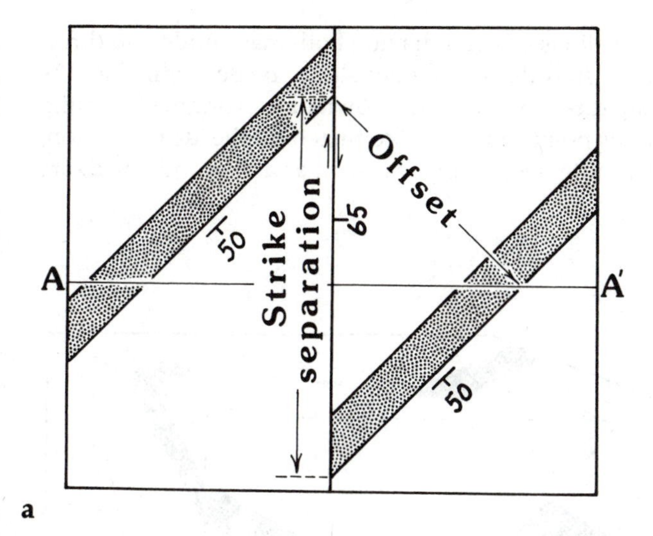

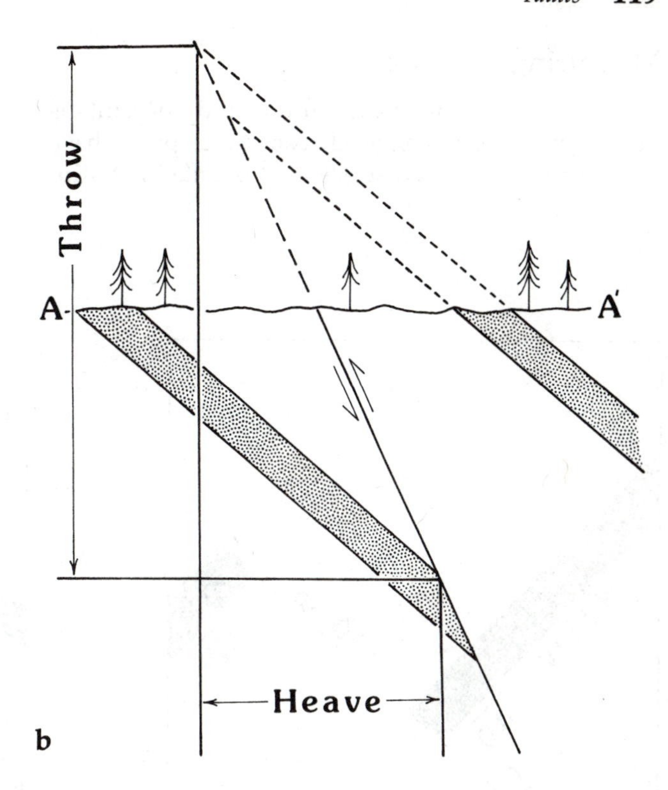

Figure 9-5
(a) Geologic map showing difference between offset and strike separation. (b) Vertical structure section showing the heave and throw components of dip separation.

Slickensides A thin film of polished mineralized material that lines some fault planes. Slickensides contain striations parallel to the direction of latest movement. Often it is not possible to tell from slickenside lineations alone in which of two possible directions movement actually occurred.

Fault trace Exposure of the fault plane on the earth's surface.

Net slip The displacement of originally adjacent points on opposite sides of the fault.

Offset Horizontal separation of a stratigraphic horizon measured perpendicular to the strike of the horizon (Fig. 9-5a).

Strike separation Horizontal distance parallel to the strike of the fault between a stratigraphic horizon on one side of the fault and the same horizon on the other side. Strike separation may be described as having either a right-lateral (dextral) or left-lateral (sinistral) sense of displacement (Fig. 9-5a).

Dip separation Horizontal (heave) and vertical (throw) distance between a stratigraphic horizon on one side of the fault and the same horizon on the other side as seen in a vertical cross-section drawn perpendicular to the fault plane (Fig. 9-5b). Dip separation has either a normal or reverse sense of displacement.

Notice that the term separation is concerned with the *apparent* displacement of some reference horizon, and the terms right-lateral, left-lateral, normal, and reverse can be used to describe the separation whether or not the actual direction of movement is known. Similarly, arrows are often drawn along faults on geologic maps to indicate the sense of strike separation, even on faults with no history of strike-slip movement. More often than not the actual slip path of a fault cannot be determined. When describing faults it is important to clearly distinguish between separation and net slip.

Measuring Net Slip

Of fundamental importance in the study of faults is the distance that two originally contiguous points have been separated. This displacement is called **net slip.** Net slip is a vector, having both magnitude and direction. In order for the net slip to be determinable, the slip direction must be known or two originally contiguous points must be recognized. In the ideal situation, two intersecting planes, such as a dike and a bed, are

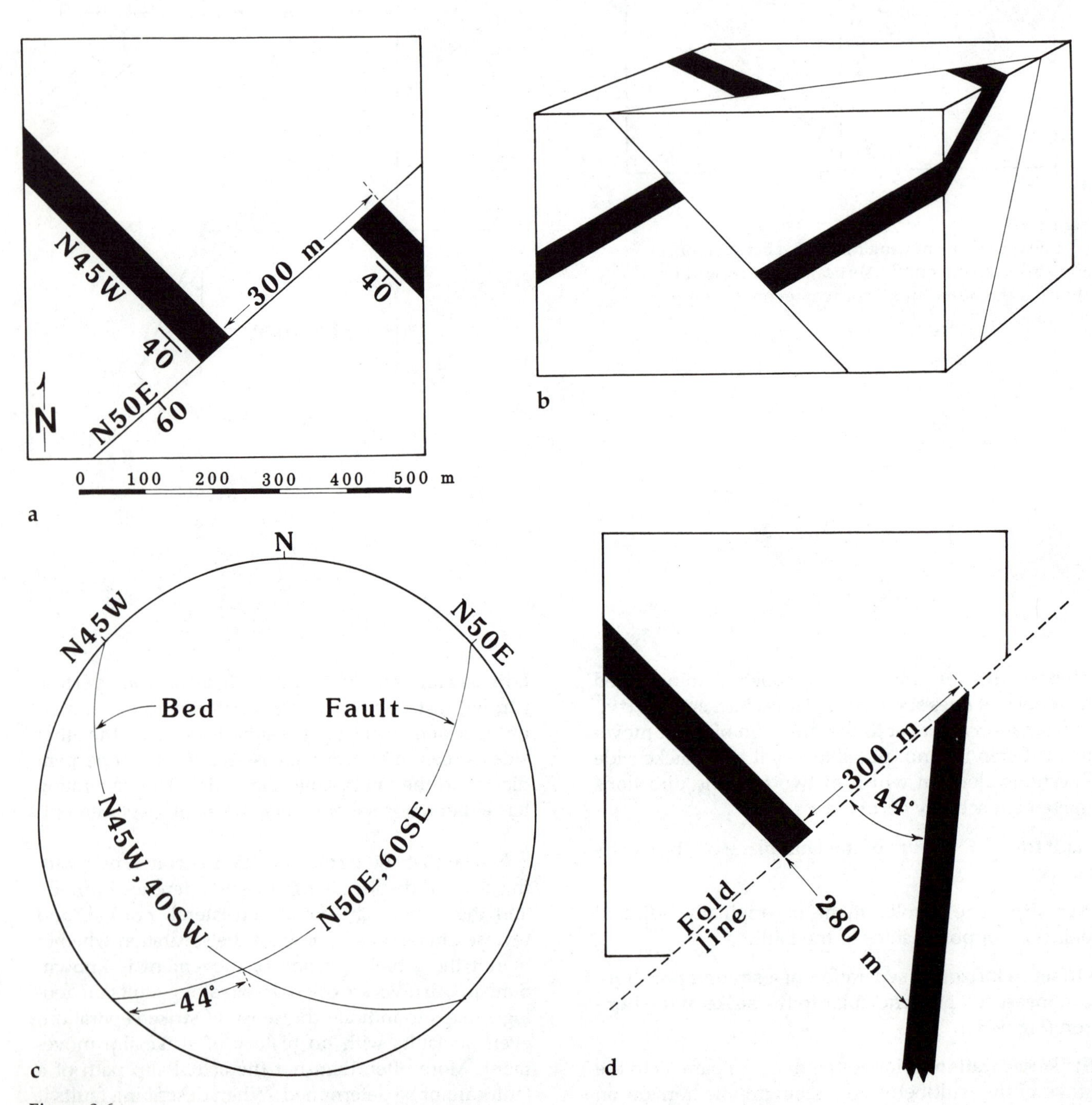

Figure 9-6
Diagrams showing solution of net slip problem. (a) Geologic map. (b) Block diagram. (c) Stereonet plot of fault plane and bedding plane. (d) Orthographic projection of fault plane showing pitch of bedding. Net slip is 280 meters in same direction as dip (direction indicated by slickenside lineations).

located on both fault blocks. The points of intersection on the hanging wall and footwall serve as the two originally contiguous points. Unfortunately, one almost never finds such a happy situation in the field. More commonly the structural geologist must use a single distinctive bed together with slickenside lineations to estimate net slip. The danger here is that the slickenside lineations indicate the orientation of only the latest movement. Some faults have complex slip paths that cannot be reconstructed from slickenside lineations.

Figure 9-6a is a geologic map showing the trace of a fault plane (N50E, 60SE) and a bed (N45W, 40SW) with 300 meters of strike separation. Figure 9-6b is a block diagram of the situation. Without further information it is impossible to know if this fault is a left-lateral strike-slip fault, a normal fault, or an oblique-slip fault. It would also be impossible to determine the net slip. The relative sense of offset may be easily visualized by the down-dip viewing method described in Chapter 6. Orient Figure 9-6a so that your line of sight is directly down the plunge of the line of intersection of the fault and the layers on one of the fault blocks. The left or east block (hanging wall) can be seen to have moved down relative to the right or west block (footwall), but this does not reveal the actual slip path.

Let us assume that the fault in Figure 9-6a is a normal fault, as indicated by slickenside lineations on the fault plane. The net slip is determined as follows:

1. On an equal-area net, draw the great circles that represent the fault plane and the plane that is offset (Fig. 9-6c).
2. Find the pitch of the offset plane in the fault plane (Fig. 9-6c). In this example the pitch is 44°.
3. Place a piece of tracing paper over the map. Draw the fault trace, and draw the offset layer on the *upthrown* block. Mark the place where the offset layer on the downthrown block intersects the fault, but don't draw it in (Fig. 9-6d).
4. The fault trace on your tracing paper is now considered to be a fold line, and the fault plane is imagined to be folded up into the horizontal plane. We know that the pitch of the offset bed in the fault plane is 44°. With the fault plane now horizontal we can draw this 44° angle on the tracing paper, showing what the offset layer looks like in the fault plane (Fig. 9-6d).
5. If we know the pitch of the net slip direction within the fault plane we can now measure the amount of net slip. Because we know that this is a normal fault, the net slip direction is 90° from the fault trace in the fault plane, that is, directly down-dip. The amount of net slip in this example is 280 meters (Fig. 9-6d). The direction of net slip is the same as the dip of the fault plane, 60, S40E.

Problem 9-1

Figure 9-7 shows the trace of a fault (N30W, 50SW) and a dike (N40E, 35SE) with 450 meters of strike separation. Assume that this is a normal fault. What is the amount of net slip?

Problem 9-2

Measure the net slip in Problem 9-1 if slickenside lineations trend northwest and have a pitch of 60°.

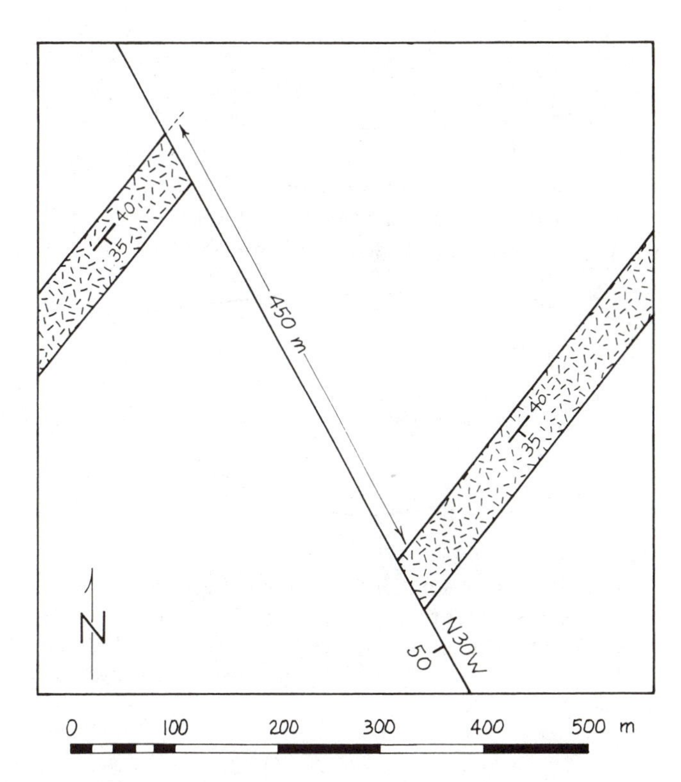

Figure 9-7
Geologic map for use in Problem 9-1.

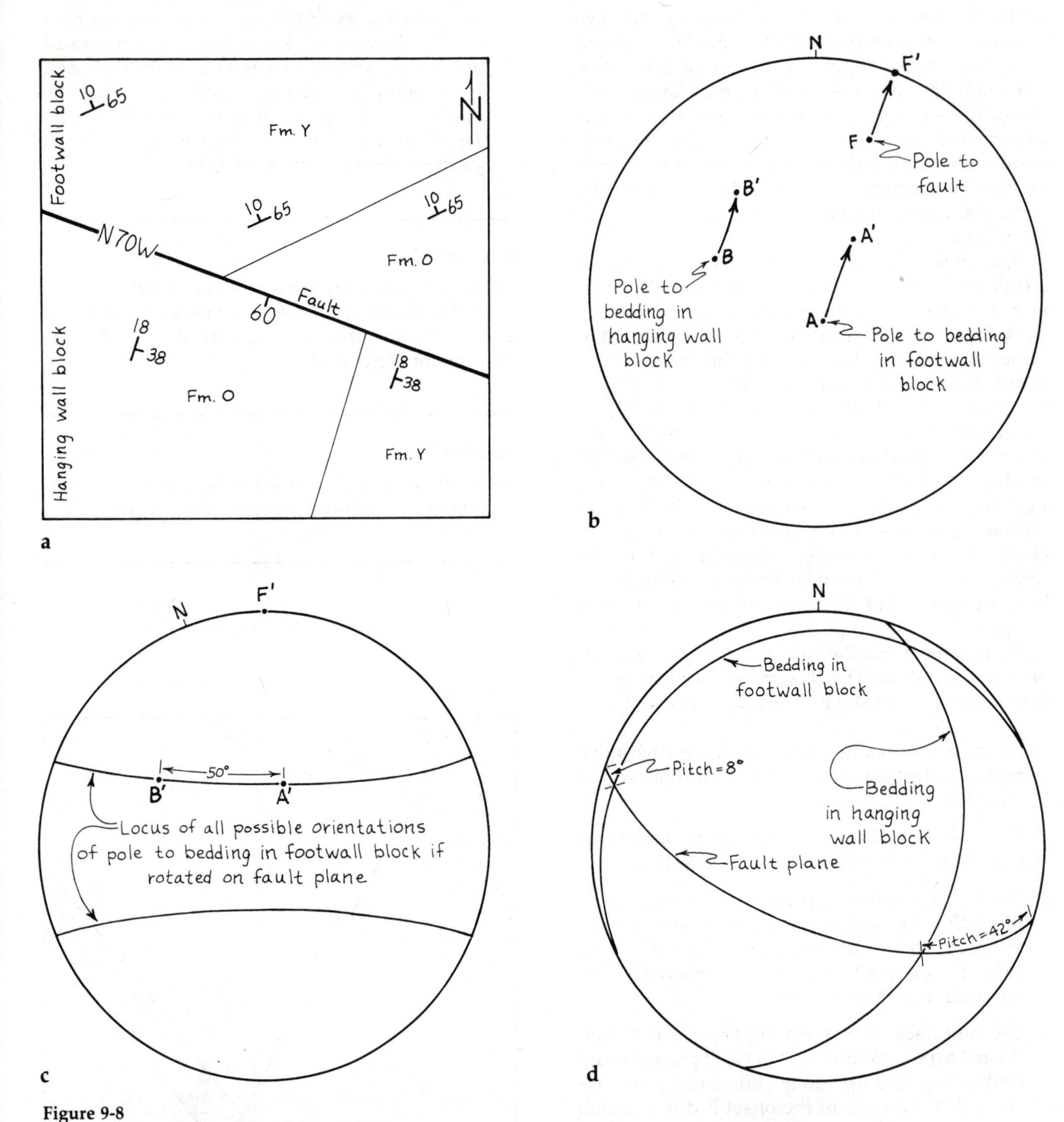

Figure 9-8
Diagrams showing solution to rotational slip problem. (a) Geologic map. (b) Stereonet plot of pole of fault (point F'), and poles to bedding in each fault block. (c) Small circles that define the locus of possible poles to bedding in footwall block if rotated on fault plane. (d) Measuring rotation by pitch of bedding on each wall of fault.

Rotational Faulting

In Figure 9-7 the beds on opposite sides of the fault have identical attitudes, indicating that all of the movement was translational. Fault movement often has a rotational component as well, which can be measured.

Consider the example in Figure 9-8a. Obviously some rotation has occurred on the fault because the beds have different attitudes on the two fault blocks. The hanging wall has rotated counterclockwise relative to the footwall. (The sense of movement on rotational faults, as on strike-slip faults, is determined by imagining yourself on one fault block looking across the fault at the other fault block.) Before we determine how much rotation has occurred in this example it will be instructive to examine the range of possible attitudes that rotation on this fault could produce. This can be done as follows:

1. Plot the pole of the fault (point F) and the poles of the bedding in the footwall block (point A) and in the hanging wall block (point B) on the equal-area net (Fig. 9-8b).
2. We now need to orient the fault so that it is vertical, so we move point F 30° to point F′ on the primitive circle. Points A and B move 30° to points A′ and B′ (Fig. 9-8b). (You may want to review Example 8 in Chapter 5 concerning rotation of lines on the stereonet.)
3. During rotational faulting the axis of rotation is perpendicular to the fault plane. Look at the geologic map (Fig. 9-8a) again and imagine rotating the fault from its 60S dip into a vertical position. Use your left hand to represent the fault and your right hand to represent the bedding in the footwall. Stick a pencil through the fingers of your right hand to represent the pole to bedding. Now, keeping your left hand vertical, rotate your hands 180° and observe the relationship between the vertical fault plane and the pole to bedding. This relationship can be plotted on the stereonet by turning the tracing paper to put the pole of the fault (point F′) at the north (or south) pole of the net. The small circle on which A′ now lies, together with its mirror image across the equator, defines the locus of all possible pole-to-bedding orientations if the footwall block is rotated about an axis perpendicular to the fault (Fig. 9-8c).

Having plotted the range of possible attitudes that rotation on this fault could produce, we can confirm that B′ is included within this set. Rotation on this fault is equal to the angle between A′ and B′, which is 50° (Fig. 9-8c).

Another approach to measuring rotation on faults involves determining the pitch of each apparent dip in the fault plane. Because the pitch on the footwall and hanging wall were identical prior to rotation, the difference in pitch equals the amount of rotation. In this problem the apparent dips are in opposite directions so the pitches are added together. The pitch on the hanging wall in the fault plane is 42° and the pitch on the footwall is 8° (Fig. 9-8d), indicating 50° of rotation.

Problem 9-3

Figure 9-9 is a geologic map showing a fault. On the east side of the fault the beds all have an attitude of N25W, 40E. West of the fault the rocks are poorly exposed, with only two outcrops, which have different attitudes.

1. Plot the hanging wall attitude on the equal-area net, and also plot the range of possible attitudes in the footwall that could result from rotation on this fault. Determine if either of the two exposures west of the fault could be the result of rotation on the fault.
2. Determine the direction and amount of rotation.

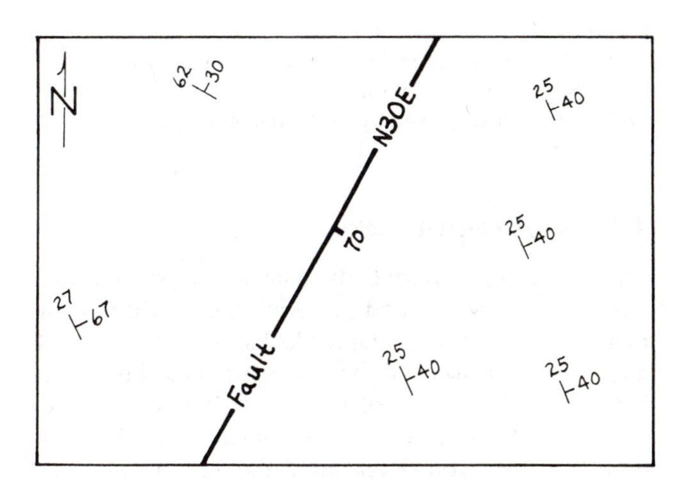

Figure 9-9
Geologic map for use in Problem 9-3.

Locating the Pivot Point

If the total movement on a fault is assumed to be rotational, the pivot point may be located by orthographically projecting the footwall and hanging wall onto separate sheets of tracing paper. Tape one sheet down and superimpose the other on top of it such that their relationship is exactly the same as in the fault plane. Hold the point of a pin on the loose sheet and rotate it. By trial and error you can locate the unique pivot point about which a horizon exposed in the footwall exactly coincides with its counterpart in the hanging wall.

It is a rare situation, however, in which you can safely assume that no translational movement has occurred. The pivot point, therefore, cannot usually be determined. In situations where bedding is consistent for long distances on each side of the fault, the rotational component, at least, can be reliably determined, as described above and shown in Figure 9-8. When faults cut across fold axes, however, purely translational faults produce field relationships that can incorrectly be interpreted as rotational in origin (Fig. 9-10).

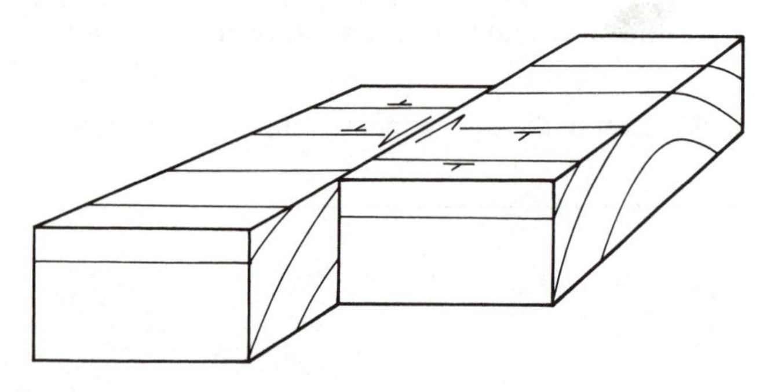

Figure 9-10
Translational fault cutting across the axis of folded beds results in field relations that could be misinterpreted as caused by rotational faulting.

Tilting of Fault Blocks

Tertiary or Quaternary beds that were deposited horizontally and then tilted provide an opportunity for measuring rotation of fault blocks. Tilting occurred after the deposition of the youngest tilted beds and before the deposition of the oldest horizontal beds within the fault block. Review, for example, Problem 5-11, in which you determined the amount of post-Rohan Tuff/pre-Helm's Deep Sandstone tilting of the northeastern fault block of the Bree Creek Quadrangle.

Problem 9-4

By stereographic projection, using the attitudes determined in Problem 3-1, determine the amount (in degrees) and direction of Neogene tilting on the fault blocks of the Bree Creek Quadrangle that contain Tertiary rocks.

	post-Rohan Tuff/pre-Helm's Deep Sandstone tilting	post-Helm's Deep Sandstone tilting
Northeastern fault block	______	______
Central fault block	______	______
Western fault block		______

The tilting determined in Problem 9-4 includes regional tilting as well as the tilting unique to each fault block. In order to determine the amount of rotation on specific faults you must first find the difference in tilt between adjacent fault blocks. This is a simple two-tilt stereonet problem.

Problem 9-5

Determine the difference in tilt between each of the following pairs of fault blocks for the stratigraphic intervals indicated. Write your answer as follows: "The southern block is tilted 15° S48W relative to the northern block."

1. Northeastern block relative to the central block, post-Rohan Tuff/pre-Helm's Deep Sandstone: ______
______
2. Northeastern block relative to the western half of the quadrangle, post-Helm's Deep Sandstone: ___
______

Finally, to convert the tilt information of Problem 9-5 into rotational movement on faults you must determine what component of the tilting lies within the fault plane. To do this, simply draw the great circles representing the relative tilt and the fault plane on the equal-area net and measure the pitch of the tilt plane in the fault plane. You will, of course, have to determine the attitude of the fault plane. When the strike is somewhat variable, as on the Bree Creek Quadrangle, estimate it as best you can.

Problem 9-6

Determine the amount and direction of rotation for the following faults and intervals. Write your answer as follows: "The northern block rotated 4° counter-clockwise relative to the southern block."

1. Southern Bree Creek/Gollum Ridge fault, post-Rohan Tuff/pre-Helm's Deep Sandstone: ________
   ____
2. Southern Bree Creek/Gollum Ridge fault, post-Helm's Deep Sandstone: ________________
   ____
3. Northern Bree Creek fault (the portion that trends northeast), post-Helm's Deep Sandstone (assume no rotation of southeastern fault block): ________
   ____

Problem 9-7

The pre-Gondor Conglomerate rotational movement on faults in the Bree Creek Quadrangle may be determined using the π-axis orientations determined in Problem 7-2 and displayed in Figure 7-14. Assume that the π-axis on each fault block was identical prior to faulting and that the Gondor Conglomerate was deposited on a horizontal surface.

1. What was the amount and direction of pre-Gondor tilting on each fault block?

   pre-Gondor tilting

   Northeastern fault block ________________

   Central fault block ________________

   Western fault block ________________

2. Convert these tilt data to rotational movement on each of the faults, as explained in the introduction to Problem 9-6, and use complete sentences as in Problem 9-6.

pre-Gondor rotation

Mirkwood fault: ________________
____
____

Gollum Ridge/South Bree Creek fault: ________
____
____

North Bree Creek fault: ________________
____
____

Problem 9-8

Using the geologic map, your structure sections, and your work in this chapter, write a succinct, complete description of each fault in the Bree Creek Quadrangle. Use complete sentences. Avoid long, rambling sentences. Include the following information for each fault:

1. Type of fault (normal, reverse, strike-slip). (If you cannot determine the sense of movement with certainty, describe the possible senses of movement and the evidence for each.)
2. Attitude of the fault plane (including geographic variation).
3. Strike separation and dip separation (heave and throw) as the data allow (sometimes only a minimum amount of separation may be determined).
4. Direction and amount of rotational movement.
5. Age of movement as specifically as the evidence permits.

Further Reading

Angelier, J., Colletta, B., and Anderson, R. W. 1985. "Neogene paleostress changes in the Basin and Range: A case study at Hoover Dam, Nevada-Arizona." *Geological Society of America Bulletin*, v. 96, 347–361. A good example of detailed analysis of faults and field criteria for determining sense of slip.

Dennison, J. M. 1968. *Analysis of Geologic Structures.* New York: Norton. A structural geology workbook emphasizing the trigonometric approach to structural relationships.

Roberts, J. L. 1982. *Introduction to Geologic Maps and Structures.* Oxford: Pergamon. A textbook in structural geology emphasizing interpretation of geology maps.

Notes

Name: ____________________

Section: ____________________

Answer sheet for Problems 9-4, 9-5, 9-6, and 9-7.

Problem 9-4 (Neogene tilting)

	post-Rohan Tuff/pre-Helm's Deep Ss tilting	**post-Helm's Deep Sand-stone tilting**
NE fault block	________	________
Central fault block	________	________
Western fault block	________	________

Problem 9-5 (relative Neogene tilting between fault blocks)

1. Northeastern block relative to central block (post-Rohan Tuff/pre-Helm's Deep Sandstone):

   ____

2. Northeastern block relative to western half of quadrangle (post-Helm's Deep Sandstone):

   ____

Problem 9-6 (rotational movement on faults)

1. Southern Bree Creek/Gollum Ridge fault (post-Rohan Tuff/pre-Helm's Deep Sandstone):

   ____

2. Southern Bree Creek/Gollum Ridge fault (post-Helm's Deep Sandstone):

   ____

3. Northern Bree Creek fault (post-Helm's Deep Sandstone):

   ____

Problem 9-7 (pre-Gondor tilting and rotation)

1. Pre-Gondor tilting

   Northeastern fault block ____________________

   Central fault block ____________________

   Western fault block ____________________

2. Pre-Gondor rotation

   Mirkwood fault:

   ____

   Gollum Ridge/South Bree Creek fault:

   ____

   North Bree Creek fault:

   ____

10

Orientation of the Stress Ellipsoid

OBJECTIVE

Determine the orientation of the three principal stresses during the formation of geologic structures

Stress is measured in units of force per unit area. Within a rock mass two types of forces are present: body forces and transmitted forces. Body forces, such as gravity, act uniformly throughout the rock mass and will be ignored in the following discussion. Transmitted forces, such as lateral or vertical compression, on the other hand, are transmitted from one point to another through the earth. These are the forces that produce folds and faults. In this chapter we explore the relationship between **stress orientation** and faulting, without being concerned about the actual magnitude of the stresses. Stress magnitude is examined in Chapter 13.

Three Principal Stresses

Imagine a hand pushing diagonally on a table top (Fig. 10-1). The stress acting on the table top can be resolved into two components: **normal stress** acting perpendicular to the surface and **shear stress** acting parallel to the surface. We use the Greek letter σ (sigma) to symbolize stress; σ_n represents normal stress, and σ_s represents shear stress. (The Greek letter tau is sometimes used to represent shear stress.)

Although both normal and shear stresses are acting on the table top in Figure 10-1, we can easily imagine a plane perpendicular to the arm in which the shear stress is zero. Within a body under stress from all directions there are always three planes of zero shear stress; these are called the **principal planes of stress.** The three normal stresses that act on these planes are called the **principal stresses.** By convention the three principal stresses are named σ_1, σ_2, and σ_3 in order

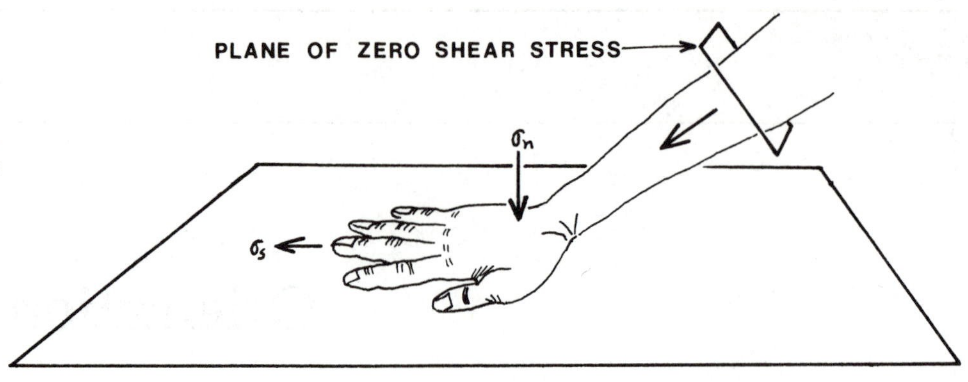

Figure 10-1
The diagonal force of a hand pushing on a table top resolved into a normal stress (σ_n) acting perpendicular to the surface and shear stress (σ_s) acting parallel to the surface. The plane of zero shear stress experiences only normal stress.

of magnitude, where $\sigma_1 \geqslant \sigma_2 \geqslant \sigma_3$ (Fig. 10-2). Together they define the **stress ellipsoid.** The normal stress acting on any plane within a stressed body cannot exceed σ_1 nor be less than σ_3. Even in situations where the crust is being extended, such as rift valleys, all three principal stresses are considered to be compressive.

Laboratory studies of rock fracturing have shown that when an isotropic body fractures under applied stress, the fracture surfaces have a predictable orientation with respect to the stress ellipsoid. As shown in Figure 10-3, there are two predicted fracture surfaces, or conjugate shear surfaces, which are both perpendicular to the σ_1–σ_3 plane. These **shear fractures** form an acute angle in the σ_1 direction and an obtuse angle in the σ_3 direction. The angle between σ_1 and each of the shear fractures is variable, depending on the difference in magnitude between σ_1 and σ_2 and on the material properties of the rock; it is always less than 45°.

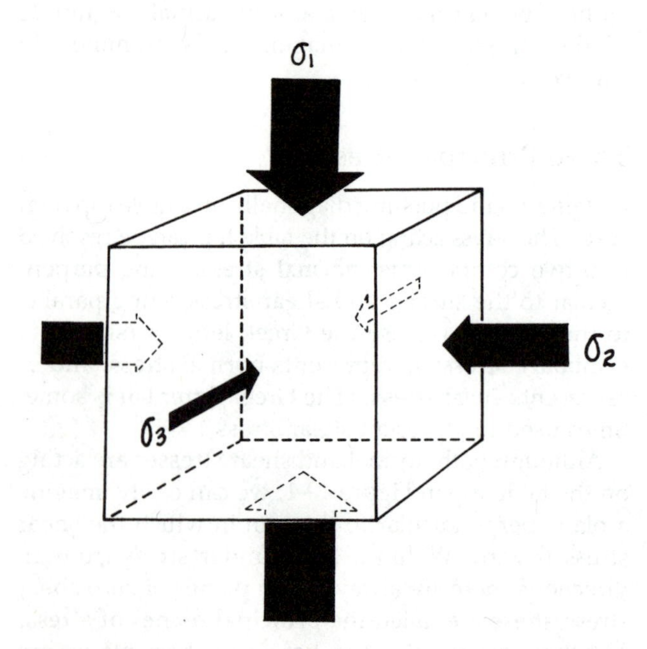

Figure 10-2
The three principal stresses that define the stress ellipsoid.

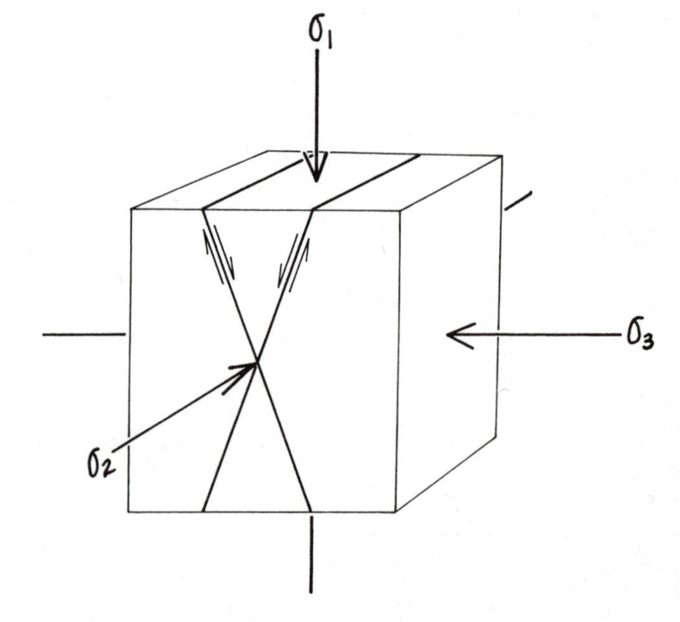

Figure 10-3
Relationship between principal stresses and conjugate shear surfaces.

Problem 10-1

Figure 10-4 shows three pairs of conjugate shear surfaces. Sketch in the three principal stress directions indicated by each.

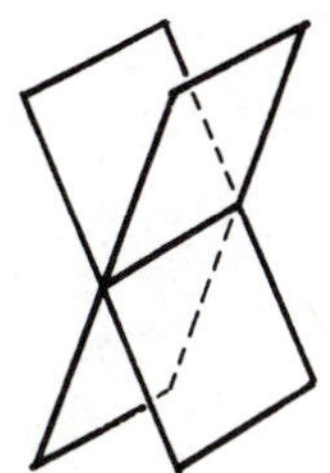

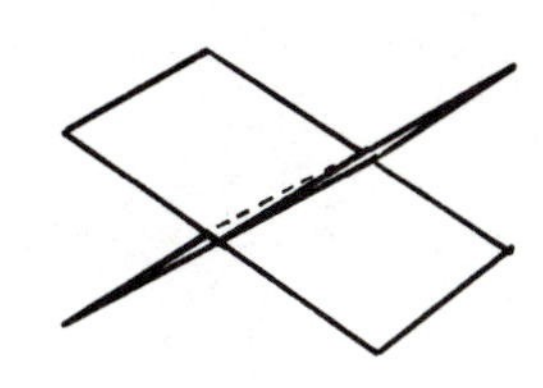

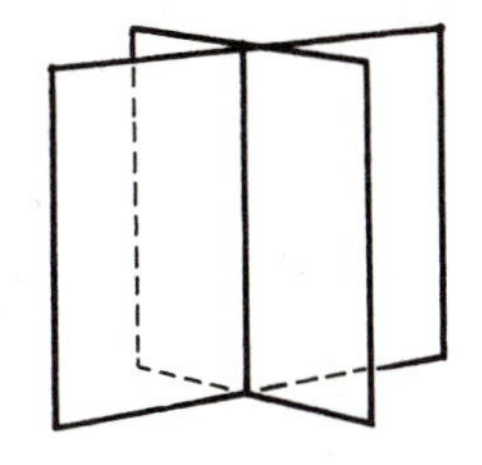

Figure 10-4
Three pairs of conjugate shear surfaces for use in Problem 10-1.

Problem 10-2 (Read page 133 first.)

In Figure 10-5 determine which of the three stress orientations indicated could have produced the folds shown and which could not have.

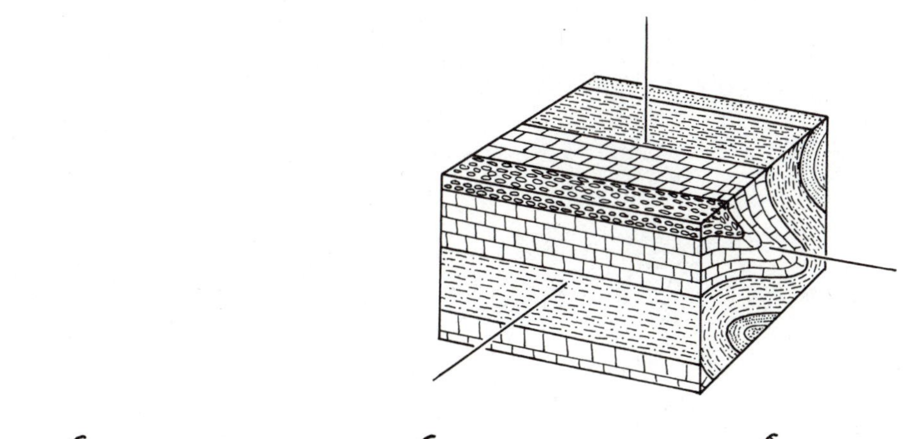

Figure 10-5
Block diagram of folded beds and three hypothetical stress ellipsoid orientations, for use in Problem 10-2.

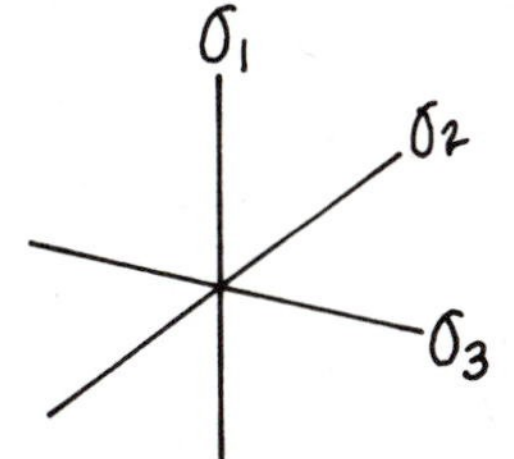

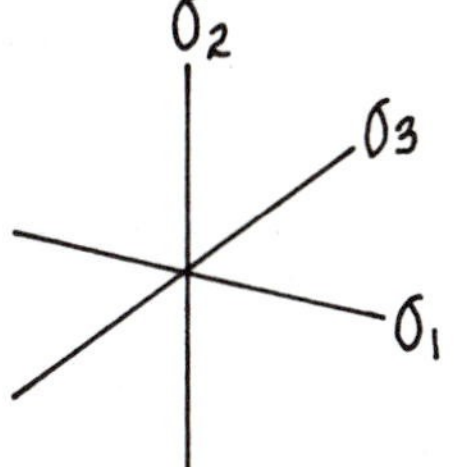

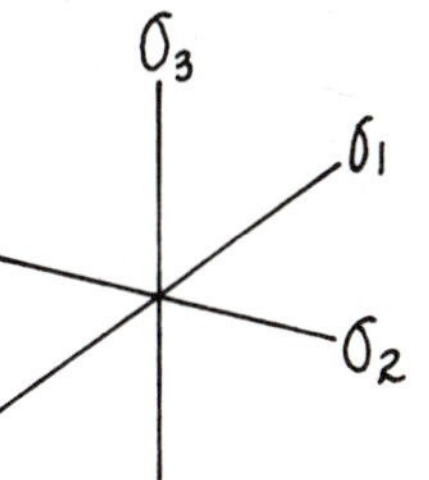

Notes

Fold Geometry and the Stress Ellipsoid

The objective of this chapter is to determine the orientation of the three principal stresses during the formation of various geologic structures. Without any prior knowledge, one might suppose that this could more easily be achieved with folds than with faults. Just the opposite is true.

In the simplest case of a flat-lying sequence of beds and σ_3 vertical, the fold axis will indeed develop perpendicular to σ_1 and parallel to σ_2 (Fig. 10-6a), but the mechanical properties and orientation of the rock layers greatly influence the ultimate geometry of the fold. Figure 10-6, for example, shows two different fold orientations resulting from the same stress ellipsoid. The orientations of σ_2 and σ_3 clearly play a less important role than does the pre-stress attitude of the layers. Notice that in Figure 10-6b even σ_1 could be rotated a considerable amount in the horizontal plane without changing the orientation of the fold that is produced.

The orientation of the stress ellipsoid can be reconstructed from some folds (see, for example, Dalziel and Stirewalt, 1975 or Burger and Hamill, 1976), but it requires a detailed analysis of fold-related fractures that is beyond the scope of this book. It *is* relatively simple, however, to differentiate folds that could have formed within a particular stress ellipsoid from those that could not have. Fold axes, for example, usually trend approximately perpendicular to σ_1.

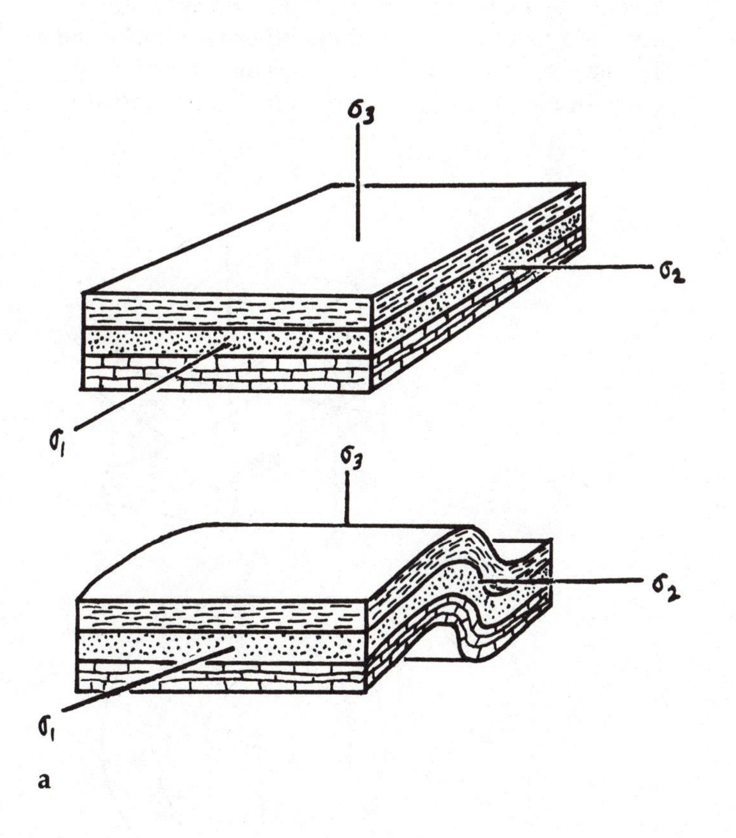

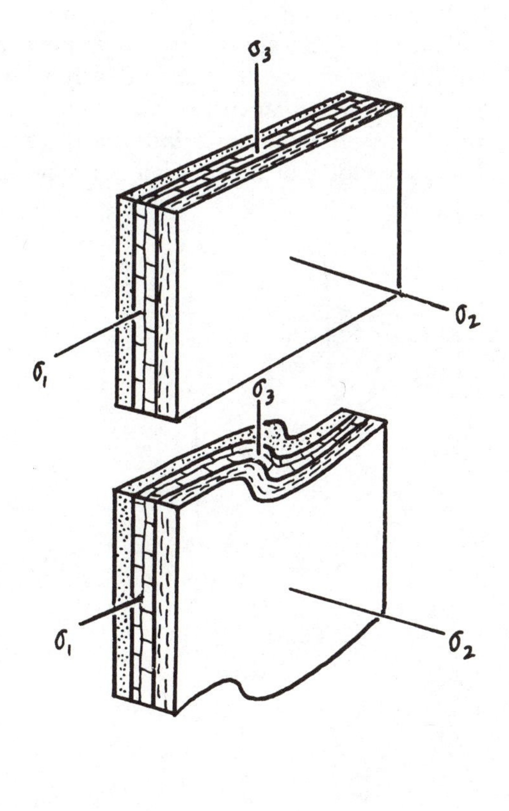

Figure 10-6
Response of horizontal (a) and vertical (b) beds to the same stress orientations. The geometry of the resultant folds is heavily influenced by the pre-fold orientation of the beds.

Fault Attitudes and the Orientation of the Stress Ellipsoid

Modern ideas about the relationship between faults and the stress ellipsoid began with the work of British geologist E. M. Anderson (1942). Anderson reasoned that because the earth's surface is an air-rock interface it must be a surface of zero shear stress and therefore a principal plane of stress. In the shallow crust (where most faulting occurs), therefore, one principal stress can be assumed to be vertical and therefore the other two must be horizontal. As a first approximation, this assumption has proved to be valid for most faults.

Anderson's assumption that one principal stress is always vertical explains the occurrence of three classes of faults: normal faults (σ_1 vertical), strike-slip faults (σ_2 vertical), and thrust faults (σ_3 vertical) (Fig. 10-7). This assumption also allows us to reconstruct the orientation of the stress ellipsoid responsible for a given population of faults.

Figure 10-7 also shows equal-area net projections of typical populations of faults of each type. Points on the fault plane great circles indicate the pitch of slickenside lineations; arrows indicate the sense of slip. The population of normal faults occurs as two subpopulations having parallel strikes and an acute angle between the fault planes. The population of thrust faults displays parallel strikes and an obtuse angle between two subpopulations. And the strike-slip fault population consists of two subpopulations with different strikes. The two subpopulations in each case represent the conjugate set of shear surfaces analogous to those observed in experimental studies (Fig. 10-3).

In order for two discrete subpopulations of faults to develop in isotropic rocks as shown in Figure 10-7, there must be a distinct quantitative difference between σ_1, σ_2, and σ_3. If σ_1 is vertical with σ_2 and σ_3 of approximately the same magnitude, for example, the resultant fault population would consist of steeply dipping normal faults with no preferred strike.

To reconstruct the orientation of the stress ellipsoid from a population of faults, draw the fault plane great circles on the equal-area net along with any available data on the orientation and sense of slip of slickenside lineations (as on Fig. 10-7). One principal stress is assumed to be vertical and the other two horizontal. The appropriate principal stresses are located by visually locating the axes of symmetry of the diagram.

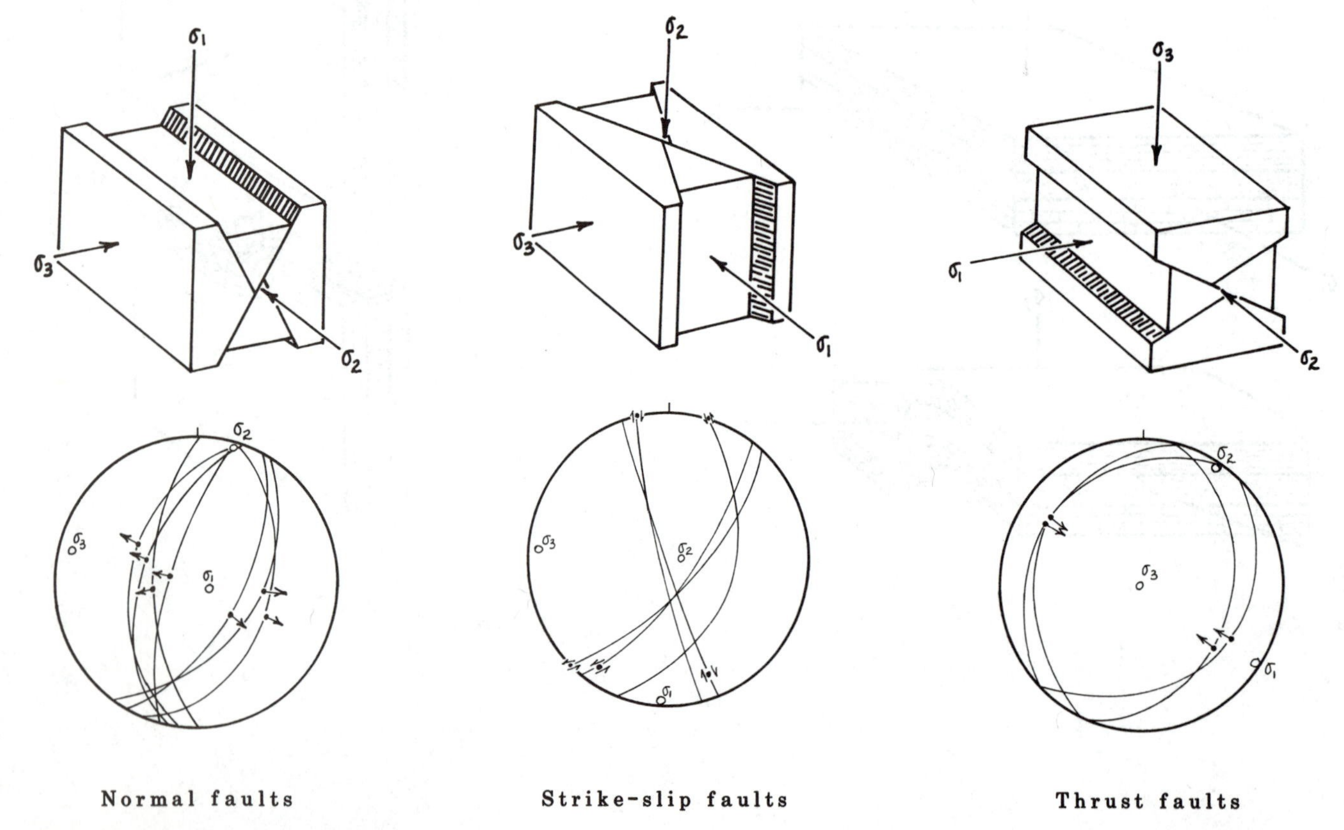

Figure 10-7
Block diagrams and equal-area plots of three classes of faults predicted by E. M. Anderson. Equal-area plots show typical data on fault and slickenside orientation. After Angelier (1979) in Suppe (1985).

Problem 10-3

The table below lists measurements from ten normal faults in a small area on the island of Crete (Angelier, 1979). Plot the data on an equal-area net and determine the orientation of the principal stresses. (In reality, 10 faults are not enough to reliably determine the orientation of the stress ellipsoid; ideally, about 40 should be used.)

Reference number	Azimuth of fault	Dip of fault	Pitch of slickensides
1	045	61S	80E
2	036	59S	80W
3	090	80N	58W
4	052	68N	78W
5	045	63N	78W
6	110	88N	59W
7	074	78N	65W
8	046	60S	80W
9	077	61N	86E
10	067	56S	88E

Problem 10-4

Figure 10-8 shows a map of a mine adit and a series of minor faults that occur in a homogeneous rock unit. Plot the fault planes on the equal-area net and determine the orientation of the stress ellipsoid.

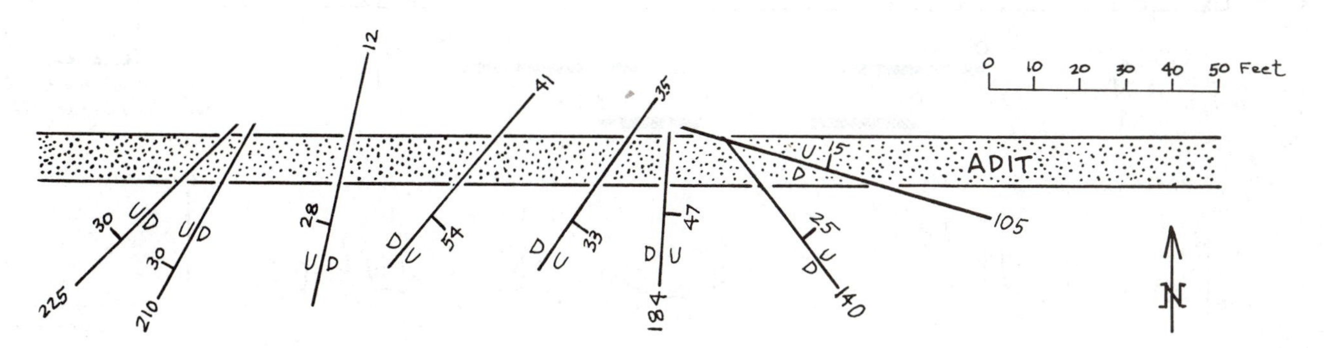

Figure 10-8
Map view of a mine adit, showing attitudes of minor faults. For use in Problem 10-4.

Problem 10-5

Figure 10-9 shows a geologic map and structure section from the Inyo Mountains of eastern California. Using complete sentences, describe the history of principal stress orientations in this area. What can you say about the specific time periods that variously oriented stresses were in effect? Is there any evidence that shear surfaces were controlled by anything other than the orientation of the principal stresses? You might want to color one or more units on the map and structure section to emphasize the geology.

Problem 10-6

In one succinct paragraph, describe the history of the principal stress orientations in the Bree Creek Quadrangle. Be as specific as possible about the time intervals during which variously oriented stress ellipsoids were in effect. Cite specific evidence to support your conclusions.

Complications Due to Pre-existing Planes of Weakness

All of the foregoing discussion assumes that the rocks being studied have no intrinsic preferred directions of shear. Of course, this assumption is often incorrect. Planes of weakness, such as bedding or cleavage planes, joints, or pre-existing faults will serve as preferred shear surfaces and will cause faults to have different attitudes than they otherwise would.

Often old faults that formed in response to one stress system will be reactivated by another one. Figure 10-10 shows an example from northern California. The San Andreas fault, a right-lateral strike-slip fault, is subparallel to a thrust fault called the Sargent-Berrocal fault. According to the principles summarized in Figure 10-7, σ_1 for the San Andreas fault should be horizontal and approximately north-south, and σ_1 for the Sargent-Berrocal fault should be horizontal and approximately northeast-southwest. Both faults are

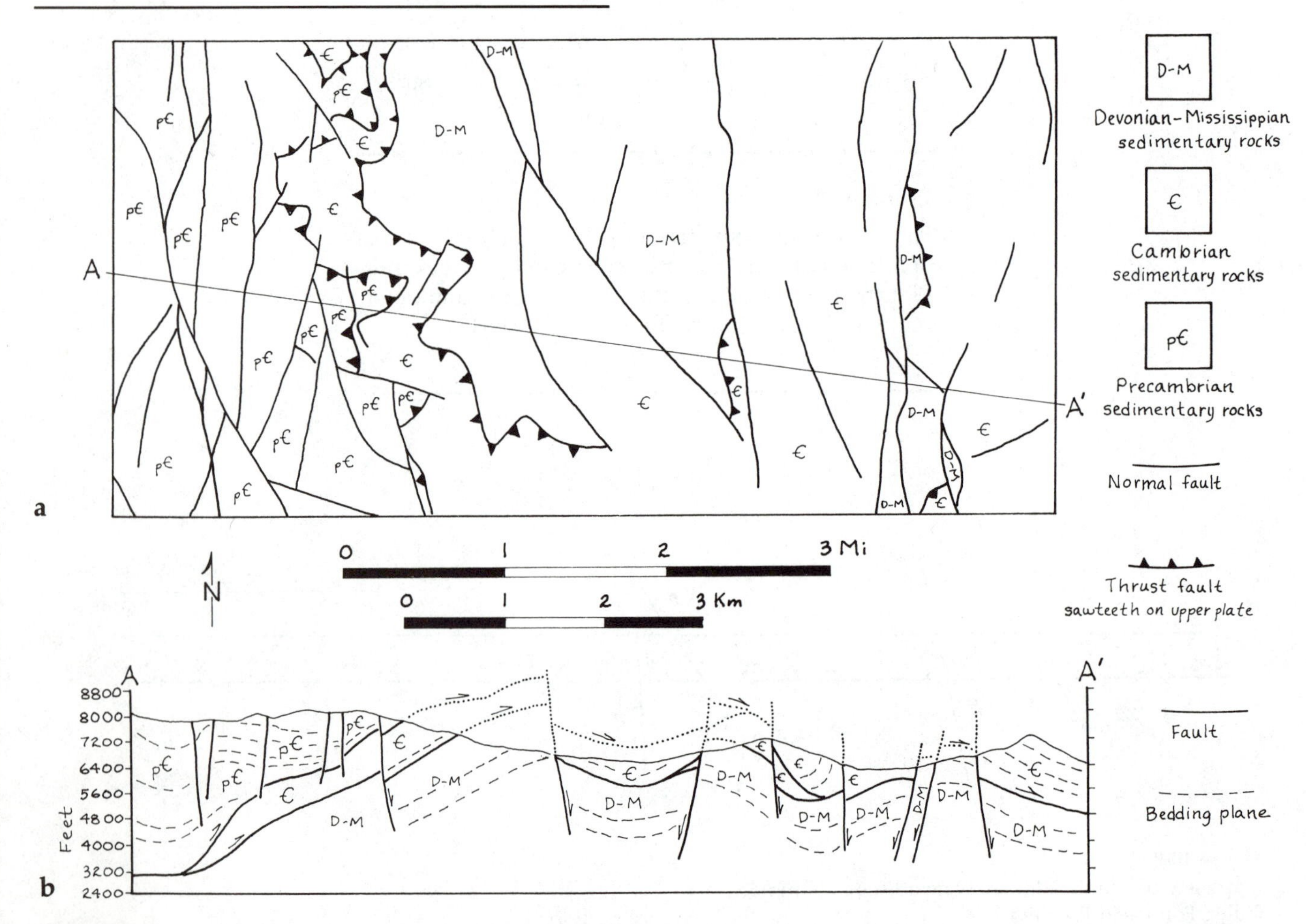

Figure 10-9
Generalized geologic map (a) and structure section (b) of a portion of the Inyo Mountains of eastern California, for use in Problem 10-5. Generalized from Nelson (1971).

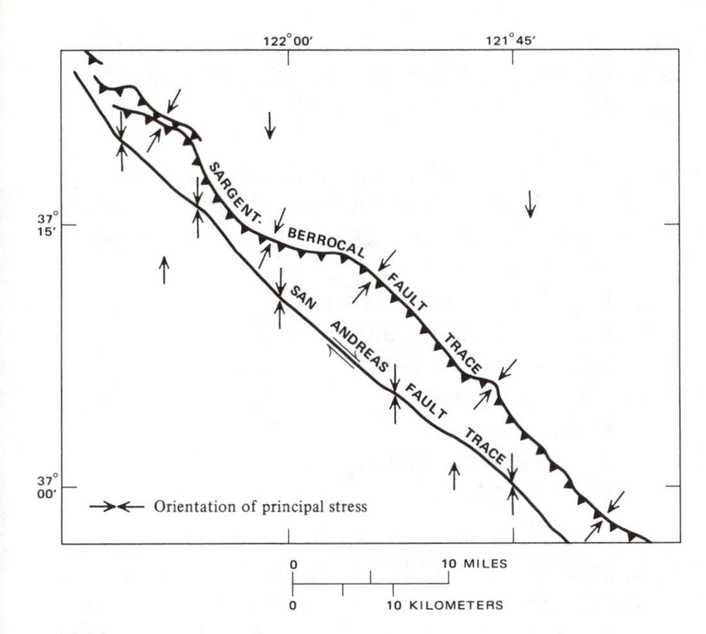

Figure 10-10
Sargent-Berrocal and San Andreas fault traces in northern California. Orientation of principal stresses based on surface trace orientation and field observations. From McLaughlin (1974).

seismically active. σ_1 cannot be oriented in two directions in the same place at the same time, so the orientation of at least one of these faults must be controlled by factors other than the present stress system. The apparent explanation is that the modern Sargent-Berrocal fault is not a product of the current stress system but is an old shear surface that developed under different stress orientations. It is an old fault that has been reactivated by new stresses (McLaughlin, 1974).

Movement on such pre-existing planes of weakness is typically oblique slip. The orientation of the stress ellipsoid cannot be determined in such cases unless there is a population of variously oriented faults with slickenside lineations. If such faults do occur, the theoretically preferred fault plane is the great circle defined by the stereographically projected slickenside lineations that occur on the pre-existing planes of weakness.

Nonuniform Stress Fields

Although Anderson's assumptions about faulting and the stress ellipsoid have proved to be extremely useful, we now know that they are not always valid. For example, instead of the two sets of conjugate faults predicted by Anderson, sometimes a rhombohedral network of four fault sets forms in isotropic rock (Aydin and Reches, 1982). And in the cases of thrust faults and strike-slip faults, usually only one of the predicted two sets of faults actually develops.

Anderson also assumed that the orientation of the stress ellipsoid does not change with depth and that the stress field causing deformation in the shallow crust is uniform over a large area. Implicit in these assumptions was the expectation that a single fault type would characterize a region and that multiple fault types require multiple deformation episodes. This expectation has turned out to be wrong. We will conclude this chapter with an examination of two examples of nonuniform stress fields occurring during a single tectonic episode. The first is from an extensional tectonic regime, the Basin and Range province of the western United States, in which the stress system has been nonuniform through time, and the second is from a compressive regime, the Himalayan-Tibetan region of Asia, in which the stress system is highly variable in space.

Problem 10-7

Figure 10-11 is a generalized map of the southwestern United States showing the Basin and Range province and bordering regions. The Basin and Range province derives its name from the north-south-trending basins and ranges that occur there. Two late Cenozoic deformational fields are recognizable in this area (Wright, 1976). Field I, including central and northern Nevada and western Utah, is characterized by listric normal faults. Field II, including southern and westernmost Nevada and eastern California, consists of a combination of normal and strike-slip faults (both dextral and sinistral).

1. On Figure 10-11 indicate the orientations of the two horizontal principal stresses (σ_2 and σ_3) in field I of the Basin and Range province.

The orientation of the stress ellipsoid in field II is more puzzling because both normal and strike-slip faults occur there. The first principles of faulting do not permit such faults to coexist during the same deformational event. In order to examine field II more carefully, we will look at the detailed fault patterns of one small area at Hoover Dam on the Arizona-Nevada border. The faults at Hoover Dam have been studied in detail by Angelier and others (1985). Figure 10-12 is a schematic summary of their results. Angelier and others (1985) recognized four faulting episodes. The first (early normal faulting) is depicted in Figure 10-12a, and the second (early strike-slip faulting) is depicted in Figure 10-12b.

2. Draw the three principal stresses on Figures 10-12a and b.

The early strike-slip stage was followed by a late normal-faulting stage, which was then followed by a late strike-slip stage. The end result is schematically shown in Figure 10-12c.

3. On Figure 10-12c, draw the three principal stresses that produce the late strike-slip stage.

In Figure 10-12b, one of the strike-slip faults is left-lateral and the other is right-lateral. Notice that one segment of the left lateral fault is reactivated in Figure 10-12c with right-lateral motion. All of the faulting depicted in Figure 10-12 took place during Miocene regional extension, when the Basin and range province developed.

4. Explain how a strike-slip fault can change from sinistral to dextral during one tectonic event. (Hint: It has to do with the geographic orientation of σ_3.)

5. The authors of this study attributed this alternation of normal and strike-slip faulting to "permutations of σ_1 and σ_2 [which] represent stress oscillations in time and space" (Angelier and others, 1985, p. 361). Bearing in mind that this was a time of active volcanism, crustal thinning, and high denudation rates, what circumstances might cause the vertical principal stress to increase and decrease in magnitude relative to the horizontal principal stresses?

Problem 10-8

Asia contains a complex array of active fault types. As shown in Figure 10-13, for example, there is a major thrust fault in the Himalaya, major left-lateral strike-slip faults (e.g., Kunlun and Altyn Tagh faults) in China, major right-lateral strike-slip faults in Soviet Central Asia (e.g., Talasso-Fergana fault) and in Indochina (e.g., Red River fault), and normal fault systems in Siberia (Baikal rift system) and China (Shansi graben system). Obviously no single stress ellipsoid orientation can account for this tectonic nightmare, yet all of these faults probably owe their existence to the collision and continued compression between India and Asia which began in the Eocene (Molnar and Tapponnier, 1975).

The India-Asia collision has been experimentally reconstructed with plasticene, producing insightful results (Tapponnier and others, 1982). Figure 10-14 shows drawings made from photographs taken during one of the plasticene experiments. The upper and lower surfaces of the plasticene were confined between two plates, preventing the development of dip-slip faults, but in spite of this limitation there is a remarkable similarity between the features in Figure 10-14c and the fault map of Asia (Fig. 10-13).

1. On the basis of fault type and orientation, on Figure 10-13 draw the orientations of the two horizontal principal stresses acting on each of the seven faults listed below.
   Himalayan frontal thrust
   Quetta-Chaman fault
   Talasso-Fergana fault
   Altyn Tagh fault
   Baikal rift system
   Shansi graben system
   Kang Ting fault

2. Place a sheet of tracing paper over the three drawings of Figure 10-14. On Figure 10-14c locate the equivalent seven faults listed above. Draw these faults on the Figure 10-14c overlay, and transfer the stress orientations from Figure 10-13.

3. On the overlays of Figures 10-14a and b draw the major faults and indicate the orientation of the horizontal principal stresses at various places on the map.

4. In one succinct paragraph describe the evolution of regional stresses during the India-Asia collision.

Further Reading

Anderson, E. M. 1952. *The Dynamics of Faulting*. London: Oliver and Boyd. A slightly revised edition of Anderson's very influential 1942 book. Mostly of historical interest.

Angelier, J. 1979. "Determination of the mean principal directions of stresses for a given fault population." *Tectonophysics*, v. 56, T17–T26. New methods for determining mean principal directions of stresses from fault orientations.

Angelier, J., Colletta, B., and Anderson, R. W. 1985. "Neogene paleostress changes in the Basin and Range." *Geological Society of America Bulletin*, v. 96, 347–361. A good example of detailed analysis of paleostress.

Dalziel, I. W. D., and Stirewalt, G. L. 1975. "Stress history of folding and cleavage development, Baraboo syncline, Wisconsin." *Geological Society of America Bulletin*, v. 86, 1671–1690. Good example of a study of the relationship between the stress ellipsoid and folds.

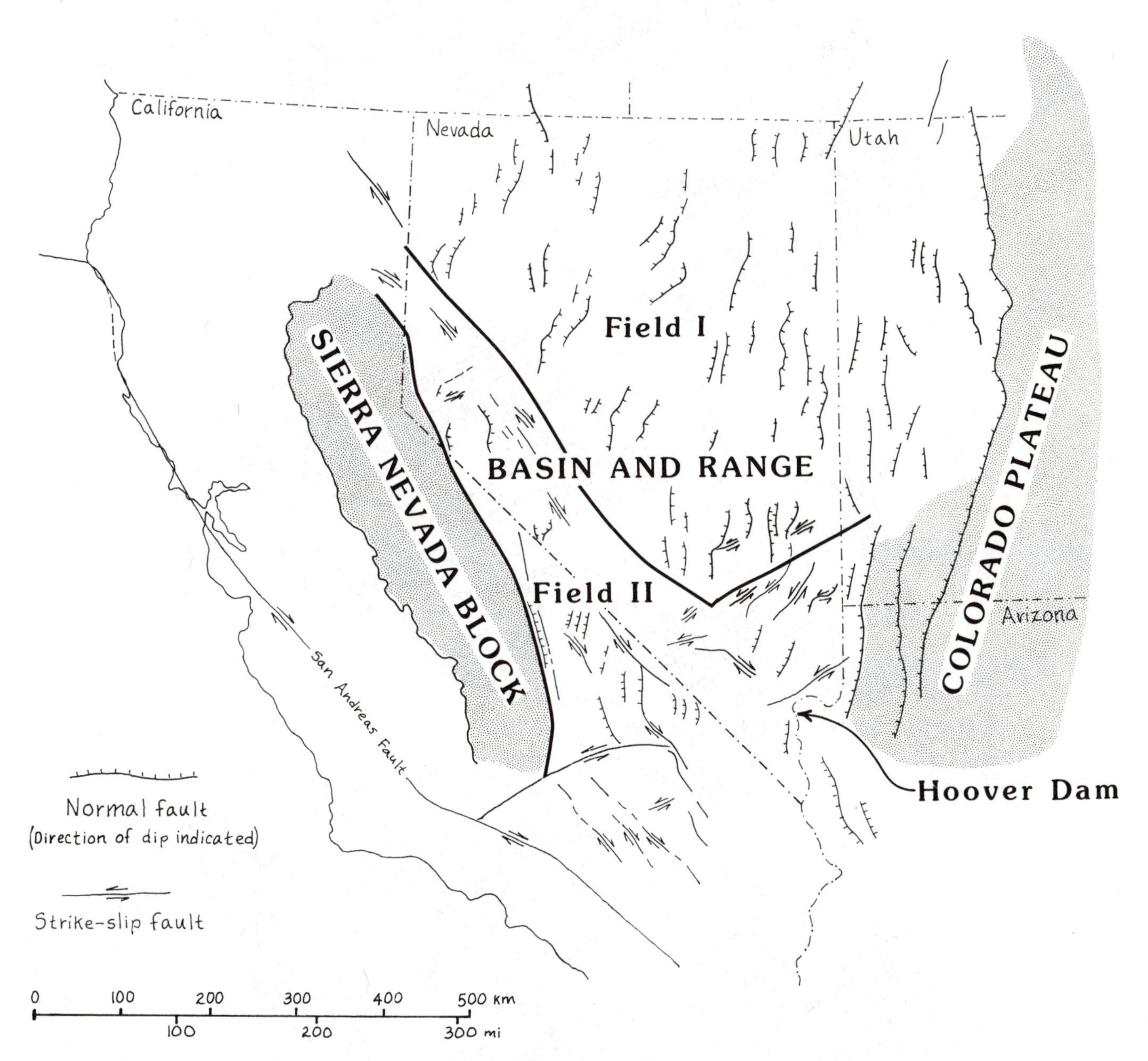

Figure 10-11
Generalized map of late Cenozoic structural features of the southwestern United States. Field I is characterized by listric normal faults. Field II is characterized by normal faults as well as dextral and sinistral strike-slip faults. After Wright (1976).

a

b

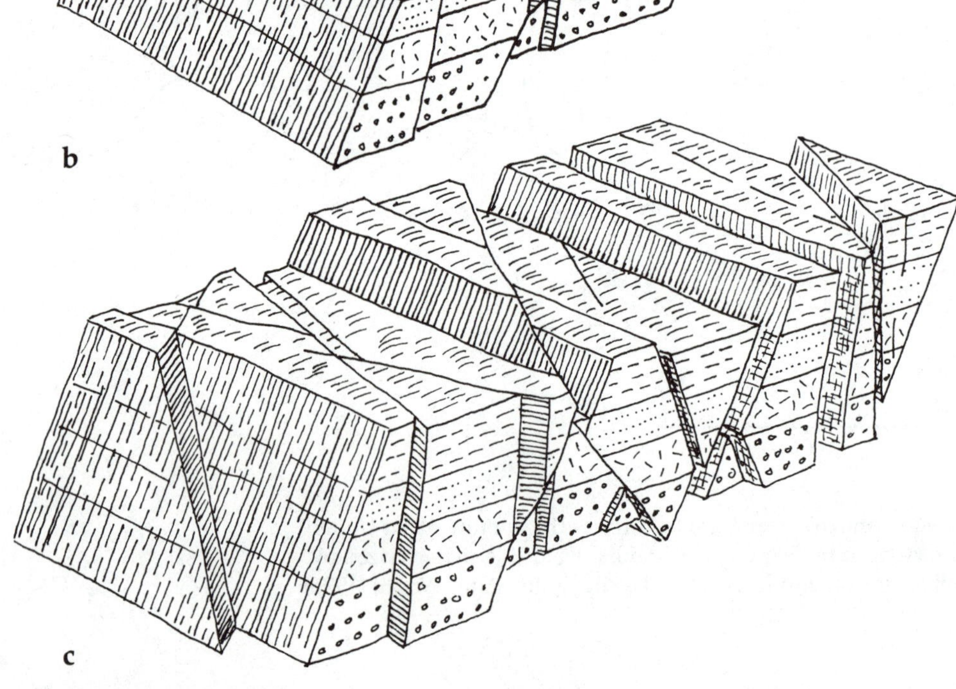

c

Figure 10-12
Schematic block diagrams showing the main characteristics of faulting at Hoover Dam. After Angelier and others (1985).

Figure 10-13
Generalized map of southern Asia showing the major active faults. After Molnar and Tapponnier (1975) and Tapponnier and others (1982).

Notes

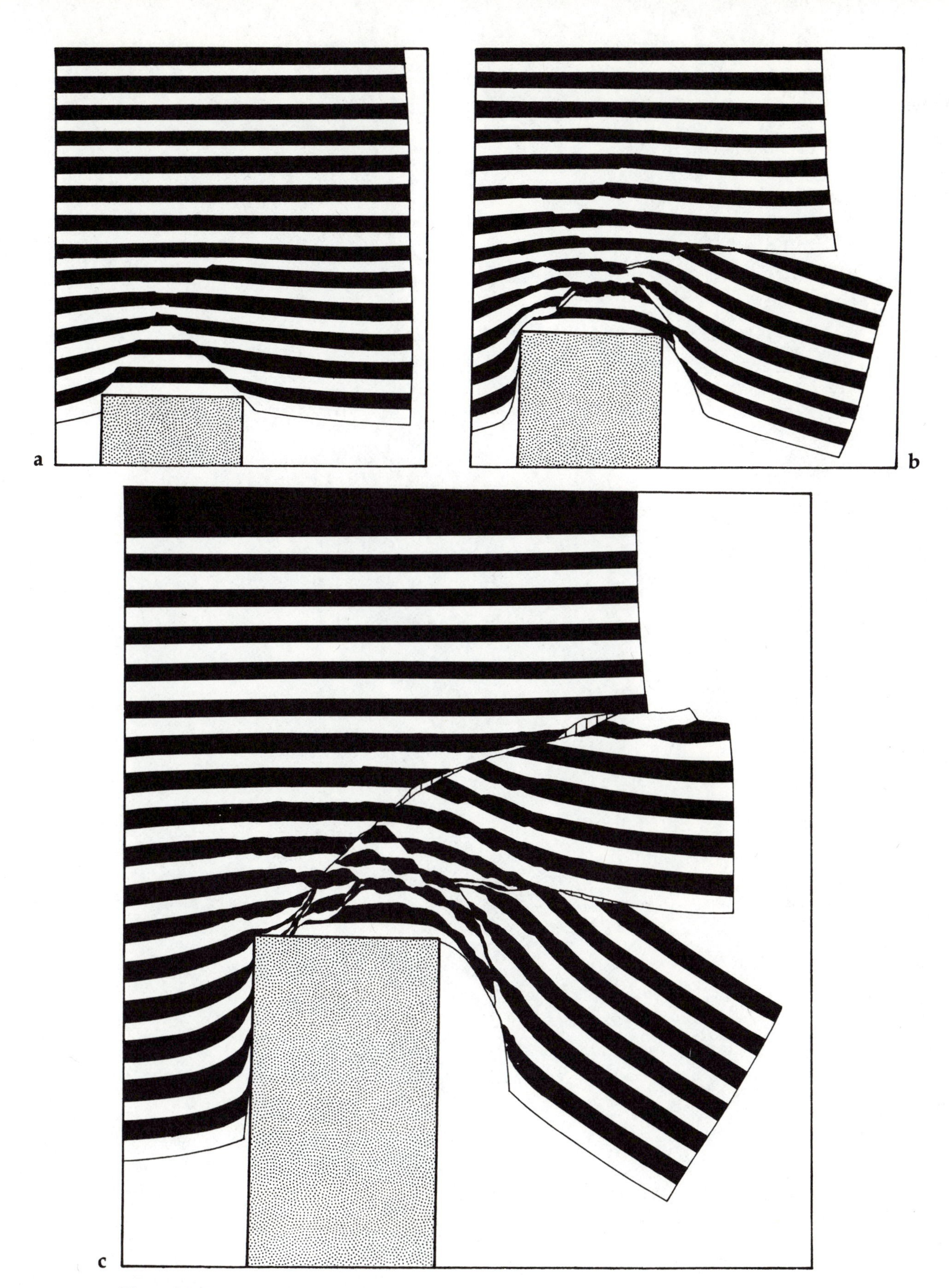

Figure 10-14
Three stages of a plasticene experiment designed to simulate the collision of India and Asia. In this particular experiment the layers of plasticene are confined at the top and on the left side but unconfined on the right side. After Tapponnier and others (1982).

Notes

11

A Structural Synthesis

OBJECTIVE

Write a professional-quality structural history of the Bree Creek Quadrangle

The ultimate objective of analyzing the structures of an area is to reconstruct the area's structural history. Even when the impetus is purely economic, a great deal of time and money is often spent on sorting out generations of deformation and compiling a detailed geologic history. Such knowledge is not just academic; it may be crucial for the successful discovery of ore bodies and petroleum reservoirs.

Structural Synthesis of the Bree Creek Quadrangle

From your work on the Bree Creek Quadrangle map you have the data to reconstruct quite a detailed structural history of that area. The task in this chapter is to do just that. Below is a list of the problems in this book that deal with the Bree Creek Quadrangle:

Problem	Task
2-1	Draw structure contours on upper surface of Bree Conglomerate
3-1	Determine attitudes of Neogene units
3-2	Determine thicknesses of Paleogene units
3-3	Determine approximate thickness of Neogene units
3-5	Construct a stratigraphic column
4-4	Draw structure sections A–A′ and B–B′
5-11	Determine amount of post-Rohan, pre-Helm's deep tilting of northeast fault block

7-2	Construct contoured pi-diagrams, profile views, dip isogons, and summary diagrams of folds; describe folds
8-3	Draw axial surface traces of superposed folds; draw structure sections C–C′ and D–D′; describe superposed folds
9-4	Determine amount and direction of Neogene tilting on fault blocks
9-5	Determine difference in amounts of tilt between adjacent pairs of fault blocks
9-6	Determine amount and direction of rotation on faults
9-7	Determine amount of pre-Gondor tilting and rotation on faults
9-8	Describe faults
10-6	Describe history of principal stress orientation

Synthesize as many of these tasks as you have completed into a cohesive summary of the structural history of the Bree Creek Quadrangle. This synthesis should consist of the following:

A. Geologic map (in an envelope at the back of the report).

B. Structure sections A–A′, B–B′, C–C′, and D–D′, neatly drawn, inked, and colored (in an envelope with the map).

C. Text (to include the following, with a *subheading* for each section):

1. Title
2. Abstract
   Write this *last,* but put it at the front of your report on a separate page. An abstract is a concise but comprehensive summary. It *is* the report, condensed and packed with concentrated information and significant results.
3. Table of contents
4. Introduction
   Briefly introduce the terrain and the structures. Since you didn't actually do the field work yourself, your introduction in this case should be one succinct paragraph.
5. Stratigraphy
   Add the Paleozoic and Mesozoic rocks to the stratigraphic column you drew in Problem 3-5 and include this complete column in your report. The approximate thicknesses of the Paleozoic units can be measured directly off your structure sections C–C′ or D–D′. Irregular plutonic units, such as the Dark Tower Granodiorite, are not assigned a thickness.

   Briefly describe the stratigraphy, paying special attention to unconformities; they often have structural significance. If you had mapped the Bree Creek Quadrangle yourself, you would include detailed rock descriptions here as well.
6. Folds
   Combine your work from Problems 7-2 and 8-3 into a detailed but readable description of the folds. Use your structure sections, contoured diagrams, and profile views to support and illustrate your descriptions. For descriptive purposes the Bree Creek Quadrangle can be conveniently divided into four subareas, each of which is a separate fault block. To allow a quick comparison of the folding in each subarea, include the page-sized reference map from Chapter 7 (Fig. 7-14).

   Prepare a *synoptic diagram* that shows the variation in the orientation of the fold axes of the folded Tertiary rocks. A synoptic diagram shows data from different subareas plotted together. Figure 11-1 is an example of such a diagram.
7. Faults
   As specifically as possible, describe the age of each fault, orientation of the fault surface, sense of movement, and amount of offset.
8. Orientation of Principal Stresses
   Review the orientation of principal stresses at various times, citing *specific* structural features to support your statements.
9. Discussion
   Discuss how faulting, folding, and stratigraphy relate to one another. Did faulting precede folding or follow folding or both? Have folded sections been rotated by faulting? Can sedimentation, erosion, intrusion, or metamorphism be related to structural events? Specifically how? Has the stress orientation of the area changed? When and in what way? Support your statements with references to the map, structure sections, and other diagrams.
10. Summary of Structural History
    Succinctly summarize the structural history that you have just discussed in detail. Begin

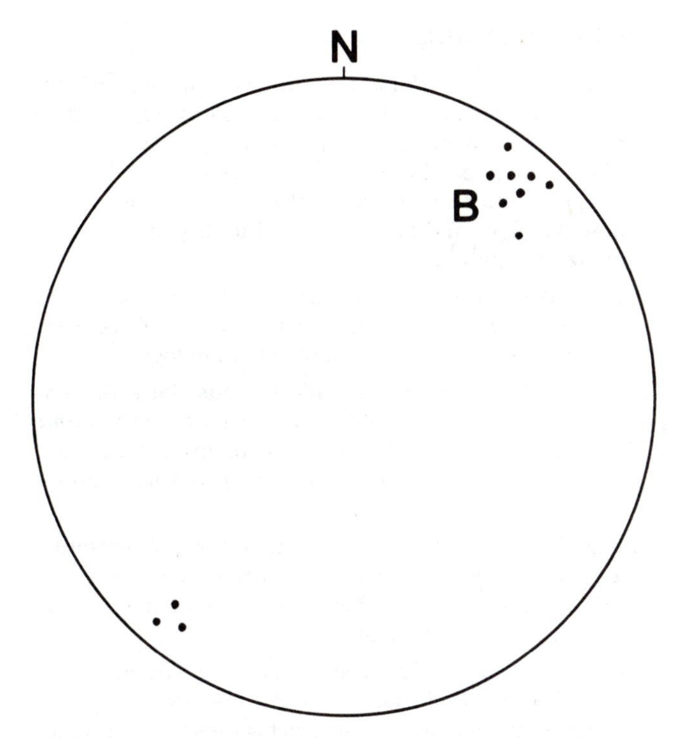

Figure 11-1
Synoptic B-axis diagram prepared from 11 subareas (after Weiss, 1954).

with the oldest events and work forward. Be very specific about the period or epoch in which an event occurred. This is your chance to tie everything together into a neat package.

11. References Cited
Give credit to your sources of information by citing references. Many students seem to have trouble learning when and how to cite references. It is worth the trouble to look at any geologic journal and pay close attention to reference citation.

 Generally speaking, any time you use someone else's observations or conclusions you need to cite the reference. This is done by providing the author and date of the publication in either of the following two ways: "Smith (1985) demonstrated that the Saddle Island fault is a low-angle normal fault." or "The Saddle Island fault is a low-angle normal fault (Smith, 1985)." The complete bibliographic reference is then provided in the "References Cited" section of the report.

 If a reference has two authors it is normal practice to list both of them in the text of the paper: (Smith and Jones, 1985). With three or more authors, the citation is: (Smith and others, 1985) or (Smith *et al.*, 1985). All authors' names must still be listed in the "References Cited" section, however. Do *not* list references in your "References Cited" section if they are not referred to in the text of the report.

Writing Style

Write with a specific reader in mind. This should be a geologist who has never seen the area you are describing. You do *not* need to explain basic geologic concepts and terms (e.g., anticlines are folds with the oldest rocks in their cores), but you *do* need to explain things that are known only to geologists familiar with the local area (e.g., the Bree Creek fault strikes north-south and dips 50° to the west).

Explain your data and conclusions in clear, simple prose. Aim for short sentences. Try reading aloud what you have written; if it doesn't flow smoothly, then it needs to be rewritten. Here is an example of a sentence that makes the reader struggle: "It should be noted that a clast of indurated crustal material perpetually rotating on its axis along the modern air-lithosphere interface is somewhat unlikely to accumulate an accretion of bryophytic vegetation."

Do not put all of your diagrams at the back of the report, where they are difficult for the reader to find. In this report, the map and structure sections must be separated from the text, but in general it is desirable to put diagrams as close as possible to the part of the text where they are most relevant. Label your diagrams "Figure 1," "Figure 2," etc., and refer to each figure in the report. Be sure each figure has a caption that explains its significance, even though this will duplicate some of the explanation in the text.

Within reason, avoid using passive voice and third-person constructions, such as: "This area was studied by the author in 1985." It is much more direct to write: "I studied this area in 1985." The old scientific convention of always referring to yourself in the third person is now considered clumsy and stiff. Furthermore, it carries a sense of trying to remove yourself from your work, as if you don't want to be responsible if it isn't quite correct.

Beware of vague qualifiers such as "rather," "somewhat," and "fairly" and "weasel words" such as "seems" and "might." Sometimes these words are right, but people often use them by reflex and then do not think through what they are saying. For example, a geologist might write: "The orientation of σ_1 seems to change clockwise during the Tertiary by a rather small amount." If the evidence is not conclusive, it is better to write: "My analysis suggests that the orientation of σ_1 rotated clockwise by less than 10° during the Tertiary, but more fault orientations should be measured to confirm the rotation."

Common Errors in Geology Reports

Here are a few errors that repeatedly appear in geology student reports:

1. The most commonly misspelled word in student reports is "occurred," which like "occurring" and "occurrence" has two Rs.
2. The most commonly misspelled geologic period is the Ordovician, with the Cretaceous a close second. Do not confuse Paleocene with Paleogene.
3. When you are describing rocks that you personally examined, use the present tense. Students often write such things as, "The Tapeats Sandstone *was* a coarse-grained quartz sandstone," because that *was* what they saw. If the rocks still exist, use the present tense.
4. Watch your use of upper/late and lower/early. Upper and lower refer to lithostratigraphic position, while early and late refer to time. The Upper Cambrian Nopah Formation is of Late Cambrian age. It was deposited during the Late Cambrian Epoch. For capitalization, check a detailed geologic time scale—the Cambrian, for instance, has three formal epochs named Early, Middle, and Late but the Cretaceous has only two.

Further Reading

Amenta, R. V. 1974. "Multiple deformation and metamorphism from structural analysis in the eastern Pennsylvania piedmont." *Geological Society of America Bulletin*, v. 85, 1647–1660. A good example of a structural synthesis of a region with several structural events. Note especially how diagrams and tables are used to display and summarize complex data.

Cluff, L. 1980. "Time and time again." *Journal of Sedimentary Petrology*, v. 50, 1021–1022. A review of the correct usage of temporal and lithostratigraphic terminology.

Cochran, W., Fenner, P., and Hill, M., eds. 1984. *Geowriting—A guide to writing, editing, and printing in earth science* (4th edition). Falls Church, VA, American Geological Institute. A very useful handbook for geologists who are writing for publication.

Murray, M. W. 1968. "Written communication—A substitute for good dialog." *American Association of Petroleum Geologists Bulletin*, v. 52, 2092–2097. An excellent guide to effective organization of scientific reports.

Norris, R. M. 1983. "Field geology and the written word." *Journal of Geological Education*, v. 31, 184–189. A detailed review of the method one instructor uses to teach report writing skills in conjunction with field mapping.

12

Rheological Models

EQUIPMENT NEEDED FOR THIS CHAPTER

rubber bands

plastic disposable syringe (available from any medical facility)

string

block of wood

Silly Putty (available at toy stores)

OBJECTIVE

Acquire a qualitative understanding of rheological models as analogs of rock deformation

Rocks respond in complex ways to stress: the same rock layer will fold under one set of conditions and fracture under another. Adjacent layers may behave differently under the same conditions. Various aspects of stress and strain are examined in Chapters 10, 13, and 14. In this chapter we will investigate idealized relationships between stress, strain, and strain rate.

Stress, symbolized by the Greek letter σ (sigma), is measured in units of *force per unit area*. Strain, symbolized by the Greek letter ϵ (epsilon), is a *change in shape or volume* measured in various ways. In addition to stress and strain, *time* is an important element in the study of deformation. The study of the relationships between stress, strain, and time is called **rheology,** from the Greek word *rheos,* which means a flow or current. A rheological model is a characteristic relationship between stress, strain, and time, exhibited by an object being deformed.

In order to gain a qualitative understanding of stress and strain it will be useful to examine three rheological models: elastic deformation, viscous deformation, and plastic deformation. We will examine these separately and in combination in an attempt to understand deformation in rocks.

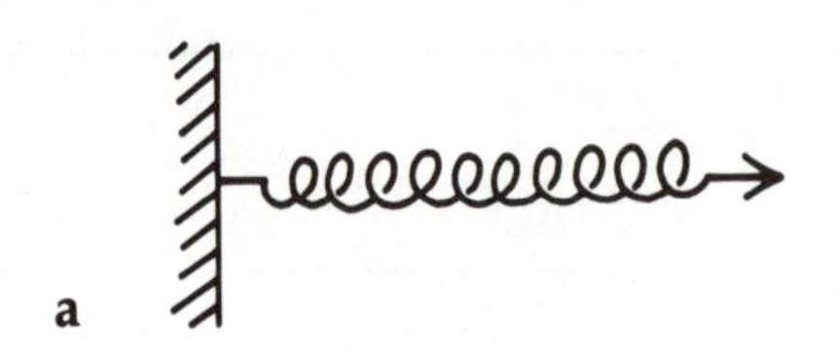

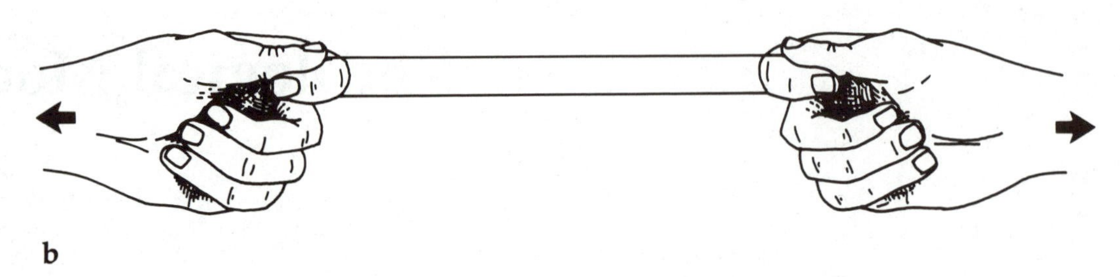

Figure 12-1
Elastic deformation. (a) Schematically represented as a coiled spring. (b) Simulated with a rubber band.

Elastic Deformation—Instantaneous, Recoverable Strain

Elastic deformation is exhibited by a rubber band or a coiled spring (Fig. 12-1). With a perfectly elastic body the strain is strictly a function of stress, and stress graphed against strain is a straight line (Fig. 12-2). Elastic deformation is described by Hooke's law, $\sigma = E\epsilon$, where E is the elasticity (Young's modulus) of the material. Objects that display perfect elastic behavior are called **Hookean bodies.** Rocks behave as Hookean bodies during earthquakes, when they transmit seismic waves.

Unlike other types of deformation, elastic deformation occurs very quickly in the earth and for our purposes will be assumed to be instantaneous. Another unique characteristic of elastic deformation is that the strain is recovered when the stress is removed, providing that the elastic limit of the material has not been exceeded.

To summarize the key features of elastic deformation: strain is directly proportional to stress, strain is (for our purposes) instantaneous, and strain is completely recovered when the stress is removed (unless the elastic limit has been exceeded).

σ

ϵ

(percent lengthening or shortening)

Figure 12-2
Stress/strain graph of elastic deformation. Slope of line varies with elasticity (Young's modulus) of the material.

Viscous Deformation—Continuous Strain under Any Stress

Some materials, such as water, flow in response to the slightest stress. Such behavior is known as viscous deformation, and materials that behave this way are called **Newtonian fluids.** The schematic analog of viscous deformation is a porous piston in a fluid-filled cylinder, together called a dashpot (Fig. 12-3a). A suitable dashpot for our experimentation is a plastic, disposable syringe common in hospitals (Fig. 12-3b).

Because viscous deformation is continuous at any stress, it is meaningless to graph stress against strain as in Figure 12-2. Here it is the strain *rate* rather than absolute strain that is significant. Strain rate is symbolized $\dot{\epsilon}$ (the first time derivative of ϵ). The strain rate of viscous material is a function of stress and viscosity: $\sigma = \eta\dot{\epsilon}$, where η (eta) is the coefficient of viscosity of the material. The greater the stress, the faster the deformation. Figure 12-4 shows σ graphed against $\dot{\epsilon}$ for a given material. Unlike elastic deformation, viscous deformation is permanent.

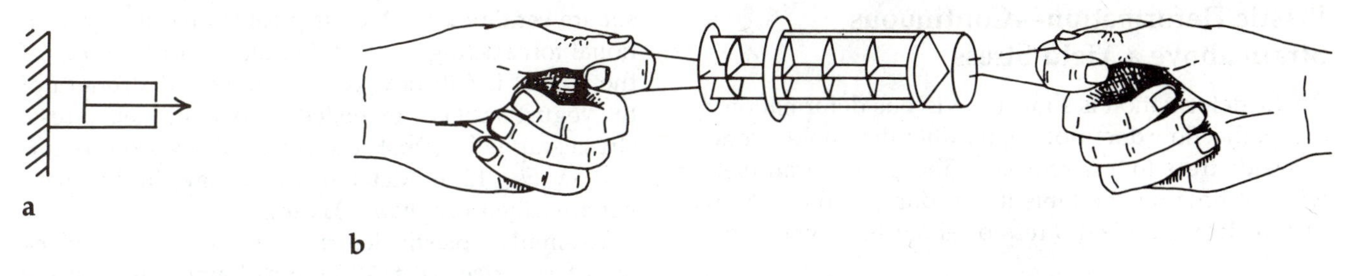

Figure 12-3
Viscous deformation. (a) Schematically represented as a leaky piston in a fluid-filled cylinder (together called a dashpot). (b) Simulated with a disposable syringe.

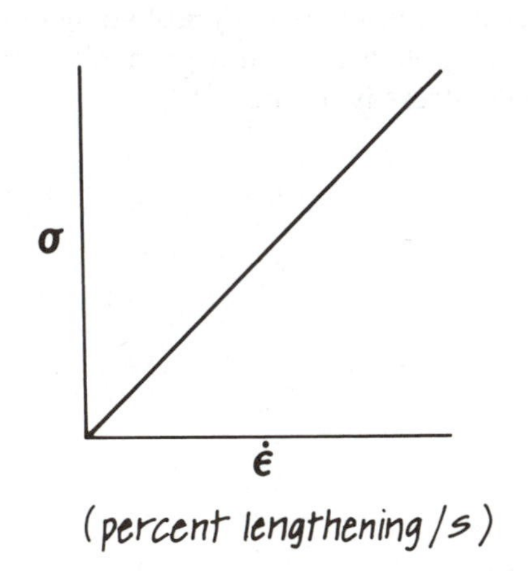

Figure 12-4
Stress/strain-rate graph of viscous deformation. Slope of line varies with viscosity of the material.

Because the total strain is partly a function of time, it is instructive to graph stress and strain separately against time, as in Figure 12-5. Examine the two graphs in Figure 12-5 carefully, and be sure that you understand how they relate to each other. We will be using such pairs of graphs throughout the rest of this chapter. In Figure 12-5a stress is shown first applied at t_1 and removed at t_2. In Figure 12-5b it can be seen that strain is continuous from t_1 to t_2, after which no more strain occurs.

Strain is commonly measured in percent lengthening or shortening per second because seconds are convenient units of time for lab experiments. For example, if an object under stress were shortened from 10 cm to 9 cm in 100 seconds the strain rate would be: $10\%/100\ s = .1/100\ s = .001/s = 1 \times 10^{-3}/s$. Strain rates in the earth's crust, of course, are many orders of magnitude slower. For example, by measuring the change in distance between points on opposite sides of the San Andreas fault, a strain rate of $1.5 \times 10^{-13}/s$ has been determined.

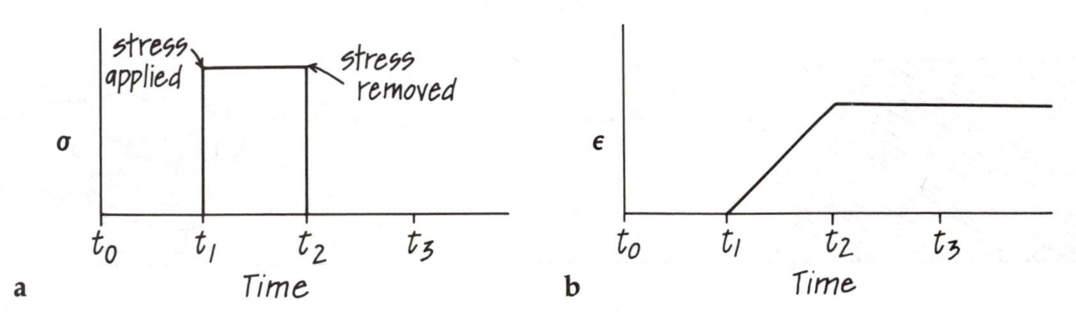

Figure 12-5
Viscous deformation experiment in which time is simultaneously graphed against stress (a) and strain (b).

Plastic Deformation—Continuous Strain above a Yield Stress

Plastic deformation is similar to viscous deformation, except that flow does not begin until a threshold stress or yield stress (σ_y) is achieved. Yogurt, for example, will not flow off the table if you dump it out of the carton. It has a yield stress of about 800 dynes per square centimeter, which is greater than the gravitational force acting on it. If you place a dish on top of the yogurt it will flow because the yield strength of the yogurt has been exceeded. Above the yield stress, stress graphed against strain rate is like viscous deformation (Fig. 12-6). Materials that behave in this manner are called **Bingham plastics.**

To simulate plastic deformation we will use a block on a flat surface (Fig. 12-7). Small amounts of stress may be applied with no movement at all. There exists a yield stress σ_y, however, that will overcome the frictional force on the stationary block. Once the yield stress is applied the frictional force is overcome, and the block begins to move and continues to move. A block on a table is not really a case of plastic deformation; a key difference is that while the block is not deformed, a plastic body is deformed. However, the similarity in the relationships between stress, strain, and time allows us to use the block as an analog of plastic deformation.

Notice that after the yield stress is applied, the amount of strain is a function of time. In Figure 12-8 stress and strain are separately graphed against time. Stress is first applied at t_1 and gradually increased until the yield stress is reached at t_2.

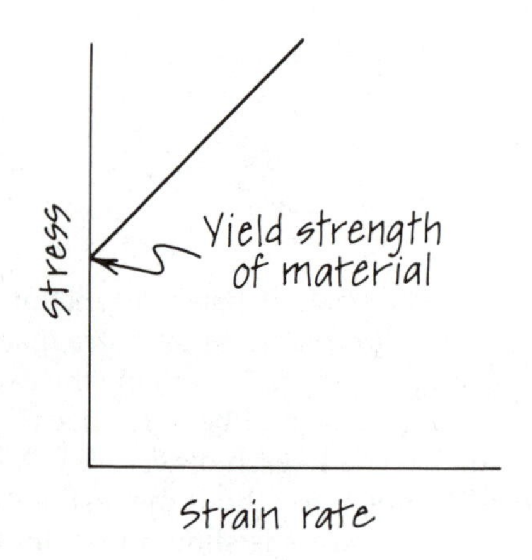

Figure 12-6
Stress/strain-rate graph of plastic deformation. Once yield strength of the material has been exceeded, behavior is viscous.

a

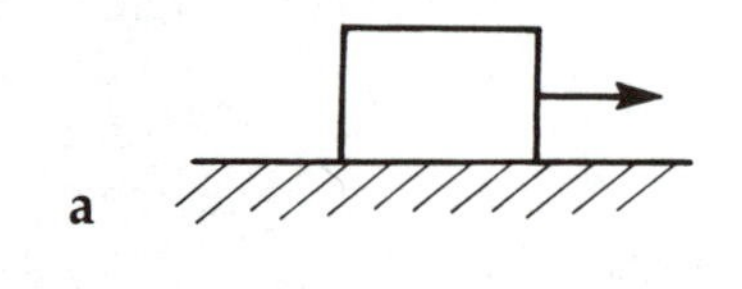

b

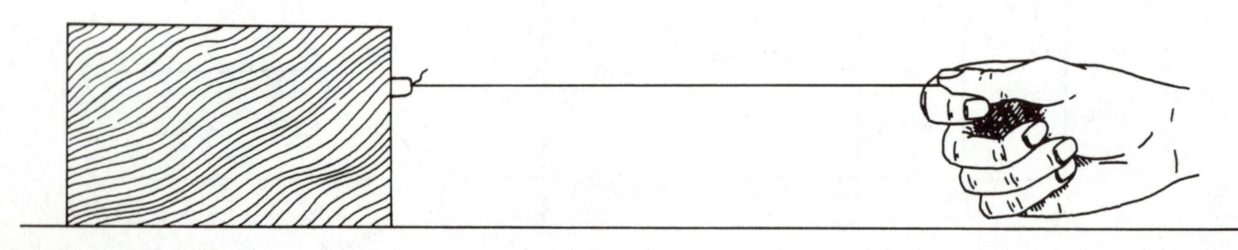

Figure 12-7
Plastic deformation. (a) Schematically represented as a block on a flat surface. (b) Simulated by pulling a wooden block with a string.

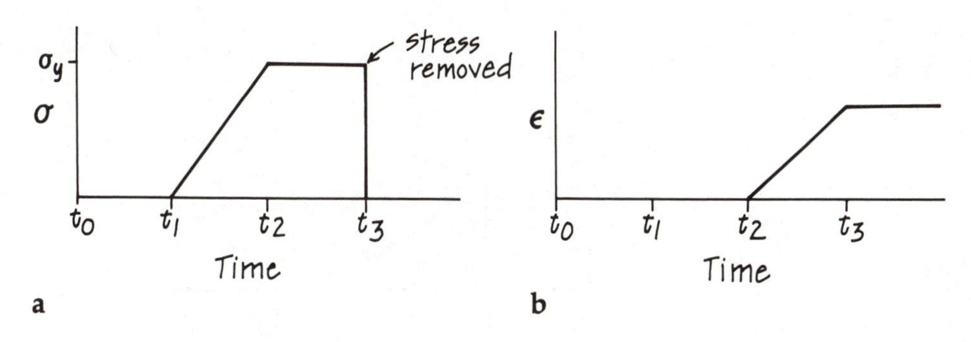

Figure 12-8
Plastic deformation experiment in which time is simultaneously graphed against stress (a) and strain (b).

Elastico-Plastic Deformation

Most materials display complex rheological characteristics that can be simulated with some combination of elastic, plastic, and viscous deformation. Attach a rubber band to a wooden block and conduct the experiment shown in Figure 12-9. Consider the behavior of the rubber band and block as a unit and carefully examine how this behavior is reflected in the σ/time and ϵ/time graph pair in Figure 12-9. Notice that the rubber band provides an elastic component and causes the strain to begin at t_1, even before the yield stress is reached. When stress is removed at t_3, however, the elastic deformation is recovered and the permanent deformation is a result of the plastic component.

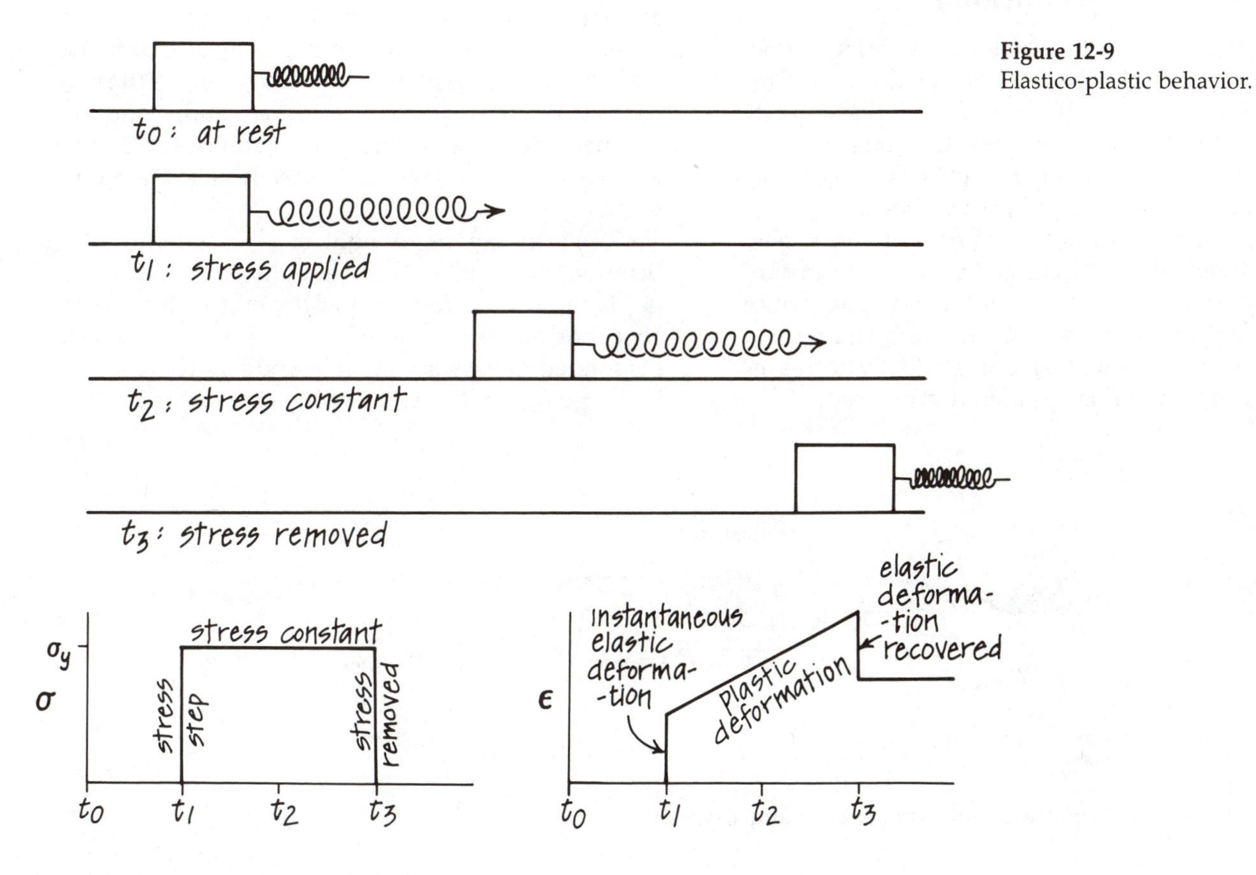

Figure 12-9
Elastico-plastic behavior.

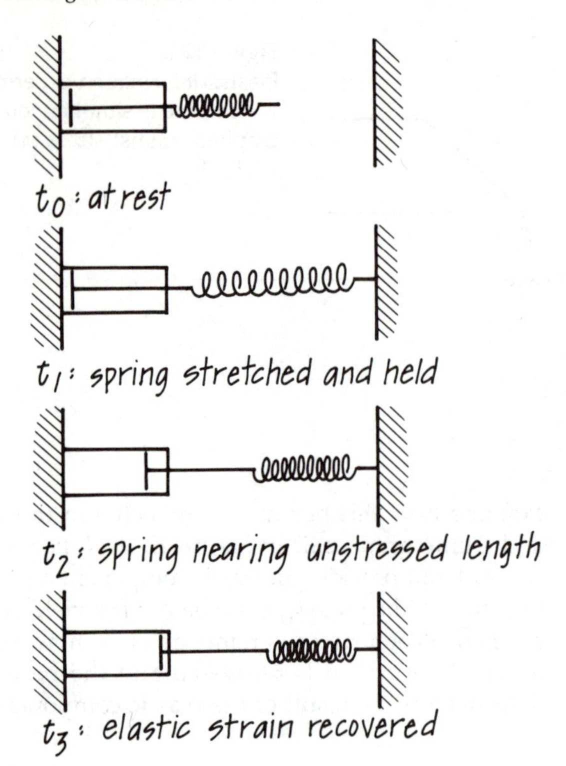

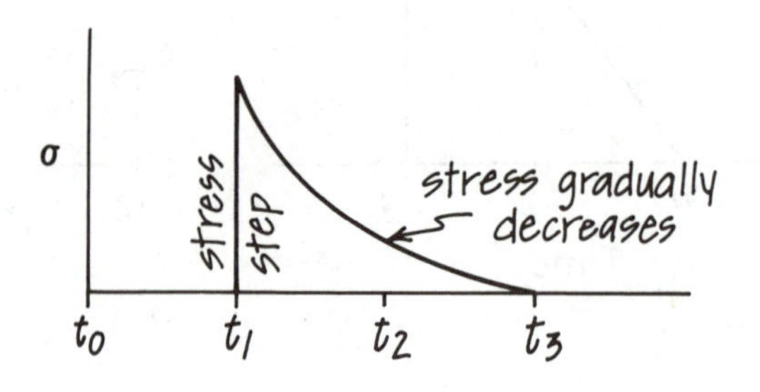

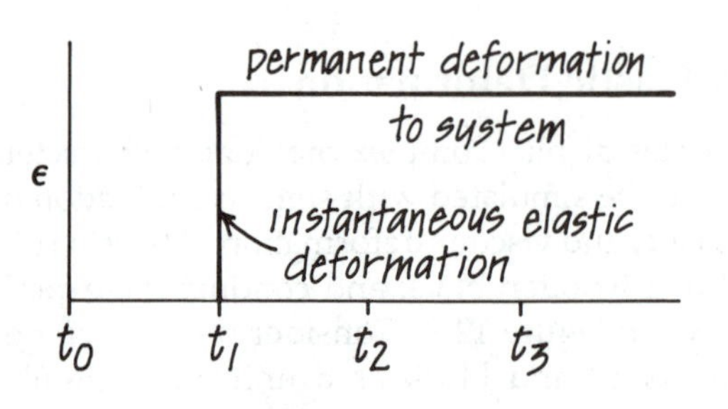

Figure 12-10
Elastico-viscous behavior.

Elastico-Viscous Deformation

Attach a rubber band to a syringe (as in Fig. 12-3b) and experiment with the behavior of the unit. This apparatus behaves just like the elastico-plastic body except that there is no yield stress that must be overcome before permanent strain begins. An object that behaves this way is called a **Maxwell body.**

In the experiment shown in Figure 12-10 the rubber band is stretched and fixed, giving the body instantaneous permanent strain. In the two graphs, notice that although the strain is instantaneous and permanent the stress is greatest at t_1 and gradually decreases until the elastic strain is completely recovered.

Firmo-Viscous Deformation

Combine a rubber band and a syringe as shown in Figure 12-11. Neither the elastic nor the viscous component can move without the other. Examine the stress and strain graphs of Figure 12-12 and note that even though the stress is constant the strain rate decreases with time as the rubber band lengthens. When the stress is removed at t_2 the strain rate jumps and then gradually decreases until all of the strain is recovered. An object that behaves this way is called a **Kelvin body.**

The earth can be thought of as a self-gravitating firmo-viscous sphere. For example, when the weight of glacial ice was removed at the end of the Pleistocene, northern portions of Europe and North America responded by isostatically rebounding. This rebound is still going on, but at a steadily decreasing rate.

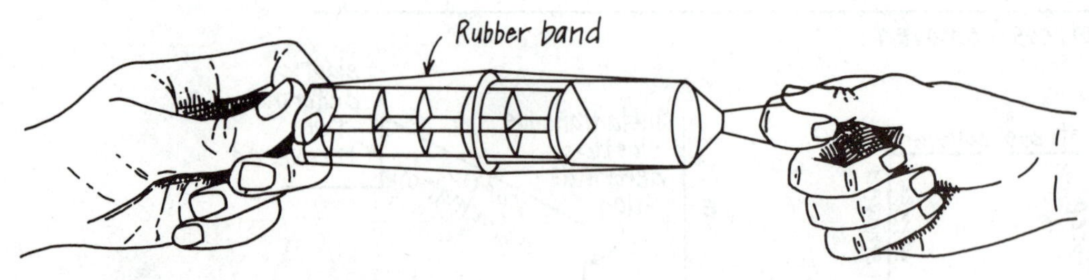

Figure 12-11
Firmo-viscous behavior simulated with a rubber band and a syringe.

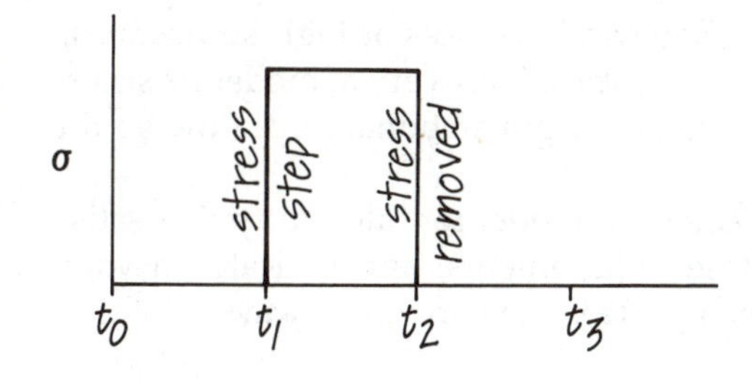

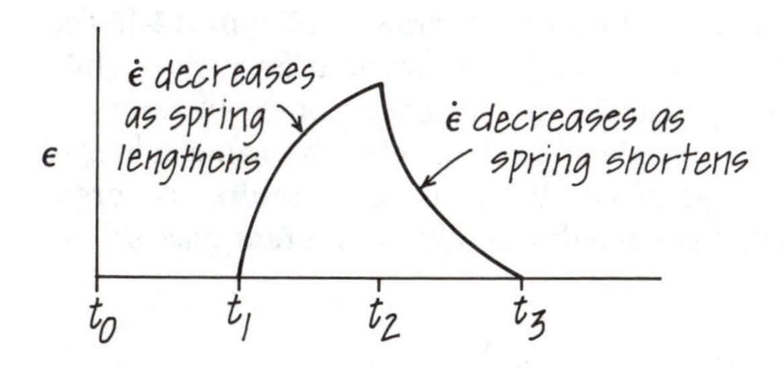

Figure 12-12
Firmo-viscous behavior.

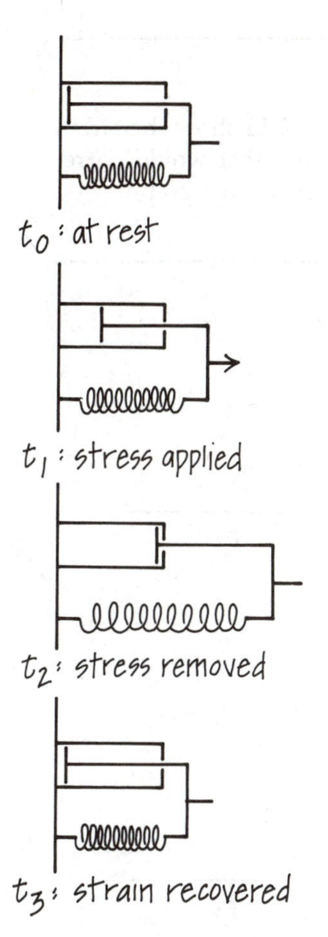

Within Every Rock Is a Little Dashpot

Under conditions of low temperatures and pressures and high strain rates, rocks behave brittlely. (The conditions under which rocks fracture are explored in Chapter 13.) In the lower crust, temperatures and pressures are such that the rocks are quite ductile. The boundary between shallow crust brittleness and deeper crust ductility is a relatively narrow zone called the brittle-ductile transition.

Even above this transition zone, at low strain rates rocks are surprisingly ductile. Over very long time intervals rocks are unable to sustain any differential stress. On a large planet with a strong gravitational field and no active tectonism there would be no mountains. They would flow like Silly Putty. An analogous situation exists on Europa, one of the satellites of Jupiter. Europa has an ice crust that is pockmarked with very few impact craters; it is the smoothest known body in the solar system. The mass of the satellite and the rheological properties of the ice combine to erase craters soon after they form. Within every rock is a little dashpot.

Many real solids behave like the rheological model shown in Figure 12-13. If stress is applied and immediately released, then the strain is elastic and is immediately recovered. But if stress is applied and held for a while, then the firmo-viscous component (dashpot and spring) becomes important. This combination of a Kelvin and an elastic body is called a **standard linear solid.** Such behavior can be seen in an old rubber band that has been wrapped around a newspaper for several weeks and is finally taken off. The limp rubber band slowly recovers some of its strain.

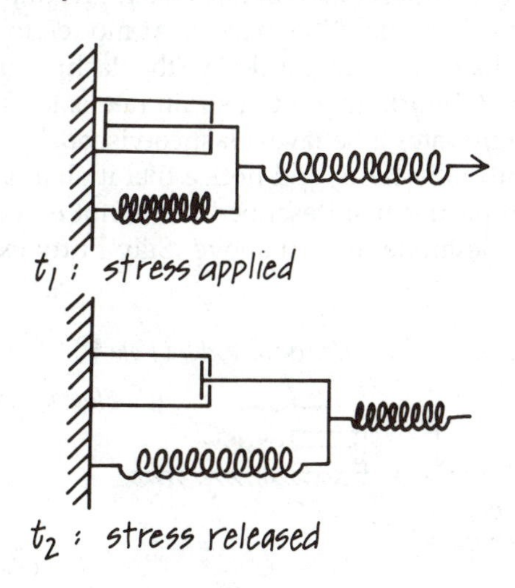

Figure 12-13
Rheological model called a standard linear solid.

Problem 12-1

On the ϵ/time graph in Figure 12-14 show the strain history of a standard linear solid that would correspond to the stress history in the σ/time graph.

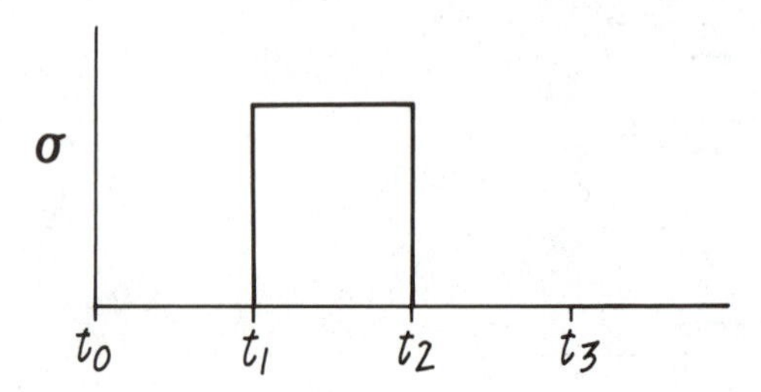

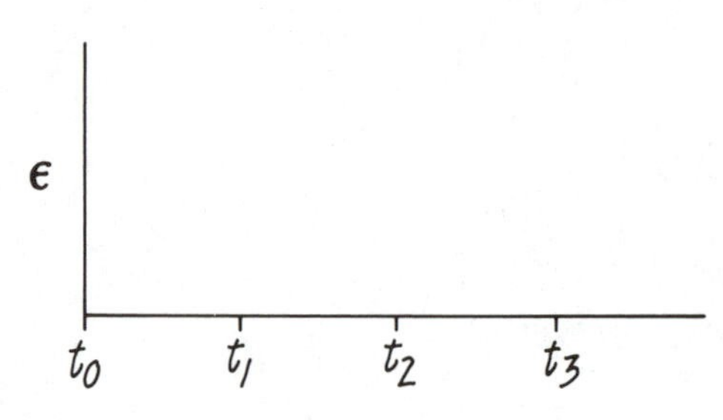

Figure 12-14
Stress/time and strain/time graphs for use in Problem 12-1. Times t_1 and t_2 correspond to t_1 and t_2 in Figure 12-13.

Problem 12-2

The rheological model shown in Figure 12-15 behaves differently at different strain rates. At high strain rates it behaves elastically ("bounces"). At moderate strain rates it behaves elastico-plastically (the dashpot doesn't have time to work unless the strain rate is low). And at low strain rates it behaves elastico-viscously. Experiment with Silly Putty and notice that it shares some of the properties just described but is not exactly the same as the model drawn above. Silly Putty exhibits the following behavior: bounces at high strain rates, stretches with slow *partial* recovery at moderate strain rates, and flows under gravitational force (low strain rates).

Draw a rheological model for Silly Putty that satisfies all of these requirements, and indicate on your drawing which parts behave in which ways.

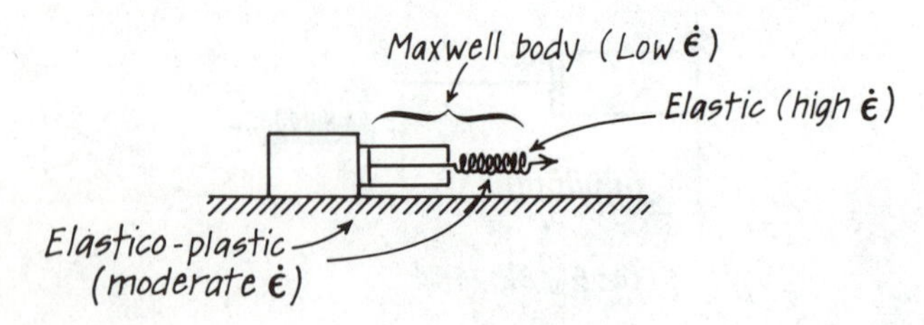

Figure 12-15
Rheological model for use in Problem 12-2.

Problem 12-3

In the deformed rock layer drawn in Figure 12-16 the limbs of the folds deformed without fracturing while fracturing occurred in the hinge zones. In terms of rheological models, explain why the same material under the same conditions of temperature and pressure could behave differently in different places?

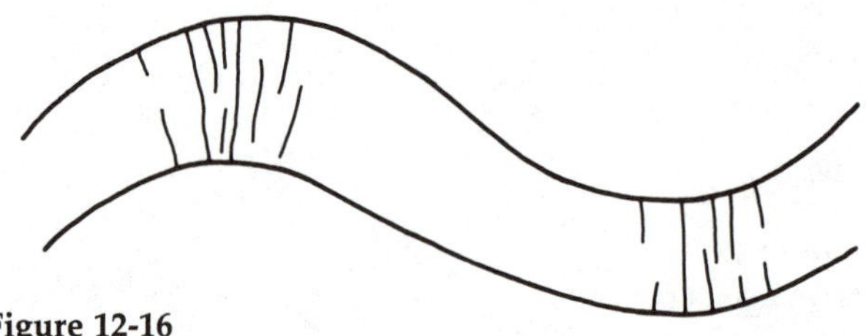

Figure 12-16
Diagram of folded and fractured rocks, for use in Problem 12-3.

Problem 12-4

The rock layers drawn in Figure 12-17 evidently folded and were then faulted. Give at least two possible reasons for the two different types of deformation.

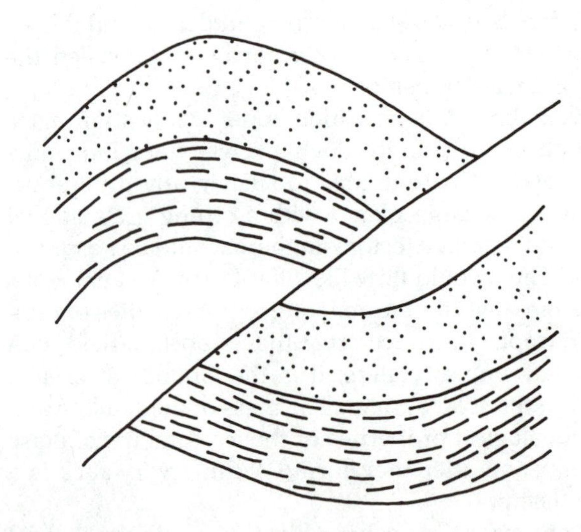

Figure 12-17
Diagram of folded and faulted rocks, for use in Problem 12-4.

Problem 12-5

Figure 12-18 shows folded, layered sediments.

1. Measure the present distance between anticlinal hinges, and also measure what the distance between these two points must have been before folding. Determine the percent shortening that these rocks have experienced.
2. If these rocks are 10 million years old, what has their average strain rate been in terms of amount of shortening per second? Show your work.

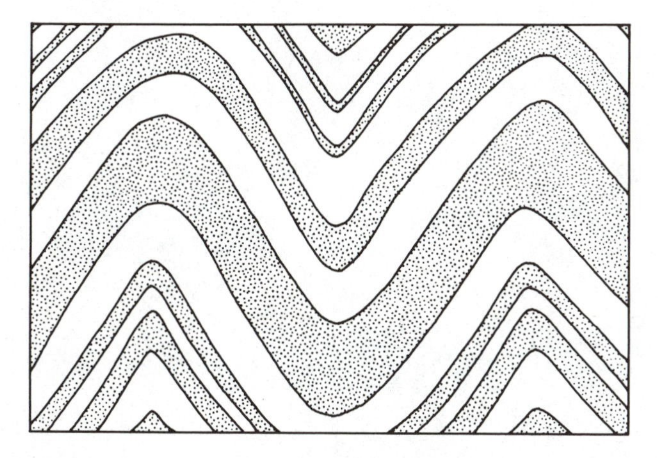

Figure 12-18
Diagram of folded layers, for use in Problem 12-5.

Further Reading

Mansfield, C. F. 1985. "Modeling Newtonian fluids and Bingham plastics." *Journal of Geological Education*, v. 33, 97–100. A very readable summary of viscous and plastic behavior, with applications to sedimentology.

Roper, P. J. 1974. "Plate tectonics: A plastic as opposed to a rigid body model." *Geology*, v. 2, 247–250. Implications of plastic behavior in plate tectonic reconstructions.

Suppe, J. 1985. *Principles of Structural Geology.* Englewood Cliffs, NJ: Prentice-Hall. A more mathematically rigorous review of rheological models is presented on pages 140–147.

Walker, J. 1978. "Serious fun with Polyox, Silly Putty, Slime and other non-Newtonian fluids." *Scientific American*, v. 329, no. 5 (November), 186–196. An insightful discussion of non-Newtonian fluid flow; not specifically related to geologic phenomena.

13

Brittle Failure

OBJECTIVE

Predict the principal stress magnitudes that will cause a given material to fracture

Chapter 10 was devoted to an examination of the orientation of the stress ellipsoid, especially with regard to faulting. Here we will investigate how the magnitude of the principal stresses influences brittle deformation. The chief objective is to be able to determine the magnitude of paleostress that caused a particular brittle failure to occur. The material in this chapter lies at the heart of engineering geology because it concerns the conditions under which rock breaks.

Quantifying Two-Dimensional Stress

Experimental rock fracturing has shown that the difference in magnitude between σ_1 and σ_3 is the most important factor in causing rocks to fracture. The magnitude of σ_2 is not believed to play a major role in the initiation of the fracture. For this reason we may profitably examine stress in two dimensions in the $\sigma_1 - \sigma_3$ plane.

If we know the orientations and magnitudes of σ_1 and σ_3, then we can determine the normal and shear stress acting across any plane perpendicular to the $\sigma_1 - \sigma_3$ plane. Consider the plane in Figure 13-1a. We want to determine the normal and shear stress acting on that plane. To simplify the situation we will isolate the plane along with two adjacent surfaces that are perpendicular to σ_1 and σ_3 (Fig. 13-1b). Viewed in the $\sigma_1 - \sigma_3$ plane, we will call these surfaces A and B and define the angle θ as the angle between the plane and the σ_3 direction (Fig. 13-1c).

If our triangle in Figure 13-1c is not moving then it must be in equilibrium. This means that the normal stress (σ_n) and shear stress (σ_s) acting on the plane must be equal to σ_1 and σ_3 acting on surfaces A and B. We will now use this equilibrium relationship to define σ_n and σ_s in terms of σ_1, σ_3, and angle θ.

Figure 13-1d shows σ_n and σ_s acting on the plane, and it also shows the horizontal and vertical components of σ_n and σ_s. The length of the line that represents the plane is $\frac{A}{\cos\theta}$ or $\frac{B}{\sin\theta}$. The vertical and horizontal forces acting on this line are indicated in Figure 13-1d. The equations of equilibrium for this plane are as follows:

$$\sigma_1 A = \frac{A}{\cos\theta}(\sigma_n \cos\theta + \sigma_s \sin\theta) \qquad (13\text{-}1)$$

$$\sigma_3 B = \frac{B}{\sin\theta}(\sigma_n \sin\theta - \sigma_s \cos\theta) \qquad (13\text{-}2)$$

Solving these equations simultaneously for σ_n and σ_s in terms of σ_1, σ_3, and θ, we derive the following equations:

$$\sigma_n = \sigma_1 \cos^2\theta + \sigma_3 \sin^2\theta \qquad (13\text{-}3)$$

$$\sigma_s = (\sigma_1 - \sigma_3)\sin\theta\cos\theta \qquad (13\text{-}4)$$

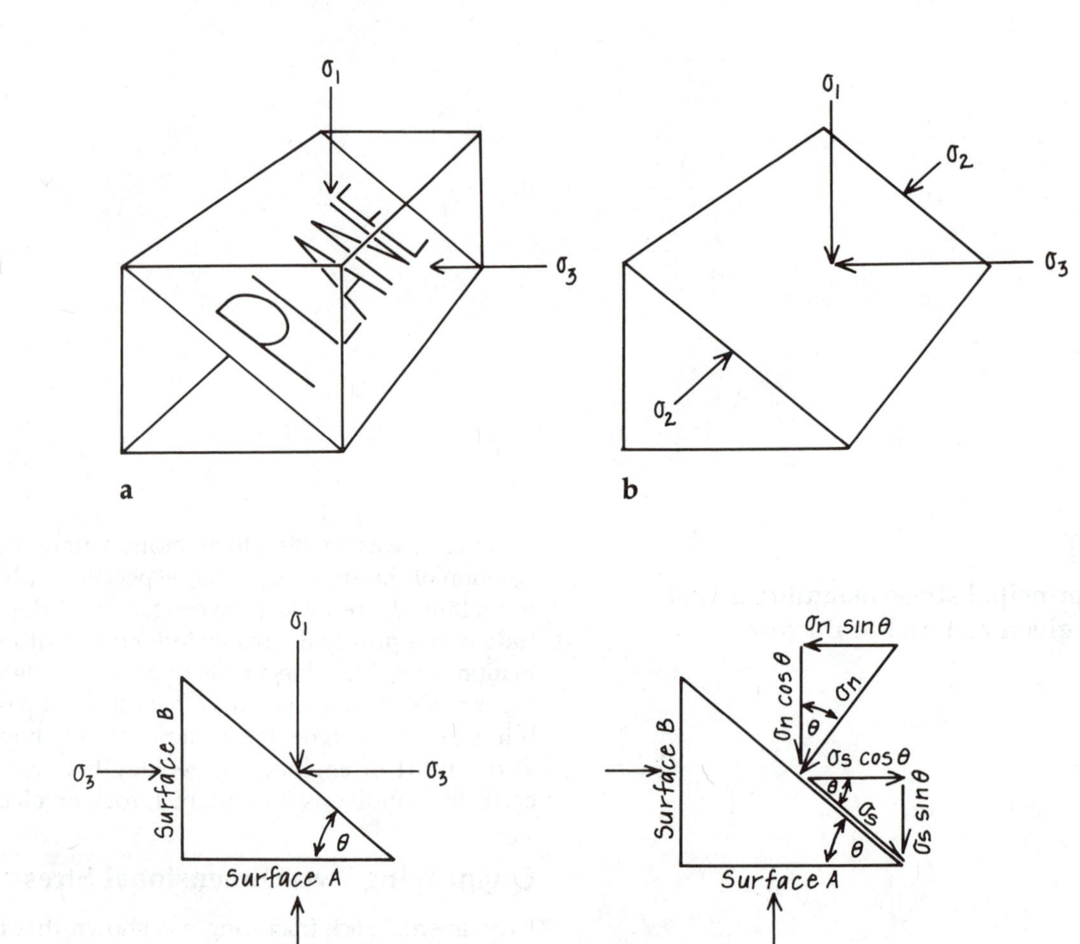

Figure 13-1
Two-dimensional relationship between a plane and its state of stress. See text for explanation.

In order to reconstitute these equations into a more useful form we can substitute the following trigonometric identities: $\sin 2\theta = 2 \sin\theta \cos\theta$, $\cos^2\theta = (1 + \cos 2\theta/2)$, and $\sin^2\theta = (1 - \cos 2\theta/2)$. The result is the following two equations:

$$\sigma_n = \left(\frac{\sigma_1 + \sigma_3}{2}\right) + \left(\frac{\sigma_1 - \sigma_3}{2}\right) \cos 2\theta \qquad (13\text{-}5)$$

$$\sigma_s = \left(\frac{\sigma_1 - \sigma_3}{2}\right) \sin 2\theta \qquad (13\text{-}6)$$

Stress is measured in units of force per unit area, for which the basic unit is the **pascal** (1 Pa = 1 newton per square meter); 10^5 Pa equals 1 bar, which is approximately equal to atmospheric pressure at sea level. The most convenient unit for most geologic applications is the megapascal (MPa), which is equal to 10^6 Pa or 10 bars. Stress within the earth's crust ranges up to about 10^3 MPa.

Using Equations 13-5 and 13-6 we can now determine the normal and shear stress acting across a plane if we know the orientations and magnitudes of σ_1 and σ_3. Suppose, for example, that in Figure 13-1 $\sigma_1 = 100$ MPa, $\sigma_3 = 20$ MPa, and $\theta = 40°$. Using Equation 13-5 the normal stress is determined as follows:

$$\begin{aligned}\sigma_n &= \left(\frac{\sigma_1 + \sigma_3}{2}\right) + \left(\frac{\sigma_1 - \sigma_3}{2}\right) \cos 2\theta \\ &= (60 \text{ MPa}) + (40 \text{ MPa})(.17) \\ &= 67 \text{ MPa}\end{aligned}$$

Similarly, Equation 13-6 can be used to determine the shear stress acting on the plane:

$$\begin{aligned}\sigma_s &= \left(\frac{\sigma_1 - \sigma_3}{2}\right) \sin 2\theta \\ &= (40 \text{ MPa})(.98) \\ &= 39 \text{ MPa}\end{aligned}$$

Problem 13-1

Given the principal stresses of $\sigma_1 = 100$ MPa (vertical) and $\sigma_3 = 20$ MPa (horizontal), determine the normal and shear stresses on a fault plane that strikes parallel to σ_2 and dips 32° (Plane 1 in Fig. 13-4).

The Mohr Diagram

In 1882 the German engineer Otto Mohr developed a very useful technique for graphing the state of stress of differently oriented planes in the same stress field. The stress (σ_n and σ_s) on a plane plots as a single point, with σ_n measured on the horizontal axis and σ_s on the vertical axis (Fig. 13-2). Such a graph is called a **Mohr diagram.**

Most stresses in the earth are compressive, so geologists, by convention, consider compression to be positive. (In engineering, tension is considered positive.) As a practical matter in structural geology, σ_n in the earth's crust is always positive and will therefore always plot on the positive (right) side of the vertical axis of the Mohr diagram.

The vertical axis of the Mohr diagram, like the horizontal axis, has a positive and a negative direction. Shearing stresses that have a sinistral (counterclockwise) sense (Fig. 13-3a) are, by convention, considered positive and are plotted above the origin. Dextral (clockwise) shearing stresses (Fig. 13-3b) are plotted on the lower, negative half of the diagram.

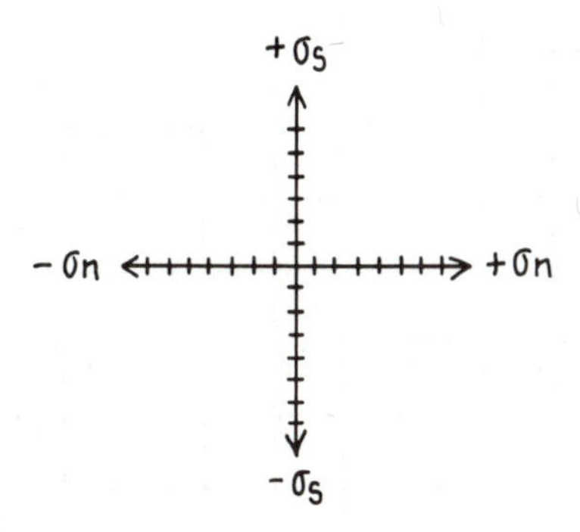

Figure 13-2
Mohr diagram for graphing the state of stress of a plane. Within a stress field consisting of a particular combination of σ_1 and σ_3, planes with different dips will experience different magnitudes of σ_n and σ_s and will therefore plot at different points on the Mohr diagram.

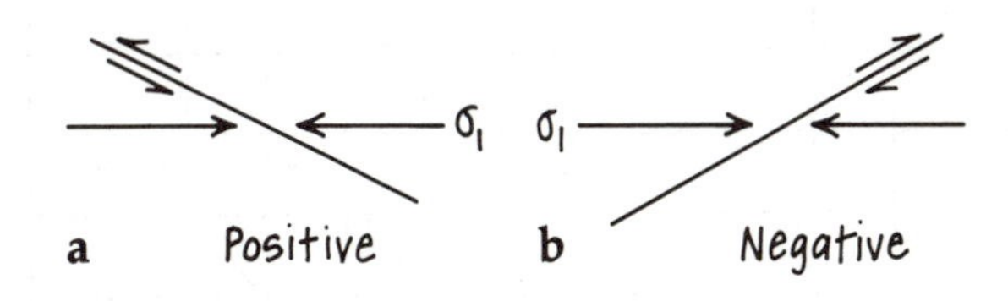

Figure 13-3
Conventional signs assigned to shearing stresses for the purpose of plotting on the Mohr diagram. Sinistral shearing (a) is considered positive; dextral shearing (b) is considered negative.

Problem 13-2

Plane 1 in Figure 13-4 has been plotted on the Mohr diagram in Figure 13-5. Determine the principal stresses on planes 2 through 5 and plot them on Figure 13-5. (Recall that trigonometric functions of angles in the second and fourth quadrants are negative, e.g., cos 180° = −1.0.)

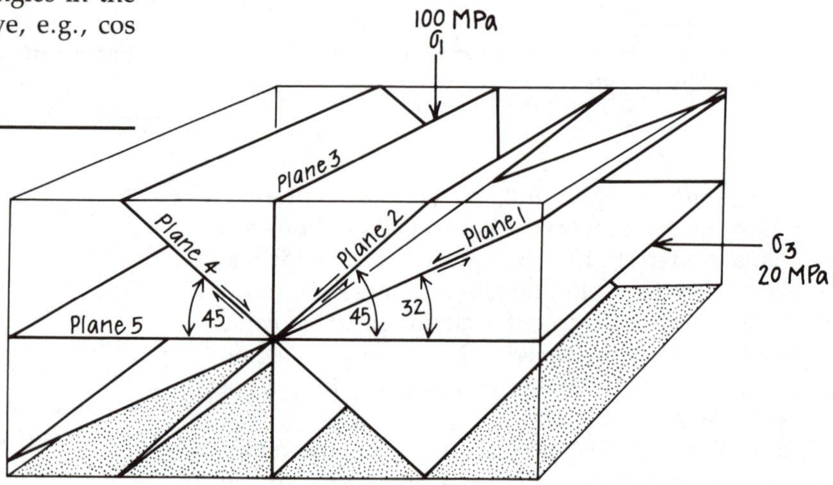

Figure 13-4
Five planes within the same stress field, for use in Problem 13-2. Stresses σ_n and σ_s acting on plane 1 were calculated in Problem 13-1.

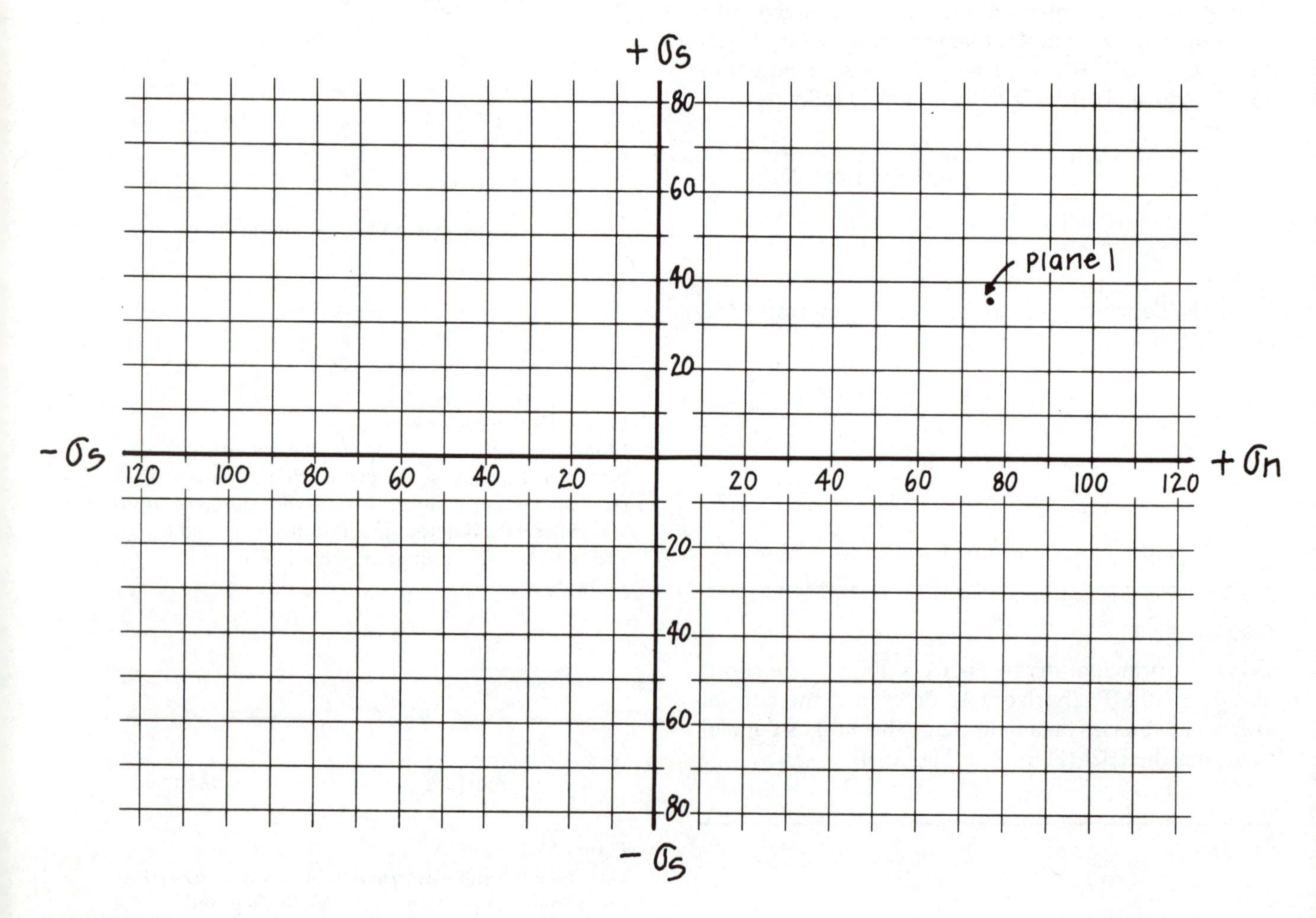

Figure 13-5
Mohr diagram for use in Problem 13-2.

The Mohr Circle of Stress

The five points plotted on Figure 13-5 should lie on a circle. A key feature of the Mohr diagram is that for a given set of principal stresses the points representing the states of stress on all possible planes perpendicular to the $\sigma_1 - \sigma_3$ plane graph as a circle. This is called the **Mohr circle.** As seen in Figure 13-6, the Mohr circle intersects the σ_n axis at values equal to σ_3 and σ_1. The radius is $(\sigma_1 - \sigma_3)/2$ and the center is at $(\sigma_1 + \sigma_3)/2$.

It is important to understand that the axes of the Mohr diagram have no geographic orientation. They merely allow the magnitudes of stresses on variously oriented planes to be plotted together. Planes perpendicular to either σ_1 or σ_3 (Planes 3 and 5 in Fig. 13-4) have no shear stress acting on them, so they plot directly on the σ_n axis. Shear stress is maximum on planes oriented 45° to the principal stress directions ($\theta = 45°$); the points representing these planes plot at the top and bottom of the Mohr circle.

Values of 2θ can be measured directly off the Mohr circle as shown in Figure 13-6. Angles 2θ corresponding to planes with positive (sinistral) shearing lie in the upper hemisphere of the Mohr circle, while those corresponding to planes with negative (dextral) shearing stresses lie in the lower hemisphere. In either case the angle 2θ is measured from the right-hand end of the σ_n axis.

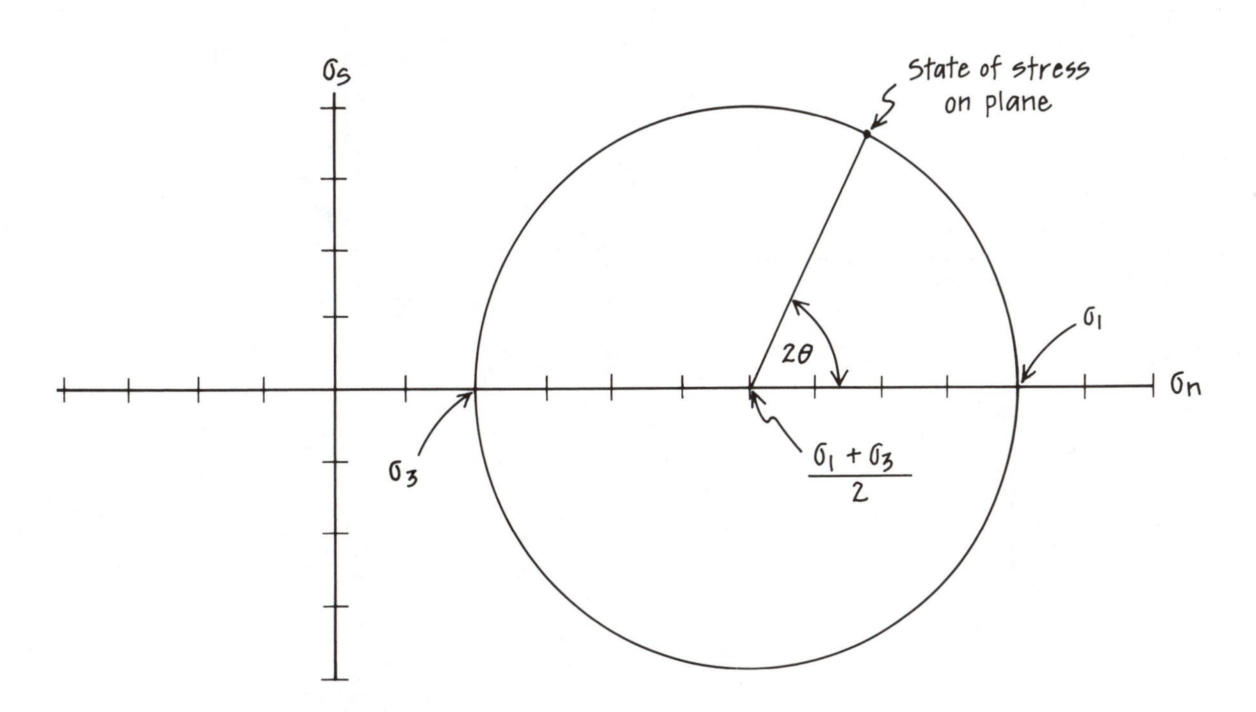

Figure 13-6
Main features of Mohr circle of stress. The Mohr circle is the set of states of stress on all possible planes in a two-dimensional stress field. The position on the circle of a given plane is determined by finding angle θ (the angle between the plane and σ_3) and plotting 2θ on the Mohr circle. Planes with sinistral shear are plotted in the upper hemisphere; planes with dextral shear are plotted in the lower hemisphere. 2θ is always measured (up or down) from the σ_1 intercept.

The chief value of the Mohr circle of stress is that it permits a rapid, graphical determination of stresses on a plane of any desired orientation. Suppose, for example, that σ_1 is oriented east-west, horizontal, and equal to 40 MPa, while σ_3 is vertical and equal to 20 MPa. We must find the normal and shear stresses on a fault plane striking north-south and dipping 55° west. The solution is as follows:

1. Figure 13-7a shows the geologic relationships. Before being concerned with the fault plane, construct a Mohr circle of stress for the given values of σ_1 and σ_3 (Fig. 13-7b).
2. Next determine the value and sign of angle 2θ for the fault plane. Angle θ is the angle between the fault plane and σ_3, which in this case is 35°. So 2θ is 70°. Shearing stresses on this fault have a dextral or negative sense, so angle 2θ is located in the lower hemisphere of the Mohr circle (Fig. 13-7b).
3. The normal and shear stress coordinates corresponding to the points thus located on the Mohr circle are read directly off the horizontal and vertical axes of the graph. In this example σ_n is 33.4 MPa and σ_s is 9.4 MPa.

Problem 13-3

If σ_1 is vertical and equal to 50 MPa and σ_3 is horizontal, east-west, and equal to 22 MPa, using a Mohr circle construction determine the normal and shear stresses on a fault striking north-south and dipping 60° east.

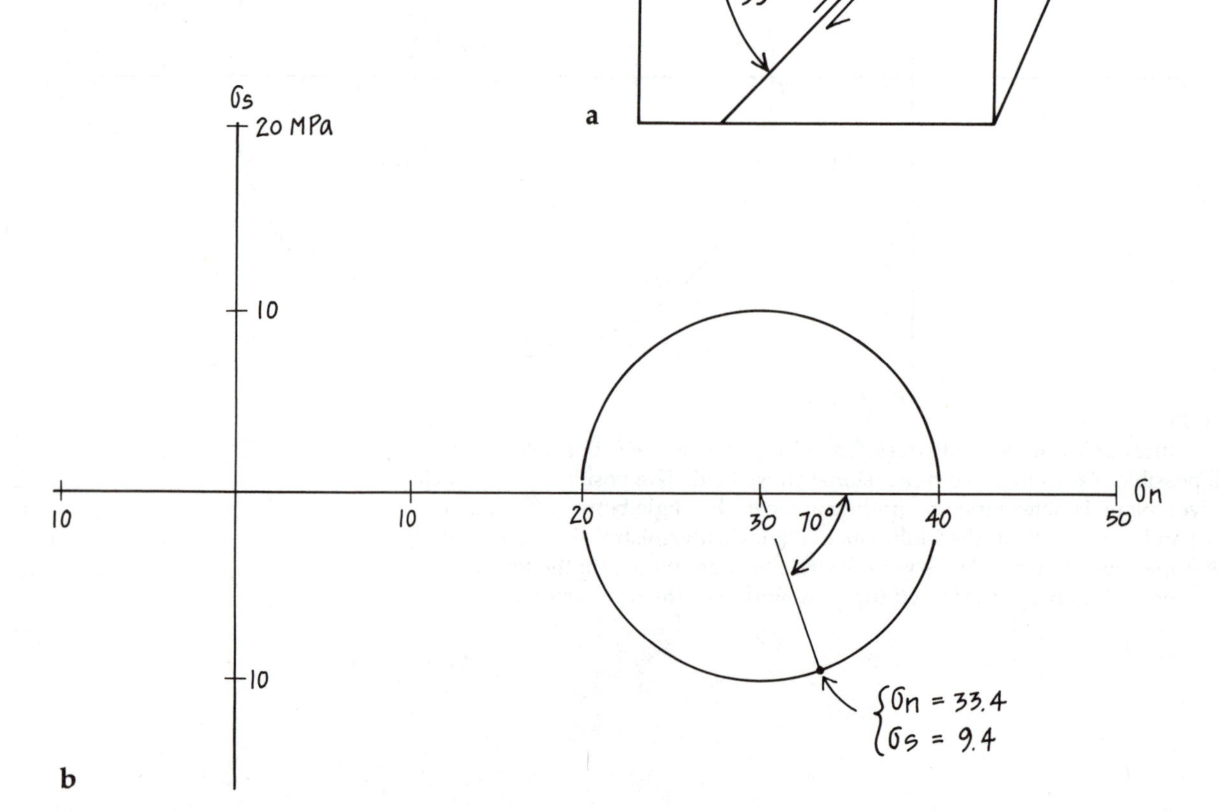

Figure 13-7
Mohr circle solution to sample problem requiring determination of σ_n and σ_s on a particular plane. (a) Block diagram. (b) Mohr circle solution.

The Mohr Envelope of Failure

Up to this point in this chapter we have examined the stresses acting on variously oriented planes. The main objective of all of this is to understand or predict the orientation and magnitude of stresses that will cause a particular rock to fracture or "fail." To begin our examination of brittle failure we will imagine an experiment in which a cylinder of rock is axially compressed (Fig. 13-8). Suppose that the radially applied **confining pressure,** σ_c, is kept constant at 40 MPa, while the **axial load,** σ_a, begins at 40 MPa and is gradually increased until the rock fails at an axial load of 540 MPa. Magnitudes of σ_a at several stages of this experiment are recorded in Table 13-1, and the corresponding Mohr circles are drawn in Figure 13-9. In this type of experiment σ_a is analogous to σ_1, and σ_c is analogous to σ_3.

As shown in Figure 13-9, a fracture experiment with constant confining pressure results in a series of progressively larger Mohr circles, all of which intersect the σ_n axis at σ_c. The **fracture strength** is the diameter of the Mohr circle ($\sigma_a - \sigma_c$) when the rock fractures. In the experiment shown in Figure 13-9 the fracture strength was determined to be 500 MPa at a confining pressure of 40 MPa.

Table 13-1
Data from a hypothetical rock fracture experiment. Mohr circles corresponding to each recorded stage are drawn in Figure 13-9.

Time	σ_a	σ_c	$\sigma_a - \sigma_c$	
t_1	40 MPa	40 MPa	0 MPa	
t_2	100 MPa	40 MPa	60 MPa	
t_3	165 MPa	40 MPa	125 MPa	
t_4	265 MPa	40 MPa	225 MPa	
t_5	413 MPa	40 MPa	373 MPa	
t_6	540 MPa	40 MPa	500 MPa	Fracture

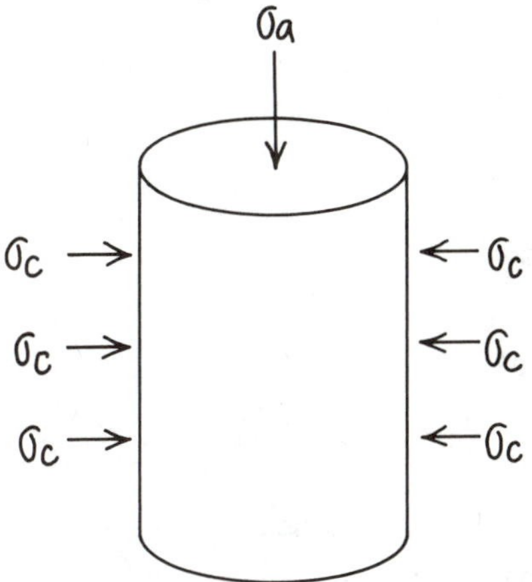

Figure 13-8
Schematic diagram of a rock-fracture experiment in which a cylinder of rock is axially compressed. Axial load (σ_a) is steadily increased while the confining pressure (σ_c) is kept constant.

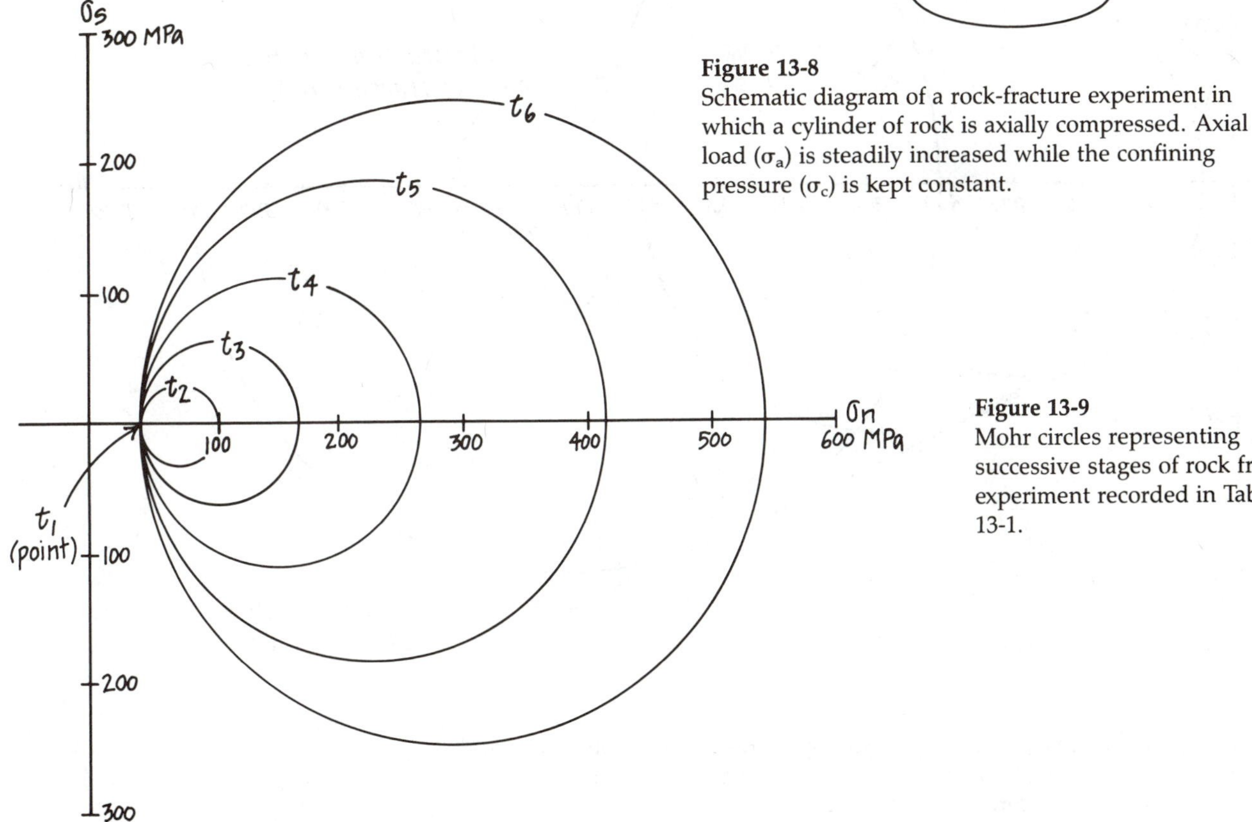

Figure 13-9
Mohr circles representing successive stages of rock fracture experiment recorded in Table 13-1.

Now, suppose we performed a series of three experiments on identical samples, but at different confining pressures. We would find that the fracture strength of the rock increases with confining pressure. Table 13-2 lists the results of our hypothetical series of experiments, with Experiment 1 being the one discussed and graphed in Figure 13-9. In Experiment 2 the confining pressure was raised to 150 MPa, and in Experiment 3 to 400 MPa. In Figure 13-10 the three resulting Mohr circles are drawn. Because each experiment in this series has a higher confining pressure than the previous one, the Mohr circles at failure become progressively larger.

Table 13-2
Data from three fracture experiments on identical rock samples. Mohr circles at failure are drawn in Figure 13-10.

Experiment No.	σ_c	σ_a at failure	$\sigma_a - \sigma_c$
1	40 MPa	540 MPa	500 MPa
2	150 MPa	800 MPa	700 MPa
3	400 MPa	1400 MPa	1000 MPa

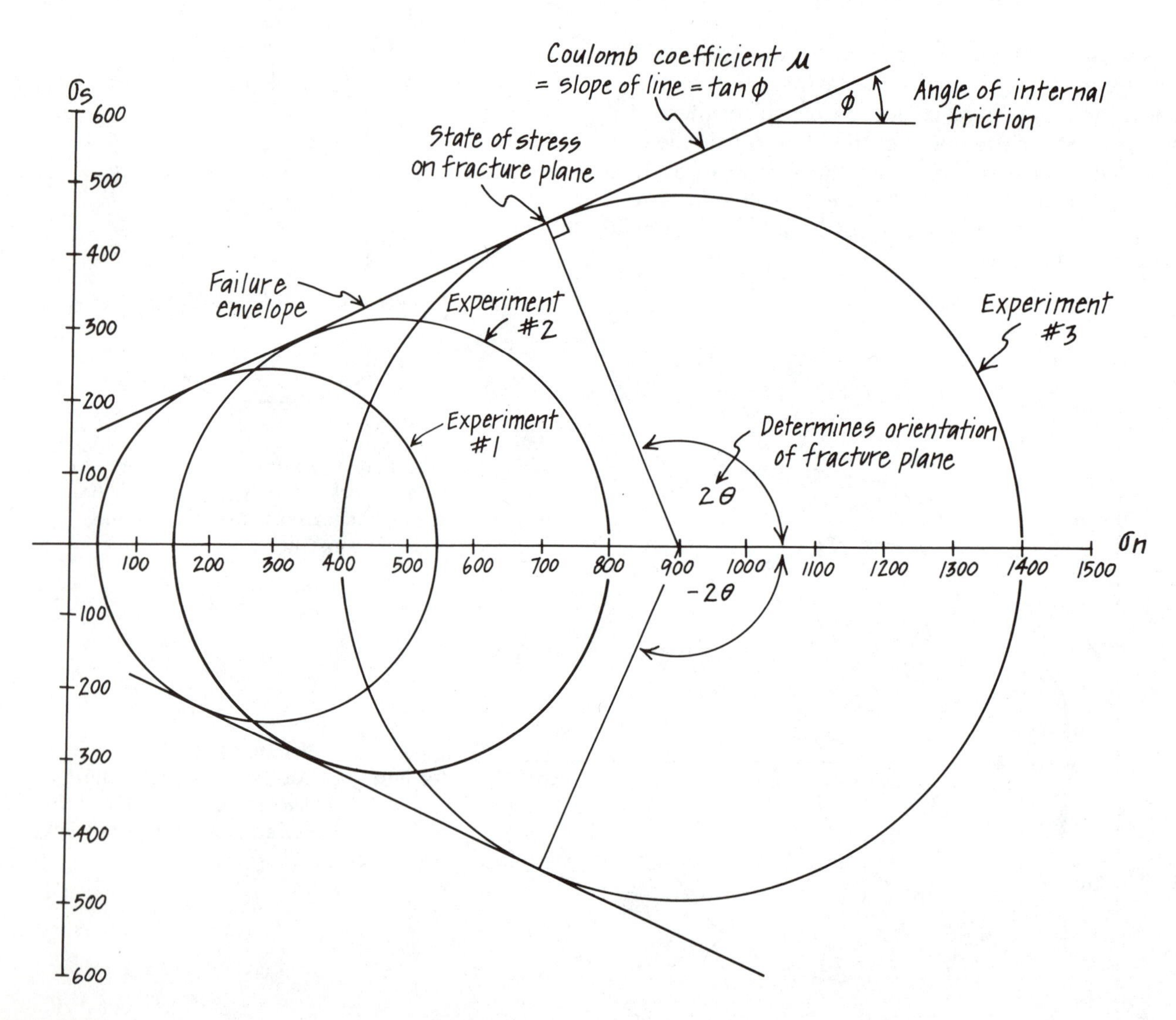

Figure 13-10
Main characteristics of a failure envelope. Envelope is defined by Mohr circles at failure of identical rock samples under different confining pressures. Data for these three envelopes are recorded in Table 13-2.

The Mohr circles at failure under different confining pressures together define a boundary called the **Mohr envelope** or **failure envelope** for a particular rock (Fig. 13-10). The failure envelope is an empirically derived characteristic that expresses the combination of σ_1 and σ_3 magnitudes that will cause a particular rock (or manmade material such as concrete) to fracture. If the Mohr circle representing a particular combination of σ_1 and σ_3 intersects the material's failure envelope then the material will fracture; if the Mohr circle does not intersect the failure envelope the material will not fracture.

The failure envelope also allows us to predict the orientation of the macroscopic fracture plane that will form when the rock fails. In an isotropic rock this plane will be the one whose state of stress is represented by the point on the Mohr circle that lies on the failure envelope (Fig. 13-10). The angle between this plane and the σ_3 direction (angle θ) can be determined by measuring angle 2θ directly off the Mohr diagram (Fig. 13-10). In the example shown in Figure 13-10, angle $2\theta = 114°$, so the fracture plane will be oriented 57° from σ_3.

At intermediate confining pressures the fracture strength usually increases linearly with increasing confining pressure, producing a failure envelope with straight lines, as in Figure 13-10. The angle between these lines and the horizontal axis is called the **angle of internal friction** ϕ (phi), and the slope of the envelope is called the **Coulomb coefficient** μ (mu):

$$\mu = \tan \phi \qquad (13\text{-}7)$$

It is helpful to develop a familiarity with the Coulomb coefficient. This is a measurable property of the rock, like specific gravity, and indicates its fracture behavior at intermediate confining pressures within the earth's crust. The Coulomb coefficient is analogous to the coefficient of friction resisting the sliding of one block over another. Consider a bottomless box sitting on top of another box (Fig. 13-11a). If the two boxes are filled with dry sand, it would be possible by pushing sideways on the upper box for a shear surface to develop between the sand in the upper box and that in the lower box. With respect to this potential shear surface, σ_n can be imagined as the force keeping the sand together, and σ_s the force trying to make the sand in the upper box slide (Fig. 13-11b). If the boxes are tilted, eventually an angle θ is reached where movement occurs on the shear surface (Fig. 13-11c). This is analogous to angle θ that we have been using for the angle between the shear plane and the σ_3 direction (Fig. 13-11d). The Coulomb coefficient is in fact sometimes called the **coefficient of internal friction.** The greater the Coulomb coefficient, the greater the resistance to fracture.

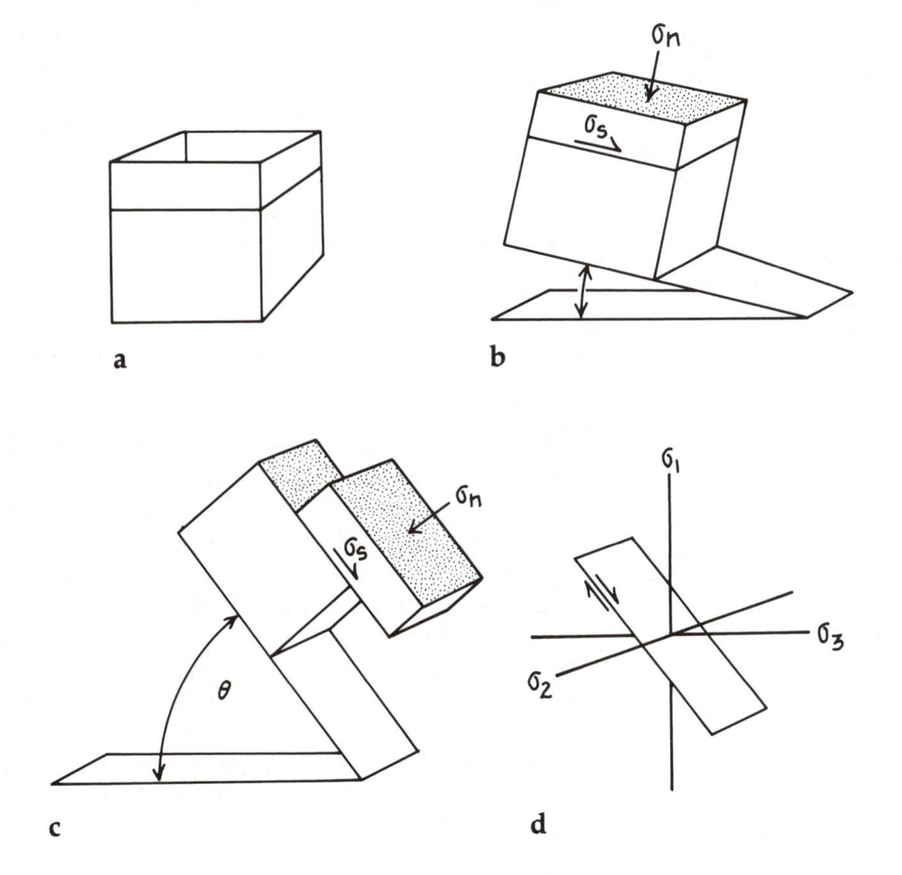

Figure 13-11
Sandbox experiment for determining cohesion properties of soil. (a) Bottomless box is placed over another box. (b) Both boxes are filled with soil and tilted. (c) Eventually an angle θ is reached at which the upper box slides. In the experiment depicted here the material in the boxes is cohesionless dry sand, analogous to a Coulomb coefficient of zero and $\theta = 45°$. (d) Orientation of shear plane with respect to principal stresses.

If the failure envelope is assumed to be straight lines, then the Coulomb coefficient can be determined from a single fracture experiment, for example, any of the three plotted in Figure 13-10. Conversely, if the Coulomb coefficient of a rock is known, the orientation of the shear surfaces relative to σ_1 and σ_3 can be predicted. It can be seen on Figure 13-10 that $2\theta = 90 + \phi$ or

$$\theta = 45 + \frac{\phi}{2} \tag{13-8}$$

Figure 13-12 summarizes the relationship between σ_1, σ_3, θ, ϕ, σ_n, and σ_s.

A material having a Coulomb coefficient μ equal to zero (analogous to cohesionless dry sand in the sandbox experiment) would have an angle of internal friction ϕ equal to zero, and $\theta = 45°$. As the value of μ increases, angle θ also increases. Measured values of μ for nine rock units are listed in Table 13-3.

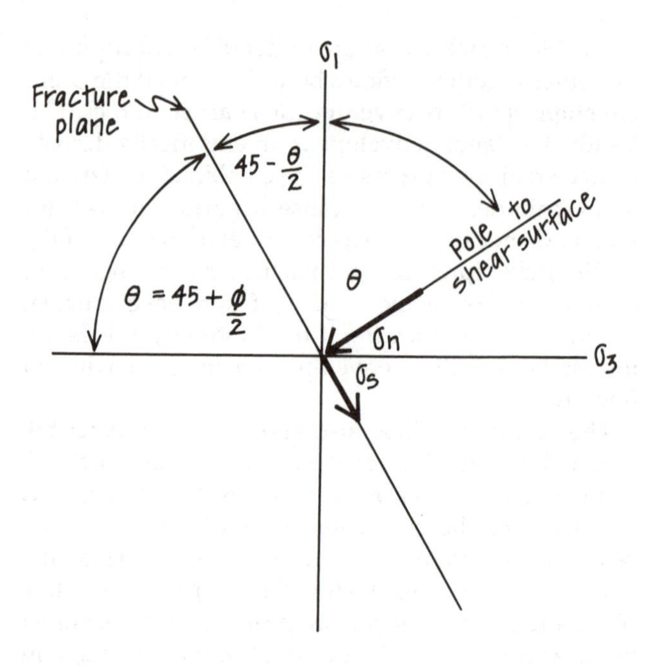

Figure 13-12
Generalized relationships between principal stresses and angles θ and ϕ.

Table 13-3
Coulomb coefficients μ of nine formations (from Suppe, 1985).

Formation	
Cheshire Quartzite	0.9
Westerly Granite	1.4
Frederick Diabase	0.8
Gosford Sandstone	0.5
Carrara Marble	0.7
Blair Dolomite	0.9
Webatuck Dolomite	0.5
Bowral Trachyte	1.0
Witwatersrand Quartzite	1.0

Table 13-4
Fracture data for use in Problem 13-4.

Experiment No.	σ_c	σ_a at failure
1	14 MPa	87 MPa
2	42 MPa	164 MPa
3	70 MPa	242 MPa
4	99 MPa	321 MPa

Problem 13-4

In Table 13-4 are recorded the results of four fracture experiments on the Rohan Tuff.

1. Draw Mohr circles for each experiment, and draw the failure envelope.
2. Determine the Coulomb coefficient of Rohan Tuff.
3. Determine the angle θ that the fracture plane is predicted to form with the σ_3 direction when a sample of Rohan Tuff fractures.

Problem 13-5

Suppose you are an engineering geologist designing a nuclear waste repository in the Rohan Tuff (see Problem 13-4). The repository is to be a large room 20 m deep within the tuff. During excavation of the repository, round pillars of tuff 5 m in diameter will be left in place to support the 20 m of overburden. Determine the maximum spacing of pillars (center to center) sufficient to support the overlying tuff. The density of the tuff is 2.0 g/cm^3. Assume that the confining pressure on the pillars is atmospheric pressure, about 0.1 MPa. Clearly show how you got your answer. (Strategy: First use your failure envelope from Problem 13-4 to find the value of σ_1 at failure when σ_3 is 0.1

MPa. Convert this compressive strength to kg/m^2 [see Appendix E]). Next determine the weight per square meter of the overburden and the area of overburden that each pillar can support. Finally, determine the closest possible spacing of pillars.)

Problem 13-6

Figure 13-13 shows a block of fine-grained limestone that was experimentally shortened about 1 percent at room temperature. Four sets of fractures developed. Fractures of sets "a" and "b" are conjugate shear surfaces. Fractures of set "c" are extension fractures that formed during loading. Fractures of set "d" are extension fractures that formed during unloading when the orientation of σ_3 became vertical in the rock-squeezing apparatus.

Determine the Coulomb coefficient μ for this rock (under atmospheric conditions).

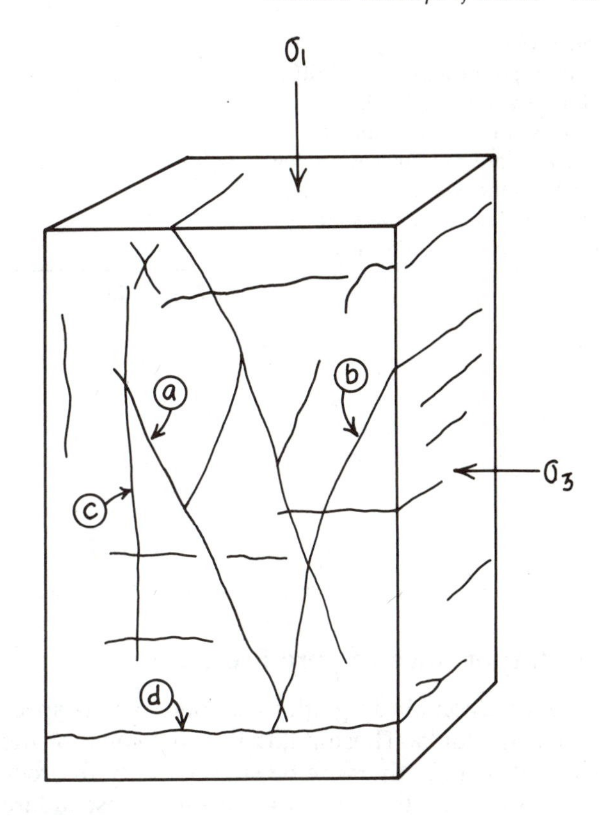

Figure 13-13
Sketch of fine-grained limestone block for use in Problem 13-6. After Hobbs and others (1976).

Problem 13-7

If rock-squeezing equipment is available (see, for example, Donath, 1970), it is desirable to carry out actual experiments and determine failure envelopes on actual rock samples. If specially designed equipment is not available, the poor man's rock squeezer shown in Figure 13-14 may be used. Using short pieces of chalk in the jaws of a pair of pliers, axial load is applied by placing stretched segments of bicycle inner tube (use large-diameter tubes) around the handles of the pliers (Fig. 13-14a) until the chalk fractures. A Mohr circle can then be drawn, with the axes of the Mohr diagram calibrated in bicycle tube units (B.T.U.s). After determining the fracture strength with zero confining pressure, place a foam pad jacket around a chalk sample, followed by successive segments of bicycle tube (Fig. 13-14b). Determine the fracture strength of the chalk under varying B.T.U.s of confining pressure. Draw a Mohr circle for each of your experiments and a failure envelope for the chalk.

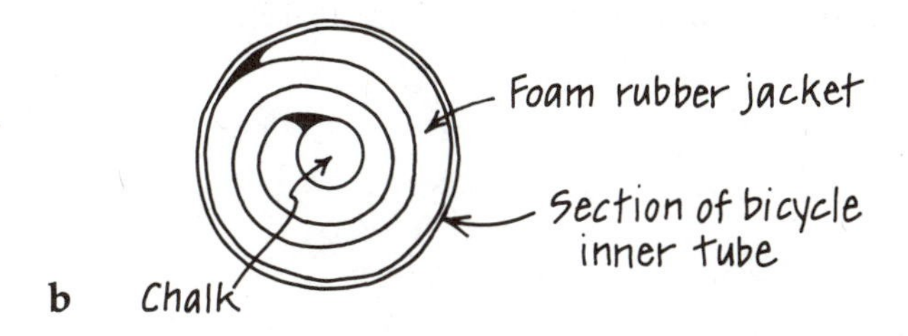

Figure 13-14
The poor man's rock press. (a) Apparatus for measuring fracture strength of chalk. Axial stress is increased by adding more sections of bicycle inner tubes. (b) Cross-section of chalk sample prior to experiment. Jacket is made from rubber insulating pad material sold in backpacking stores. Confining pressure is increased by adding more sections of inner tube.

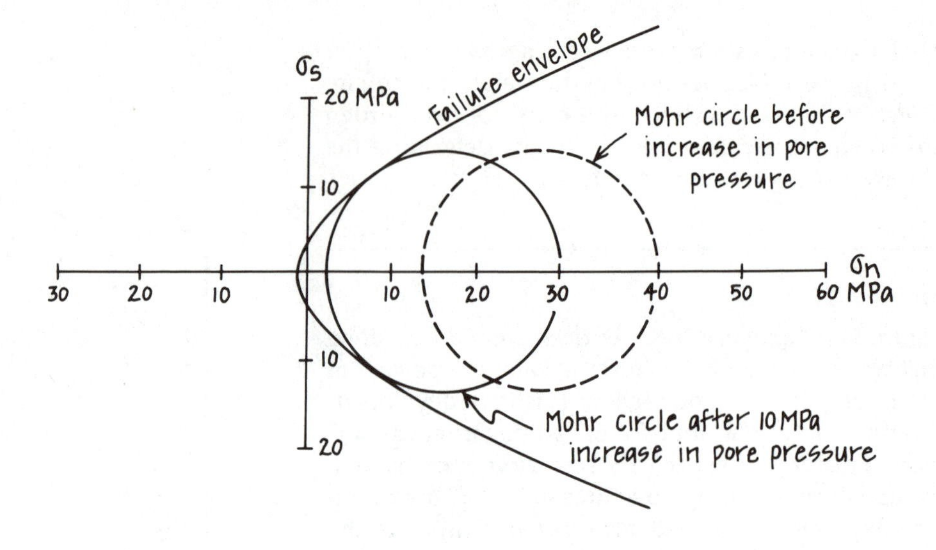

Figure 13-15
Effect of pore pressure on brittle failure. Dashed Mohr circle is based on measured principal stresses. Pore pressure effectively translates the Mohr circle to the left, as indicated by the solid Mohr circle of effective stress.

The Importance of Pore Pressure

Many rocks contain a significant amount of pore space filled with fluids. These fluids support some of the load that would otherwise be supported by the rock matrix. Consider the porous sandstone whose failure envelope is shown in Figure 13-15. This sandstone is subject to the following principal stresses: $\sigma_1 = 40$ MPa and $\sigma_3 = 13$ MPa. The dashed Mohr circle in Figure 13-15 represents this state of stress. Now, suppose we add 10 MPa of pore pressure to the rock. This has the effect of lowering the principal stresses by 10 MPa. Fluid pressure is hydrostatic (equal in all directions), so all principal stresses are equally affected. The Mohr circle remains the same size; it merely moves to the left on the horizontal axis a distance equal to the increase in pore pressure (Fig. 13-15).

The reduction of principal stresses by pore pressure is expressed through the term **effective stress.** The effective stress acting on the rock is the total (regional) stress minus the pore pressure.

Notice that in Figure 13-15 the solid Mohr circle (representing effective stress) intersects the failure envelope. The increase in pore pressure caused this rock to fracture. This phenomenon, called hydraulic fracturing, is routinely used in the petroleum industry to create fractures in low permeability rocks.

In addition to triggering the formation of new fractures, fluid pressure can control movement and earthquakes on pre-existing faults. This was first demonstrated in the 1960s when the U.S. Army accidentally triggered some earthquakes near Denver by injecting wastewater into the ground. A controlled experiment was subsequently conducted by the U.S. Geological

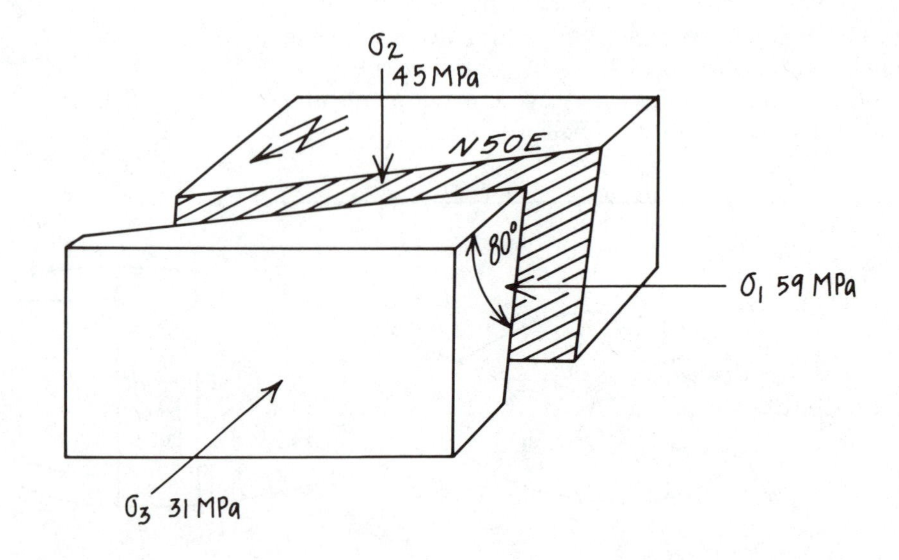

Figure 13-16
Block diagram showing oblique-slip fault that was experimentally activated at Rangely, Colorado, by increasing the pore pressure within the rocks. Principal stress magnitudes were determined from fluid pressure measurements made during hydraulic fracturing. The fault plane is subject to a normal stress of 35 MPa and a shear stress of 8 MPa. Data from Raleigh and others (1972).

Survey at Rangely, Colorado (Raleigh and others, 1972). The geologic setting and principal stresses of this experiment are schematically depicted in Figure 13-16. A pre-existing oblique-slip fault in the Weber Sandstone was successfully activated when water was injected into the ground. Pore pressure was experimentally raised and lowered while seismicity was monitored.

Mohr circles for the Rangely experiment are shown in Figure 13-17. Because movement was occurring on a pre-existing fault in this case, the failure envelope is different from the normal envelope for intact rock. The failure envelope in Figure 13-15 is actually derived from experiments with unfractured Weber Sandstone. The failure envelope in Figure 13-17 is derived from experiments with previously cut samples.

In order to determine the state of stress on the fault shown in Figure 13-16, the principal stress magnitudes ($\sigma_1 = 59$ MPa, $\sigma_3 = 31$ MPa) were resolved onto the fault plane in the direction of slip, yielding a normal stress of 35 MPa and a shear stress of 8 MPa. This state of stress is indicated on the Mohr circle in Figure 13-17a. The injection of water into the rock created a pore pressure of 27 MPa, thereby reducing σ_1 and σ_3 to effective stresses of 32 MPa and 4 MPa, respectively. The Mohr circle of effective stress, shown in Figure 13-17b, is 27 MPa to the left of the pre-injection Mohr circle.

Notice in Figure 13-17b that the point on the Mohr circle that represents the state of stress on the fault plane has crossed the failure envelope. Movement on the fault did indeed occur at this level of pore pressure. When the pressure was reduced 3.5 MPa the earthquakes stopped. This is in impressive agreement with Figure 13-17b, which indicates that if the Mohr circle is translated 3.5 MPa to the right the state of stress of the fault lies directly on the failure envelope.

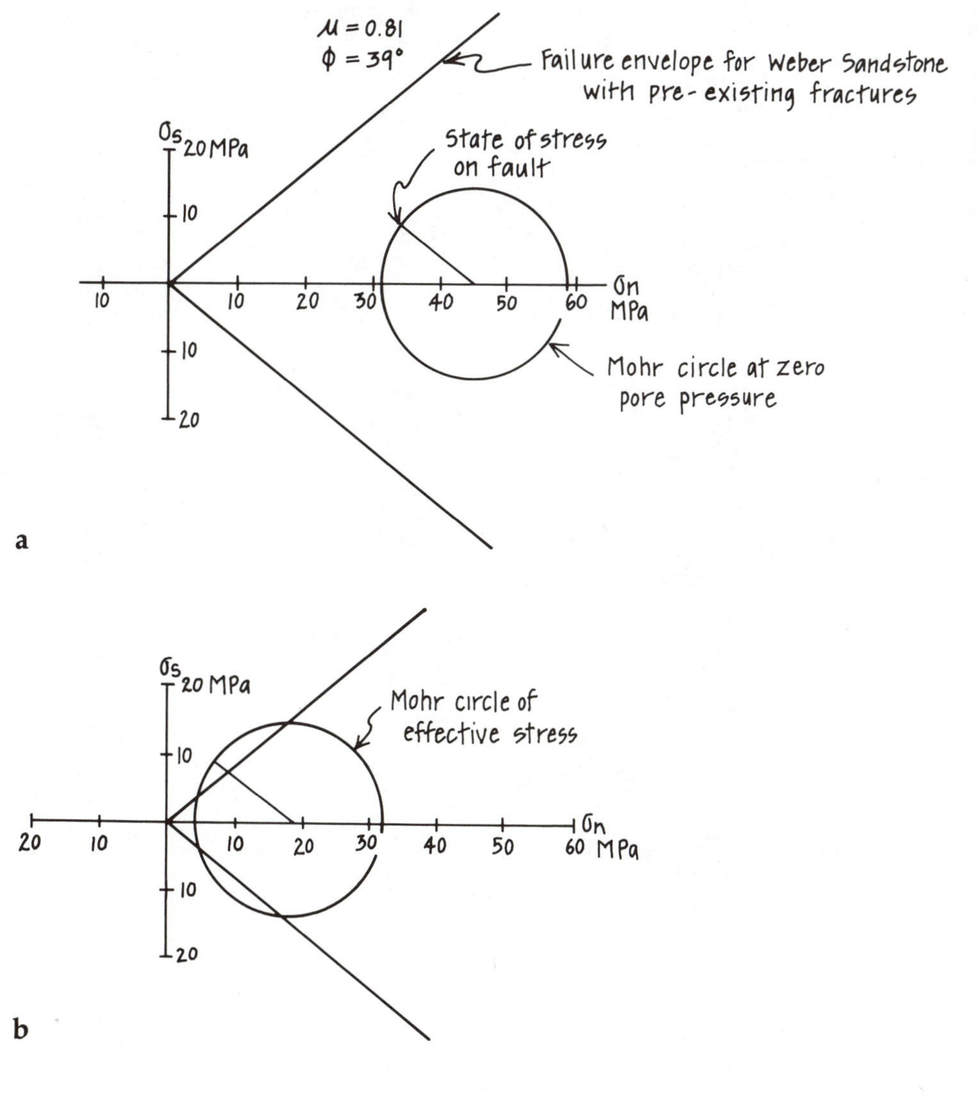

Figure 13-17
Mohr circles and failure envelope for Weber Sandstone at Rangely, Colorado. (a) Assuming zero pore pressure. (b) Mohr circle of effective stress after injection of 27 MPa of fluid pressure, which triggered a series of earthquakes. Earthquakes ceased when Mohr circle of effective stress was moved 3.5 MPa to the right.

Problem 13-8

Figure 13-19 shows the failure envelope of a "tight" (low permeability) sandstone, which is a petroleum reservoir rock. If $\sigma_1 = 72$ MPa and $\sigma_3 = 42$ MPa, determine the amount of pore pressure that would be necessary to hydraulically fracture this reservoir.

Problem 13-9

Figure 13-18 is a map of the San Francisco Peninsula showing the San Andreas fault, a vertical, strike-slip fault. Hypothetical orientations and magnitudes of σ_1 and σ_3 are indicated on the map. Figure 13-20 shows a hypothetical failure envelope (assuming pre-existing fractures) for the rocks adjacent to the fault.

Rather than waiting for the "big one," suppose San Franciscans wanted to release accumulating strain on the San Andreas fault through controlled earthquakes triggered by injection wells. Determine the pore pressure required for fault slip to occur, and estimate the annual premium for malpractice insurance for the geologist in charge.

Further Reading

Bartley, J. M., and Glazner, A. F. 1985. "Hydrothermal systems and Tertiary low-angle normal faulting in the southwestern United States." *Geology,* v. 13, 562–564. Addresses the long-standing paradox of low-angle normal faults and the role of pore pressure in their development.

Davis, G. H. 1978. "Experiencing structural geology." *Journal of Geological Education,* v. 26, 52–59. Reviews an integrated lab and field course in structural geology in which fracture strength experiments are combined with field analysis of structures.

Donath, F. A. 1970. "Rock deformation apparatus and experiments for dynamic structural geology." *Journal of Geological Education,* v. 18, 3–13. Review of equipment and procedures necessary to carry out fracture strength experiments.

Raleigh, C. B., Healy, J. H., and Bredehoeft, J. D. 1972. "Faulting and crustal stress at Rangely, Colorado." *In:* Heard, H. C., and others, eds., *Flow and Fracture of Rocks,* American Geophysical Union, Geophysical Monograph 16, 275–284.

Secor, D. T. 1965. "Role of fluid pressure in jointing." *American Journal of Science,* v. 263, 633–646.

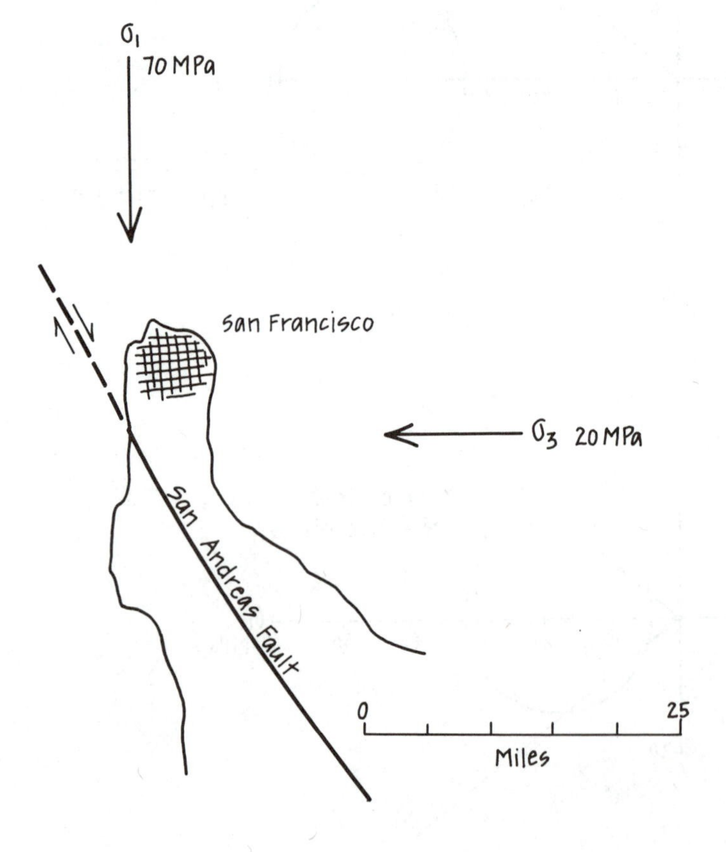

Figure 13-18
Map of San Francisco peninsula and the San Andreas fault trace. Magnitudes and orientations of principal stresses are hypothetical. For use in Problem 13-9.

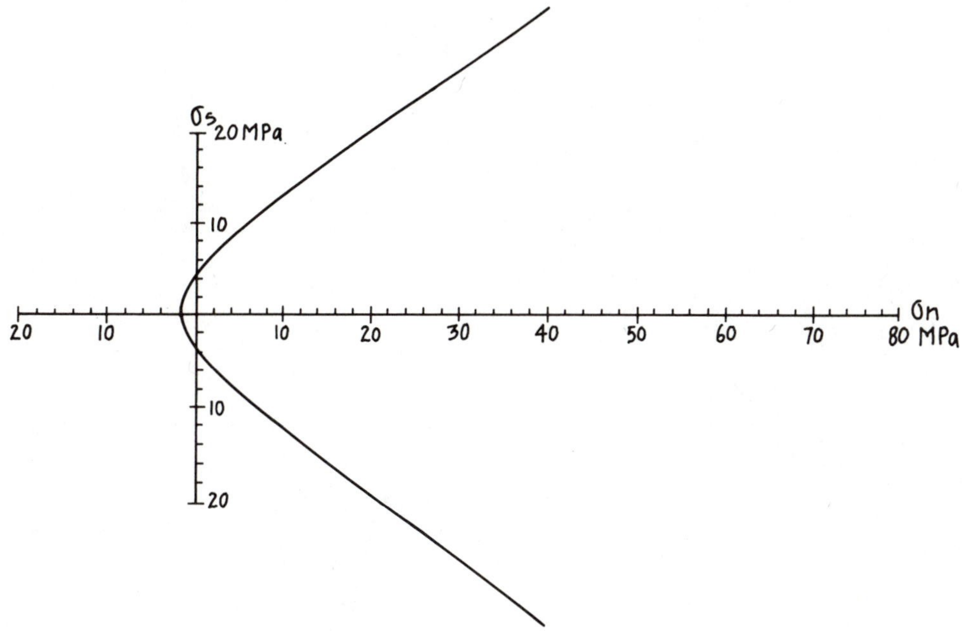

Figure 13-19
Failure envelope of a sandstone, for use in Problem 13-8.

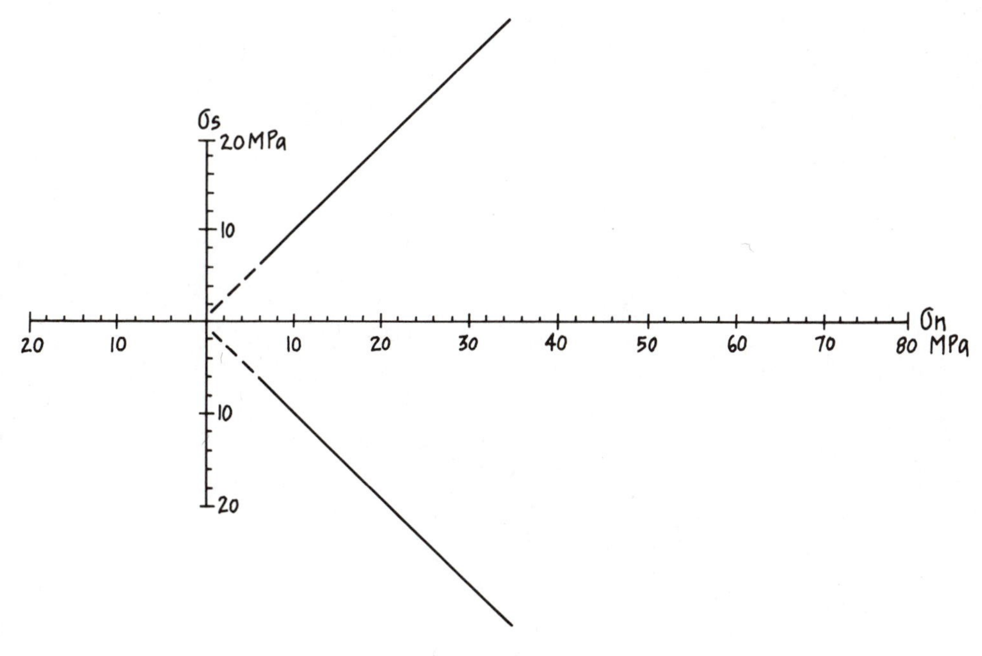

Figure 13-20
Hypothetical failure envelope for prefractured rocks adjacent to the San Andreas fault. For use in Problem 13-9.

Notes

14

Strain Measurement

OBJECTIVES

Measure longitudinal and shear strain from deformed objects

Determine the orientation and relative dimensions of the strain ellipse from deformed objects

Determine in which of three strain fields a particular structure developed

Determine the orientation of the strain ellipsoid in which a particular deformed feature developed

Determine the dimensions of the strain ellipsoid

Strain is a change of shape or volume, or both. One fundamental aspect of structural geology is the study of how rocks deform under different stresses. Our observations, however, are limited to the end product of rock deformation. Except when rocks are experimentally subjected to extremely high strain rates, we can measure only the deformation itself, not the stress that caused it. Once the strain is accurately measured, however, some inferences about the stress may sometimes be made.

The ability to deduce the stress history of a rock from measured strain is a major objective of structural geology, so careful measurement of strain is very important. In this chapter we will measure strain several different ways, starting with changes in the lengths of lines (longitudinal strain) and the angles between intersecting lines (shear strain).

For the experiments in this chapter you will need the following equipment:

play dough*

computer cards (a stack about 5 cm thick)

protractor

metric ruler

*Recipe for play dough: Mix together 1 cup flour, 1 cup water, 1 tablespoon cooking oil, ½ cup salt, 1 teaspoon cream of tartar, and food coloring as desired. Cook over medium heat until mixture pulls away from sides of pan and becomes doughlike. Knead until cool. Keeps 3 months unrefrigerated.

Longitudinal Strain

If the original length of a line is known, then a comparison can be made between the original length (l_0) and the deformed length (l_1). This value is called the **extension** (e) of the line. It is the proportional change in unit length.

$$e = \frac{l_1 - l_0}{l_0}$$

Notice that if l_1 is greater than l_0 then e will be positive, and if l_1 is less than l_0 then e will be negative. For a line that is stretched to twice its original length, $e = 1.0$ (100% has been added). For a line that is contracted to half its original length, $e = -0.5$ (50% has been eliminated). Throughout this chapter it will be assumed that undeformed lines have a length of 1 unit. After extension their length may be defined as $1 + e$.

Shear Strain

If the original shape of a deformed object is known, then changes in angular relationships can be measured. **Angular shear,** symbolized by the Greek letter ψ (psi), is the angular change after deformation of two lines that were originally perpendicular (Fig. 14-1). **Shear strain,** symbolized by the Greek letter γ (gamma), is the tangent of angular shear:

$$\gamma = \tan \psi$$

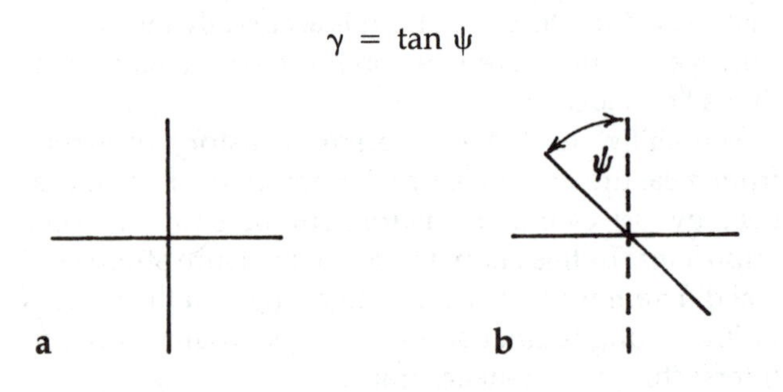

Figure 14-1
Shear strain measured as angular shear (ψ). (a) Before deformation. (b) After deformation.

Problem 14-1

Figure 14-2 shows a diagrammatic brachiopod shell before deformation (Fig. 14-2a) and after deformation (Fig. 14-2b).

1. Determine the extension e of the hinge line.
2. Determine the angular shear ψ and the shear strain γ of the shell.

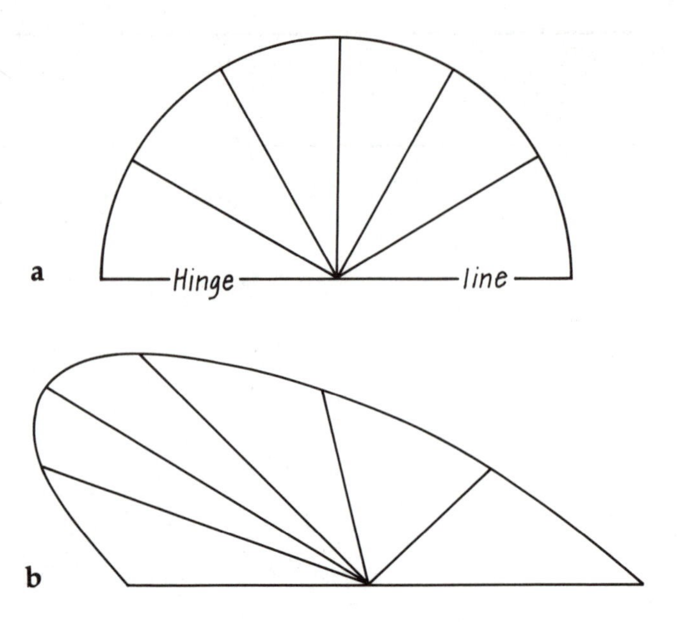

Figure 14-2
Undeformed (a) and deformed (b) brachiopod for use in Problem 14-1.

The Strain Ellipse

Deformation in rocks is described in terms of the change in shape or size of an imaginary sphere. During homogeneous deformation the imaginary sphere within the rock becomes an ellipsoid. Before considering three-dimensional deformation, however, it is instructive to examine deformation in two dimensions.

Imagine a plane containing a circle. Upon deformation, the circle becomes an ellipse (Fig. 14-3). An ellipse formed in this way is called a **strain ellipse,** and its orientation and dimensions characterize the deformation of the plane in which it lies. Figure 14-3a contains a circle from which the strain ellipse develops; it is always, by convention, given a radius of 1 arbitrary unit. Figure 14-3b contains the strain ellipse representing the deformed circle.

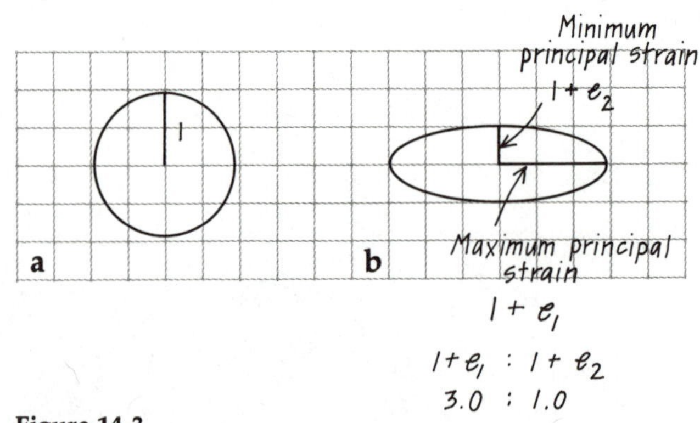

Figure 14-3
The strain ellipse. Beginning with a circle with radius of 1 unit (a), strain ellipse develops with maximum and minimum principal strain axes (b).

The strain ellipse is described in terms of the two **principal strains,** which correspond to the semi major and semi minor axes of the strain ellipse. The lengths of the maximum and minimum principal strains are $1 + e_1$ and $1 + e_2$, respectively. The shape of the ellipse is described by the ratio of the principal strains, which in this example is 3.0:1.0. Now to relate the strain ellipse to strain in rocks, do problem 14-2 (p. 179).

Three Strain Fields

Strain ellipses may occur in a variety of shapes. In Figure 14-4 are seven circles of radius 1, and the strain ellipse that has developed from each. In Figure 14-5 is a graph in which $1 + e_2$ is plotted against $1 + e_1$. The undeformed circle is shown at $1 + e_2 = 1.0$ and $1 + e_1 = 1.0$. Before reading further, plot the letter of each of the seven strain ellipses of Figure 14-4 onto its appropriate position on Figure 14-5. You will probably have difficulty understanding the following discussion if you do not take the time to do this.

The seven strain ellipses whose positions you have plotted on Figure 14-5 represent seven generalized classes. Notice that no ellipse can ever be plotted above the diagonal line on the graph, because $1 + e_1$ is always greater than or equal to $1 + e_2$. The diagonal line is the locus of all strain ellipses that are not ellipses at all but circles. "Ellipse" A, which exhibits equal elongation in all directions, and "ellipse" G, which exhibits equal contraction in all directions, both plot on this line.

The graph of Figure 14-5 can be divided into three fields with the $e_1 = 0$ and $e_2 = 0$ lines acting as dividers, as shown in Figure 14-6. Field 1 includes all ellipses in which both principal strains have positive extensions, such as ellipse B on Figure 14-4. Field 2 includes ellipses in which e_1 is positive and e_2 is negative, such as ellipse D on Figure 14-4. And field 3

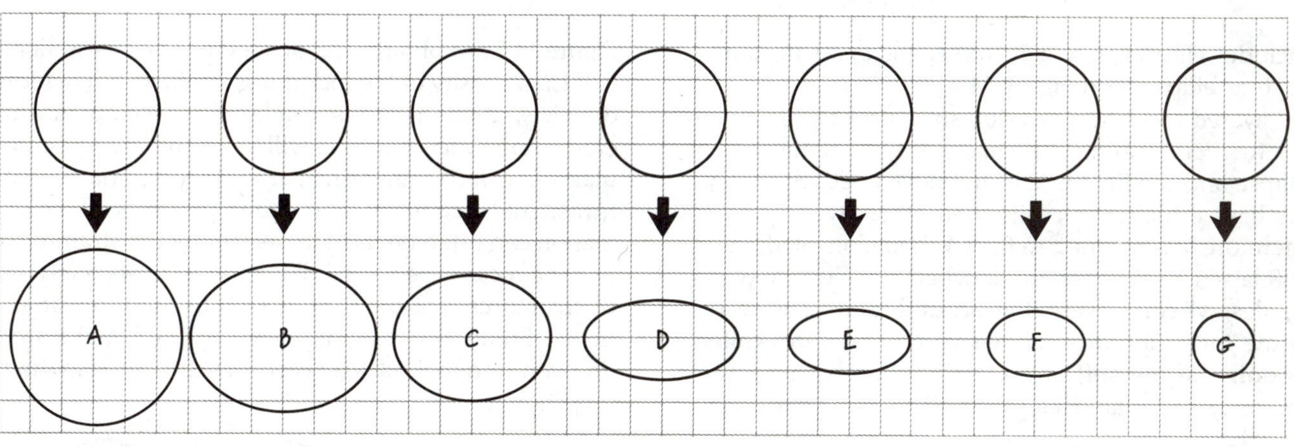

Figure 14-4
Seven circles and corresponding strain ellipses. The seven strain ellipses can be plotted on Figure 14-5.

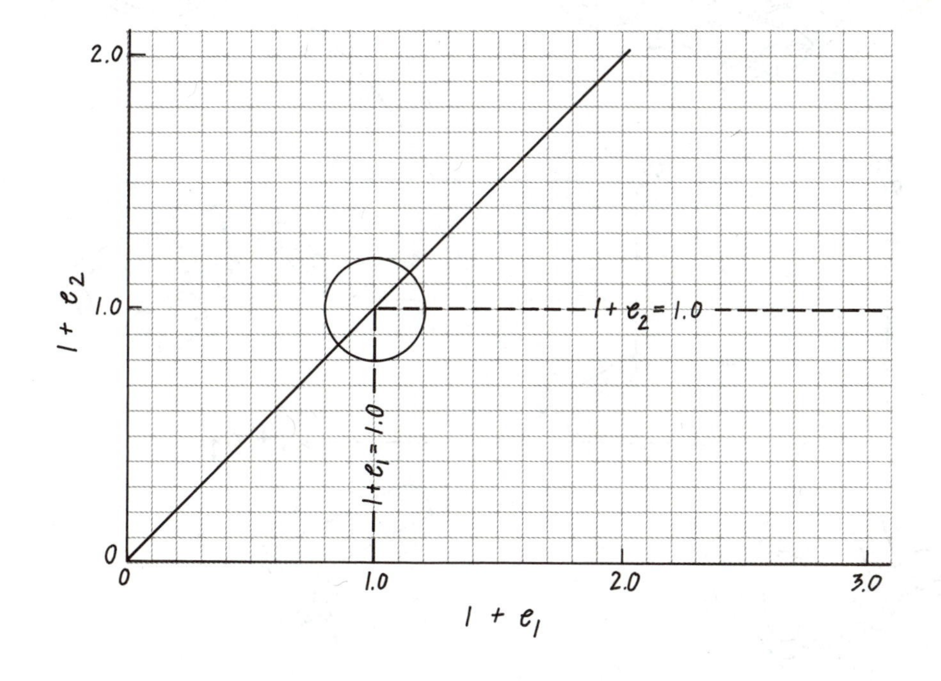

Figure 14-5
Graph on which $1 + e_2$ is plotted against $1 + e_1$ for a given strain ellipse.

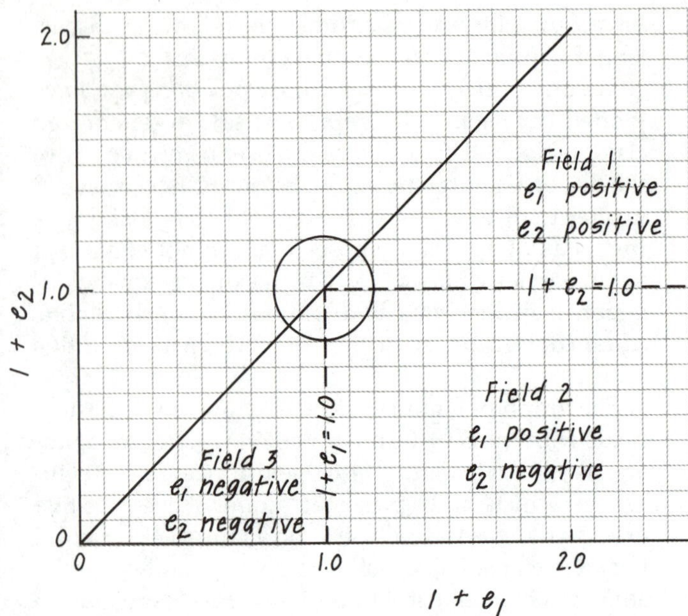

Figure 14-6
Graph for $1 + e_2$ versus $1 + e_1$ showing three fields.

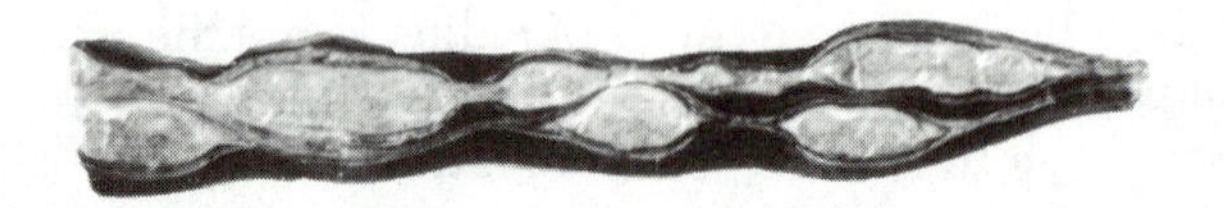

Figure 14-7
Boudinage in cross-section. From the collection of O. T. Tobisch.

includes ellipses in which both e_1 and e_2 are negative, such as ellipse F on Figure 14-4.

Layered rocks may develop structures that are useful for determining the characteristics of the strain ellipse and the field in which it developed. Some layers have a higher viscosity than other layers and are therefore less inclined to flow. When elongated, such stiff layers break up or are stretched into clumps, while the less viscous layers flow around them. The result is the formation of sausage-shaped structures called **boudins** of the stiff layers surrounded by the lower viscosity material. This process is called **boudinage.** Figure 14-7 is a photograph of boudins in cross-section.

If the viscosity of the rock does not allow it to deform ductilely, then fractures commonly develop during elongation. Such fractures will be oriented perpendicular to the maximum principal strain. Boudinage, fractures, fold geometry, and other products of deformation can often be used to determine the strain field that the rocks experienced. Figure 14-8 shows the types of structures that develop in each of the three strain fields. Study this diagram carefully, and make sure that you understand why each structure exists where it is shown.

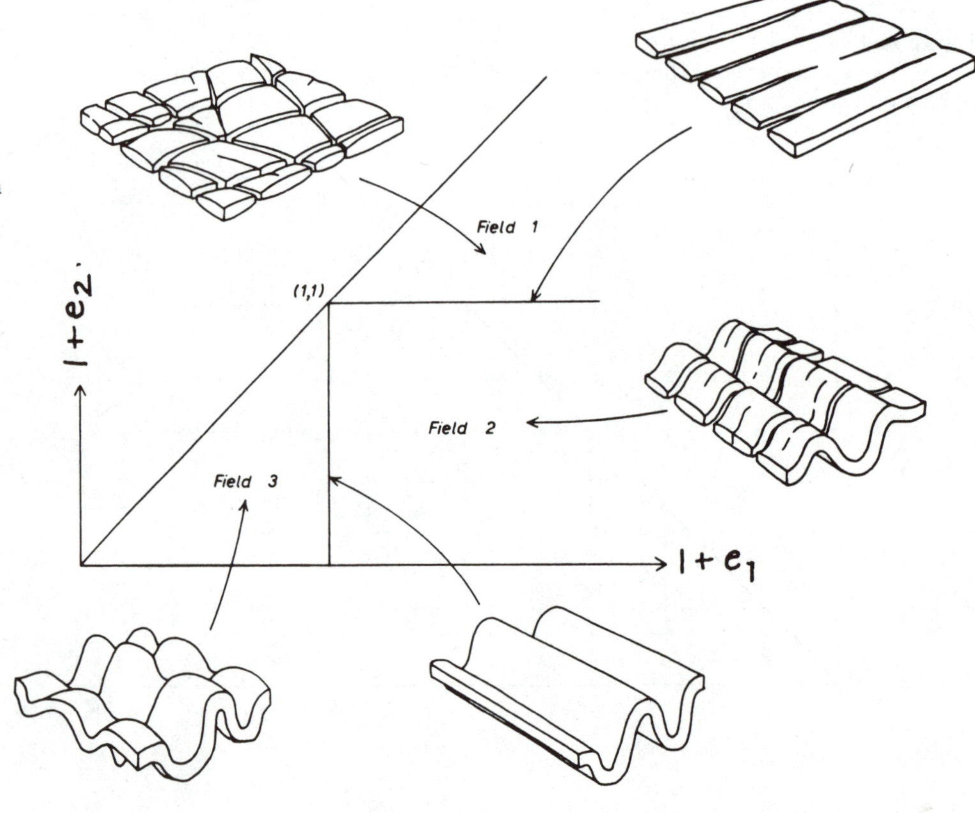

Figure 14-8
Three fields as in Figure 14-6 showing types of structures predicted to occur in each strain field. From Ramsay (1967). Reproduced with permission.

Problem 14-2

Name: ______________________

Section: ______________________

Figure 14-9a is a photograph of a slab of undeformed breccia from the Alps. Figures 14-9b, c, and d are photographs of this same breccia from nearby localities where it has been deformed. The scale is the same in all photographs.

In the lower right corner of each photograph sketch an approximate strain ellipse for the rock, and determine the $1 + e_1 : 1 + e_2$ ratio of your ellipse.

a

$1 + e_1 : 1 + e_2 = 1.0 : 1.0$

b

$1 + e_1 : 1 + e_2 =$:

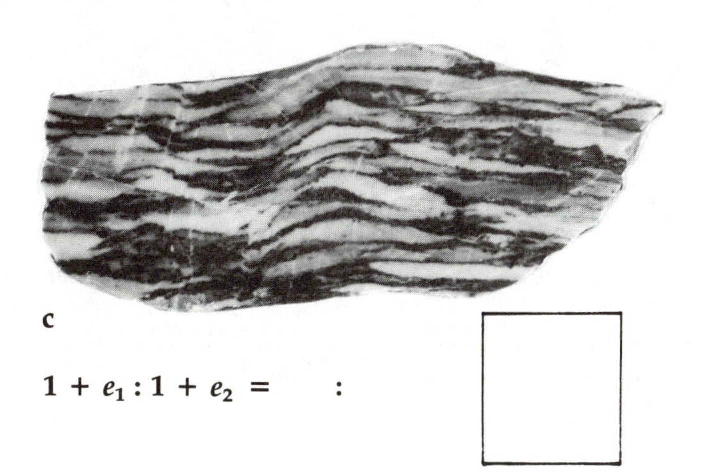

c

$1 + e_1 : 1 + e_2 =$:

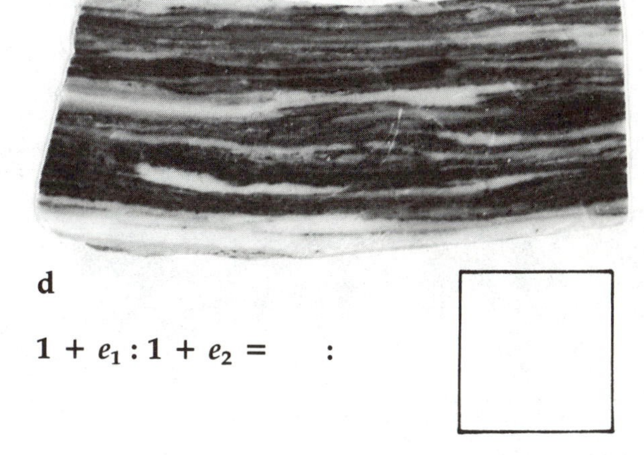

d

$1 + e_1 : 1 + e_2 =$:

Figure 14-9
Breccia from the Alps. Undeformed (a) and in various stages of deformation (b, c, and d). For use in Problem 14-2. From the collection of O. T. Tobisch.

The boomerang-shaped object in this rock slab is not a boudin but a cross-section through a dome.
Field No. ____ because ______________________________
____

This rock has a lumpy surface. Next to the main rock are two cross-section views cut at different angles.
Field No. ____ because ______________________________
____

This is another sample of the deformed breccia seen in Figure 14-9.
Field No. ____ because ______________________________
____

Figure 14-10
Three photographs of deformed structures for use in Problem 14-3. From the collection of O. T. Tobisch.

Problem 14-3

For each of the three photographs in Figure 14-10:

1. Decide which field the strain ellipse lies in, and give your reasons.
2. The circle next to each photograph represents the strain ellipse prior to deformation. Superimpose an approximation of each rock's strain ellipse on the circle.

The Coaxial Deformation Path

Up to this point we have viewed strain ellipses as the final products of deformation. It is important to see how the strain ellipse progresses as deformation proceeds. We will limit our examination to strain ellipses that lie in field 2, because these are the most common. Even with this restriction, however, strain ellipses may develop in an infinite number of ways. We will examine in some detail only the two simplest.

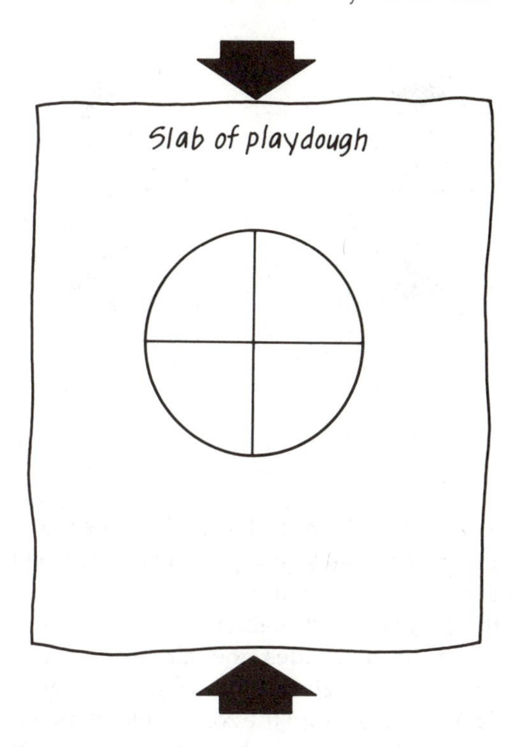

Figure 14-11
Circle and lines impressed into play dough at the beginning of Experiment 14-1.

Experiment 14-1: Coaxial Strain in Play Dough

The simplest possible strain ellipse forms by compressing a circle. An ellipse made this way in play dough is convenient for study. Record and graph the results of this experiment on the sheet provided on page 193.

Flatten a slab of play dough and impress into it a circle several centimeters in diameter. A jar lid or drinking glass can be used as a circle press. With a straightedge indent two perpendicular lines through the center of the circle, as in Figure 14-11. These two perpendicular lines are to be the lines of principal strain as the strain ellipse develops. Measure and record the radius of the circle. This measurement is l_0 for both axes of the strain ellipse. You need to know l_0 to compute e, but the radius of the circle is arbitrarily given a length of 1.0.

Compress the slab a small but measurable amount parallel to one of the two lines. Measure the lengths of the semimajor axis and semiminor axis of the resultant ellipse and determine e_1 and e_2. Proceed to fill in Table 14-1 as you deform the play dough in small increments. After measuring the dimensions of six such ellipses, graph the deformation path on Figure 14-33.

For the purpose of discussing the evolution of the strain ellipse, it will be useful to distinguish between *material* lines, such as the lines you pressed into the play dough, and *geometric* lines, such as the boundaries between the zone of compression and the zone of elongation.

The strain exhibited by the play dough strain ellipse is called **coaxial** strain because the principal axes of strain do not change their orientation with respect to the material being deformed. In **noncoaxial** strain the incremental strain ellipses rotate with respect to the material being deformed. An example of noncoaxial strain will be examined later in the chapter.

The type of deformation path you observed in Experiment 14-1 is sometimes referred to as **pure shear,** which is defined as coaxial strain with no change in volume. You could check to see if your play dough ellipse maintained its surface area by comparing the area of the undeformed circle with that of the ellipse.

Now that you have graphed a coaxial deformation path, we will take a closer look at other properties of coaxial strain.

Experiment 14-2: Lines of No Incremental Longitudinal Strain

Re-form your play dough slab, and impress a circle into it once again. As before, impress on the circle two perpendicular lines that will be the lines of principal strain. Now impress two more perpendicular lines on the circle so that they make 45° angles with the first

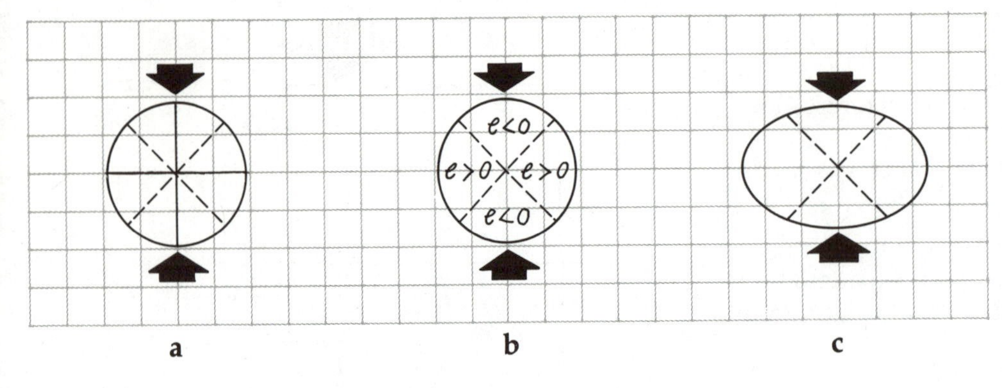

Figure 14-12
Coaxial strain, in which the principal strain axes do not rotate during deformation. (a) Solid lines are strain axes. (b) Dashed lines represent both material lines impressed into the play dough and geometric lines that divide the zone of compression from the zone of extension. (c) During deformation the geometric zone boundaries do not move, while the material lines (not shown) rotate into the zone of extension.

pair, as shown in Figure 14-12a. Deform the play dough as in Experiment 14-1, and pay close attention to the fate of the second pair of lines.

At the onset of deformation the second pair of perpendicular lines divides the circle into zones of compression and elongation (Fig. 14-12b). These material lines rotate into the zone of elongation during deformation, but the zone boundaries themselves do not move during the evolution of the strain ellipse. As shown in Figure 14-12c, the percentage of the ellipse's area in the elongation zone increases at the expense of the percentage in the shortening zone, but the boundaries of the two zones remain perpendicular to one another and at 45° to the principal strain axes.

The development of the strain ellipse is a continuous process, but it is analyzed in small increments called **incremental strain ellipses.** The geometric lines that separate the zone of elongation from the zone of shortening are called **lines of no incremental longitudinal strain** (Fig. 14-13) because for all incremental strain ellipses, $e = 0$ along these two lines. As the ellipse progressively develops, the material lines that occupy the lines of no incremental longitudinal strain in one incremental strain ellipse pass into the zone of elongation in the next increment.

Your play dough ellipse should now look something like the one in Figure 14-14a. Impress two new lines onto your strain ellipse at the positions of no incremental longitudinal strain, then impress two more lines across the center of the ellipse close to these lines, but in the zone of shortening, as shown in Figure 14-14b. This last pair of lines has up until now experienced only shortening. As you further deform your play dough, however, see how these shortened lines become lines of no incremental longitudinal strain for an instant and then begin to elongate. During deformation some lines undergo continuous elongation, some shorten and then elongate, and some shorten continuously. Such behavior explains many features seen in deformed rocks.

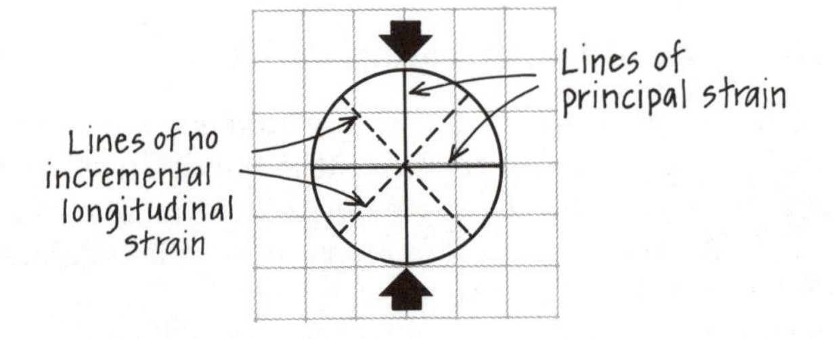

Figure 14-13
Incremental strain ellipse.

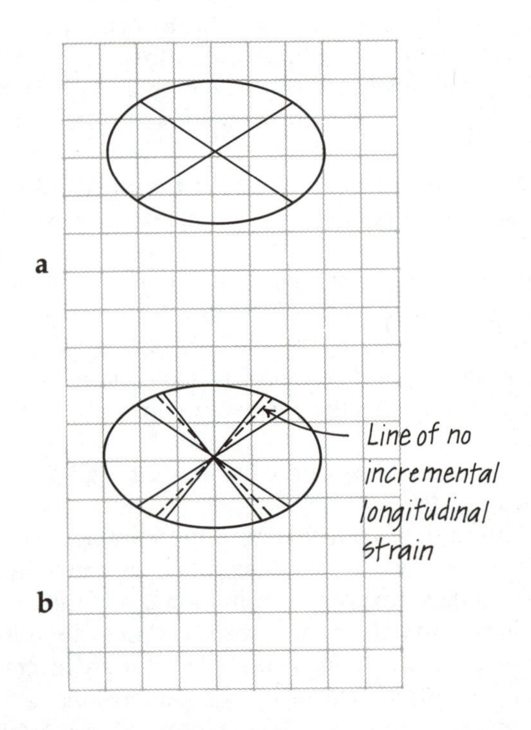

Figure 14-14
Shortening and lengthening of material lines. (a) Deformed play dough ellipse with two sets of originally perpendicular material lines. (b) Same ellipse with two additional sets of material lines.

The Coaxial Total Strain Ellipse

Once deformation has ceased, we can call the strain ellipse a **total strain ellipse.** Imagine impressing a unit circle into play dough and deforming it into an ellipse. Now imagine superimposing another unit circle on top of the ellipse, as in Figure 14-15. Notice that two axes of the ellipse are also diameters of the circle. Regardless of how these lines might have shortened and elongated during the development of the strain ellipse, the net result is that they are the same length as they started out. These are called **lines of no total longitudinal strain.** All axes of the strain ellipse that fall within the unit circle have undergone net shortening, while all of those that extend beyond the circle have undergone net elongation.

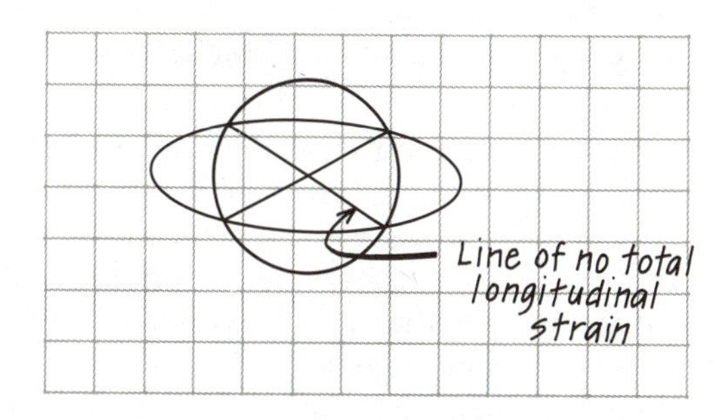

Figure 14-15
Total strain ellipse with lines of no total longitudinal strain.

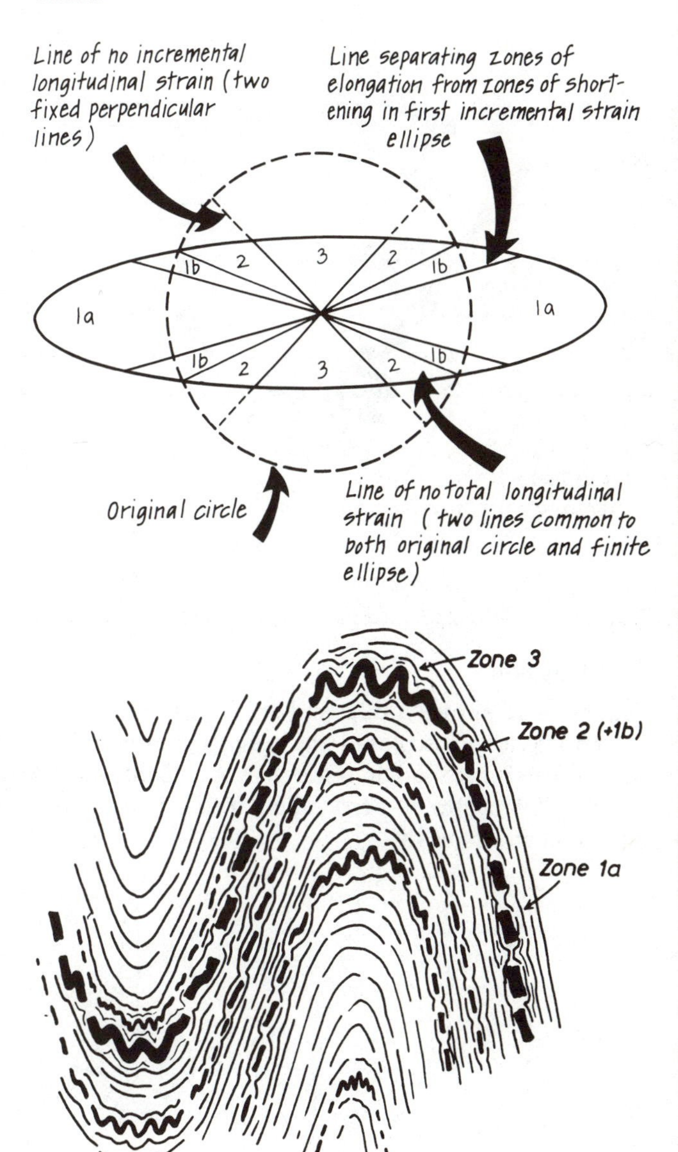

Figure 14-16
Coaxial total strain ellipse with four zones, each of which has a different deformation history. After Ramsay (1967).

Zone		Structures in competent beds
1a	Lines that have been elongated only	Boudinage
1b	Lines that underwent early shortening followed by elongation (net lengthening)	Remnants of disrupted folds and isolated fold hinges
2	Lines that underwent early shortening followed by elongation (net shortening)	Folds that are becoming unfolded and boudinaged
3	Lines that have been shortened only	Folds with large amplitude and short wavelengths

Now we can divide our coaxial total strain ellipse into four zones that characterize the deformation history of lines. Figure 14-16 shows these zones and lists the structures that are predicted to form in each. Zone 1a, which includes lines that have only elongated, produces boudinage in competent beds. Zones 1b and 2 include lines that underwent early shortening followed by elongation: those in 1b ended up long; those in 2 ended up short. In these two zones, folds which formed during the shortening stage become disrupted or unfolded during the elongation stage. Zone 3 includes lines that have been only shortened, producing folds with large amplitude and short wavelengths. Figure 14-17 shows a fold containing structures from all four zones.

Figure 14-17
Fold developed by coaxial strain showing structures in each zone. Maximum principal strain axis is vertical. From Ramsay (1967). Reproduced with permission.

Experiment 14-3: Superimposed Total Strain Ellipses

Before setting the play dough aside, one more observation needs to be made. Impress a circle on a smooth slab of play dough and deform the circle into a distinct ellipse as before. Now squeeze the slab from another direction. You will see that the total strain ellipse from the first strain regime becomes deformed into a different shaped ellipse under a differently oriented strain field. Any strain ellipse, therefore, may be the product of any number of strain episodes of varying orientation. Things may be more complicated than they appear.

Figure 14-18
Diagram for use in Problem 14-4.

Problem 14-4

Figure 14-18 shows a dike and sill complex in which a competent rock, colored black, has intruded into a schist. The horizontal lines are cleavage. Consider the cleavage planes to be perpendicular to the minimum principal strain axis.

1. What has been the approximate extension e perpendicular to the cleavage?
2. What has been the approximate extension e parallel to the cleavage?
3. Using the two extensions just determined and the structures seen in the dikes and sills, draw a properly proportioned and properly oriented strain ellipse for this rock. Label the zones within the strain ellipse, and give the $1 + e_1$:$1 + e_2$ ratio.

Noncoaxial Strain

Experiment 14-4: The Card-Deck Strain Ellipse

Noncoaxial strain, in which the principal axes of strain do change their orientation with respect to material lines, is easily demonstrated with a stack of computer cards about 5 centimeters or more in thickness. Ideally, a wooden box such as the one shown in Figure 14-19 should be constructed to hold the cards during the experiment.

First, draw a circle on the edge of the cards. Using a straightedge, produce a uniform shear in one direction, thereby deforming the circle into an ellipse (Fig. 14-20). Deck-of-cards-type deformation is referred to as **simple shear.** Simple shear is noncoaxial, constant-volume, two-dimensional deformation with no flattening perpendicular to the plane of slip. Notice that the ellipses produced by shearing the cards must all be of equal area because the component chords that lie on each computer card are of constant length.

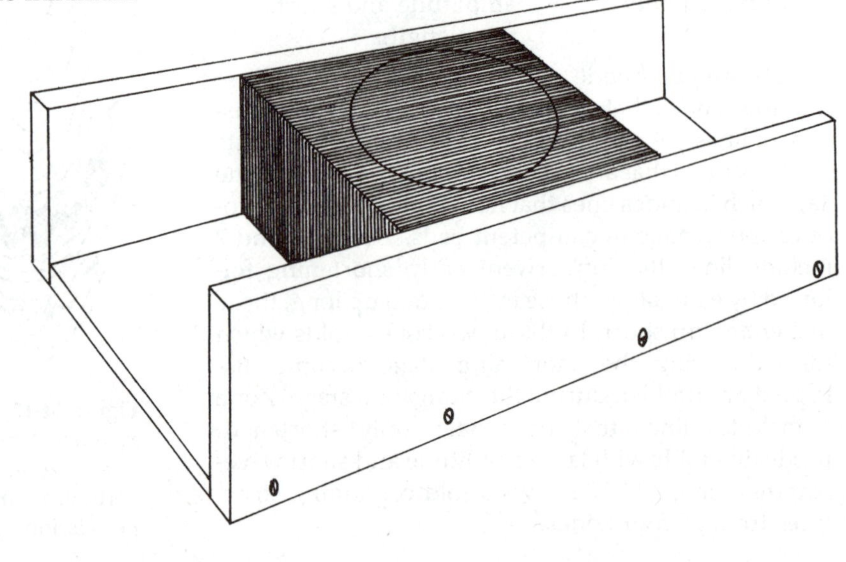

Figure 14-19
Wooden box apparatus to hold computer cards during noncoaxial strain experiments. After Ramsay and Huber (1983), Figure. 1.1.

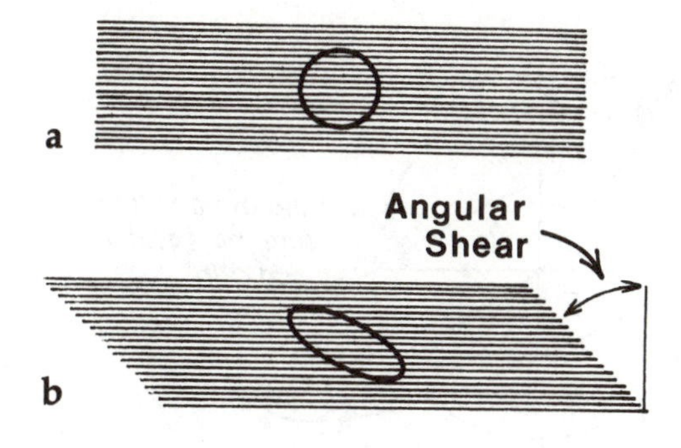

Figure 14-20
Noncoaxial strain demonstrated with computer cards. (a) Before deformation. (b) After deformation, showing angular shear.

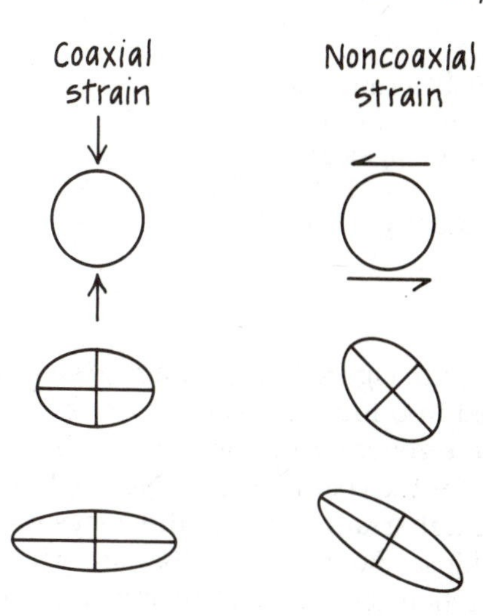

Figure 14-21
Comparison of coaxial and noncoaxial strain.

An important aspect of simple shear, and noncoaxial strain in general, is the amount of rotation. In this experiment you will measure the angle of rotation ψ as shown in Figure 14-20b.

Deform the deck in small increments of about 10° of rotation, and record the pertinent data on Table 14-2 (p. 195). Then graph the deformation path of the ellipse on Figure 14-34, as you did with the coaxial ellipse on Figure 14-33.

Compare the simple shear deformation path of the card-deck ellipse with the deformation path of the play dough ellipse. In fact there should be very little difference in the two deformation paths. The two processes are, however, quite different. As summarized in Figure 14-21, the principal strain axes rotate within the stress field during noncoaxial strain; in coaxial strain they do not. This is no trivial difference, and the structure in the rocks reflects the process.

The Noncoaxial Total Strain Ellipse

We will now examine some details of noncoaxial strain. As with the coaxial ellipse, this ellipse has two fixed, perpendicular boundaries between zones of shortening and elongation. These are the lines of no incremental longitudinal strain, and they are parallel and perpendicular to the edges of the cards. Recall that these are geometric and not material lines. As with coaxial deformation, as the ellipse deforms, the area of the ellipse in the elongating zones increases while the area in the shortening zones decreases (Fig. 14-22).

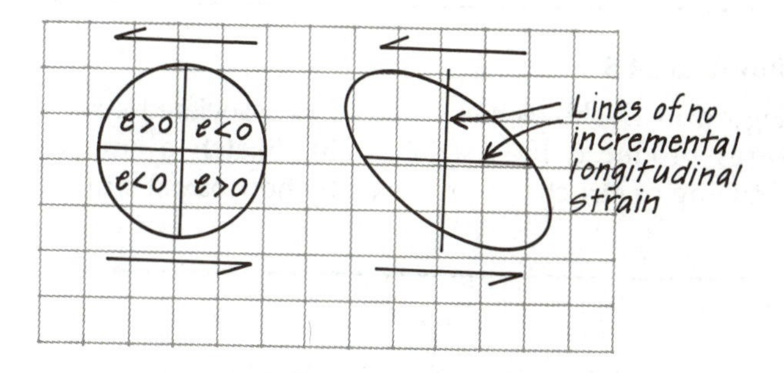

Figure 14-22
Noncoaxial strain ellipse showing zones of shortening and elongation and lines of no incremental longitudinal strain. (a) Before deformation. (b) After deformation.

Experiment 14-5: The Asymmetrically Zoned Noncoaxial Strain Ellipse

Square up your card deck and draw the lines of no incremental longitudinal strain on the undeformed circle (Fig. 14-22). Deform the circle into a distinct ellipse, and draw a circle equal in diameter to the original circle symmetrically on the ellipse. Finally, draw in the lines of no incremental longitudinal strain once again and the lines of no total longitudinal strain, as in Figure 14-16.

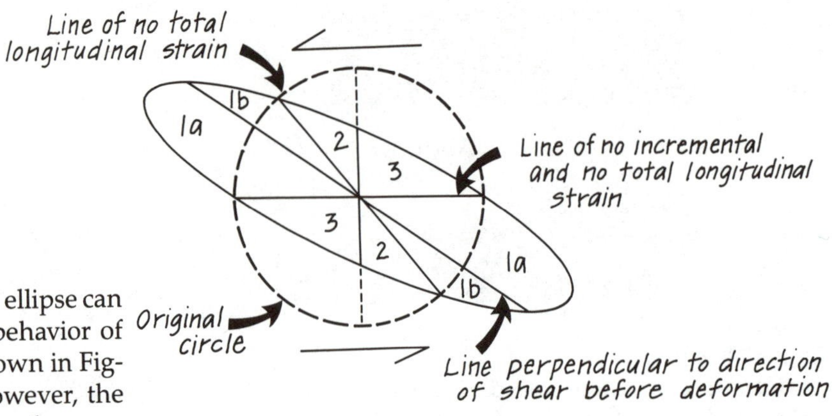

Figure 14-23
Noncoaxial total strain ellipse showing four zones. Notice asymmetric arrangement of zones compared with those in coaxial ellipse (Fig. 14-16). After Ramsay (1967).

As shown in Figure 14-23, the noncoaxial ellipse can be divided into four zones based on the behavior of lines. These zones correspond to those shown in Figure 14-16 for the coaxial ellipse. Notice, however, the asymmetrical arrangement of the zones on the noncoaxial ellipse. Because lines that are parallel to the shear direction are lines of no incremental *and* no total longitudinal strain, zones 1b and 2 only occur on one side of zone 1a in the noncoaxial ellipse. This phenomenon is useful in attempting to determine whether or not rotation has been involved in the development of certain structures.

Problem 14-5

Figure 14-17 shows some folds and associated structures produced by coaxial strain. Sketch a similar drawing of structures produced by noncoaxial strain.

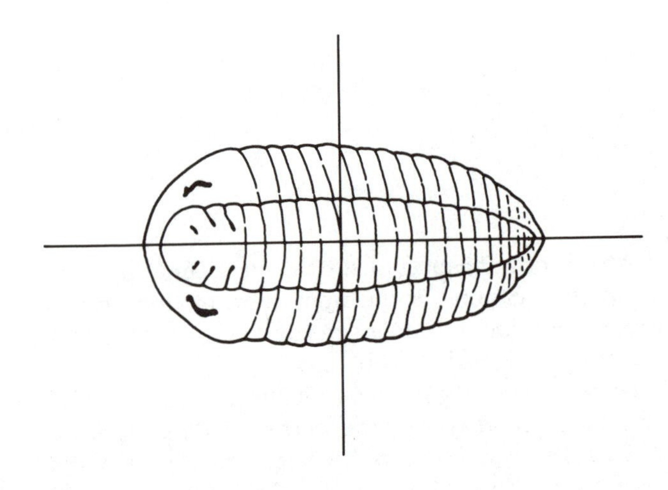

Figure 14-24
Undeformed trilobite.

Deformed Fossils as Strain Indicators

Many rocks are lithologically homogeneous and do not contain structures such as folds and boudins that reveal the strain. In such rocks fossils can sometimes be used as strain indicators. If the undeformed size and shape of the fossil are precisely known, then the problem is merely one of measuring extensions and angular shear in different directions, as in Problem 14-1. Usually, however, it is difficult or impossible to reliably determine what the lengths of lines were before deformation. This problem can be partially overcome when several variously oriented individuals are present.

Fossils that are especially suitable for strain measurement are bilaterally symmetrical ones such as brachiopods and trilobites. Trilobites, for example, have a central axis that divides the animal into mirror-image right and left sides (Fig. 14-24).

A technique developed by Wellman (1962) provides an elegant approach to the use of such fossils in strain analysis. Imagine five randomly oriented undeformed trilobites on a slab of mudstone. These are represented by five sets of perpendicular lines in Figure 14-25a. Points A and B are arbitrarily located points

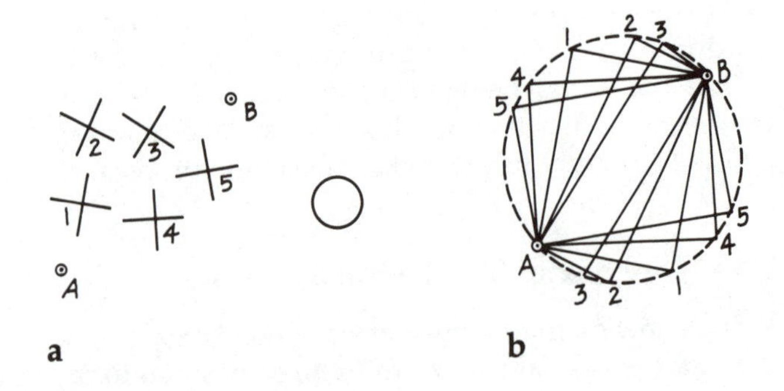

Figure 14-25
Wellman technique for determining strain ellipse. In this example the rock is not deformed, and fossils are represented by perpendicular lines, as in Figure 14-24. (a) Perpendicular lines representing five fossils. Points A and B are arbitrary points located on the rock slab. (b) Strain ellipse (dashed line) passes through corners of squares formed by extending perpendicular sets from points A and B.

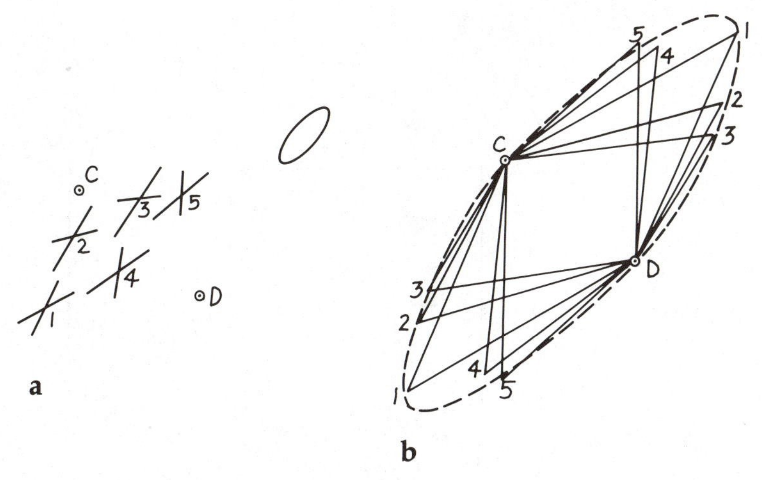

Figure 14-26
Using the Wellman technique on deformed fossils. (a) Five nonperpendicular lines representing deformed fossils. (b) Strain ellipse (dashed line) drawn through corners of parallelograms.

on the slab. Without changing the orientation of the lines, imagine translating each set of perpendicular lines to points A and B and lengthening the lines until they meet to form a rectangle. If this is done for all five sets of lines, the corners of the five rectangles lie on a common circle (Fig. 14-25b). This circle represents the strain ellipse prior to deformation.

Now imagine five deformed trilobites on a slab, represented by five pairs of nonperpendicular lines (Fig. 14-26a). Points C and D are arbitrarily located on the slab, and each pair of lines is translated to each point and extended, resulting in five parallelograms. The corners of the parallelograms define the strain ellipse (Fig. 14-26b).

To determine the axial ratio and orientation of the strain ellipse from a group of deformed fossils:

1. Draw two lines on each fossil. These lines represent perpendicular lines prior to deformation.
2. Place two points several inches apart on the photograph or drawing. The line between these two points should not be parallel to any of the lines you have drawn on the fossils.
3. Place tracing paper over the photograph, transfer the two points to the tracing paper, and proceed to transfer the pairs of lines to the two points without rotating the tracing paper with respect to the photograph.
4. Sketch the ellipse that most closely fits the corners of the parallelograms.

Do problem 14-6 on page 189.

Strain in Three Dimensions

Because rocks are three-dimensional objects, consideration must be made for the third dimension in describing and measuring strain. Undeformed rocks are imagined to contain a sphere which becomes an ellipsoid during homogeneous deformation. This is the **strain ellipsoid,** and its dimensions and orientation describe the strain in the rock.

The strain ellipsoid has three principal axes, the maximum, intermediate, and minimum principal strain axes. Their lengths are $1 + e_1 \geqslant 1 + e_2 \geqslant 1 + e_3$, respectively, with the original sphere having a radius of 1. In Figure 14-27a is an equidimensional, undeformed block. In Figure 14-27b is the same block after deformation, showing the principal strain axes.

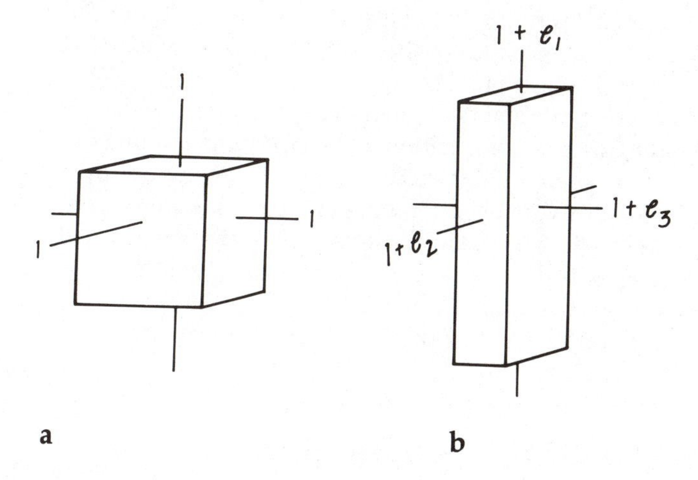

Figure 14-27
Three principal strain axes. (a) Undeformed. (b) After deformation.

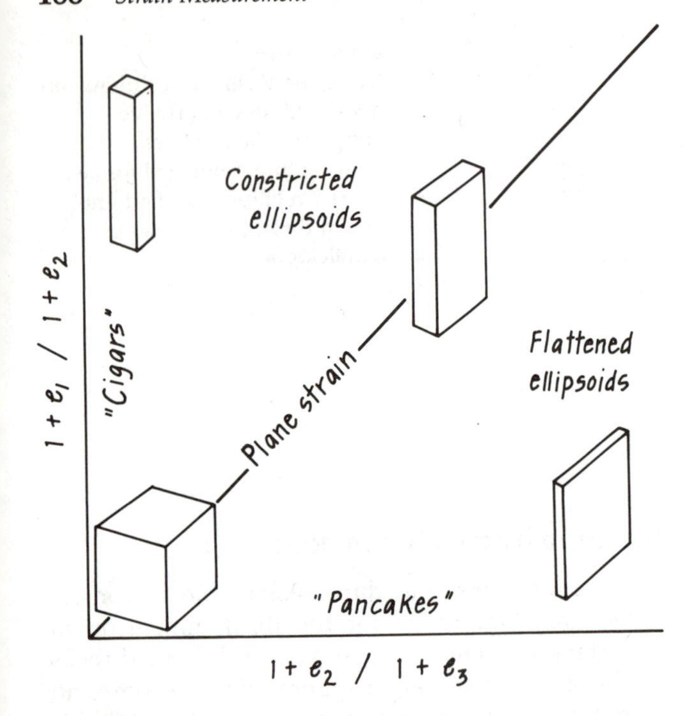

Figure 14-28
Flinn diagram. Two strain fields are divided by a plane strain line along which $e_2 = 0$ and volume is preserved. Pure shear and simple shear both fall on the plane strain line.

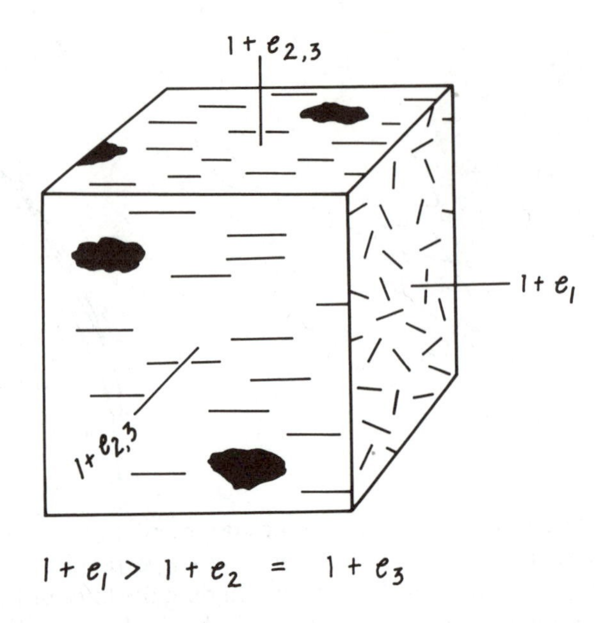

Figure 14-29
Deformed rock containing platy mineral. The consistent lineation indicates the orientation of the maximum principal strain axis. The random orientation of the flat surface indicates that $1 + e_2 = 1 + e_3$.

The graph in Figure 14-28, in which $1 + e_1/1 + e_2$ is graphed against $1 + e_2/1 + e_3$, displays the range of strain ellipsoids. This graph, called a Flinn diagram, is divided into two fields by the plane strain line. **Plane strain** is deformation in which volume is preserved and extension is zero on the intermediate principal strain axis. Above the plane strain line the ellipsoids are constricted, the ultimate being cigar-shaped ellipsoids in which $1 + e_1 > 1 + e_2 = 1 + e_3$. Below the plane strain line the ellipsoids are flattened, the ultimate being pancake-shaped ellipsoids in which $1 + e_1 = 1 + e_2 > 1 + e_3$.

The fabrics of deformed rocks may often be used to diagnose the orientation and shape of the strain ellipsoid. The rock depicted in Figure 14-29, for example, contains variably oriented, platy minerals with a common axis. The strain ellipsoid is cigar-shaped in which $1 + e_1 > 1 + e_2 = 1 + e_3$. The principal strain axes are shown on the drawing.

Do problem 14-7 on page 190.

Quantifying the Strain Ellipsoid

A variety of techniques has been devised for quantifying strain. Most are beyond the scope of an introductory course. Ramsay and Huber (1983) is the source to be consulted for a more complete discussion of this topic. One technique that is, in principle, very simple involves the measurement of deformed objects in the rock. Pebbles in deformed conglomerate, for example, may be used this way, but lack of original sphericity complicates the measurements.

A rock type that is particularly well suited to strain measurement is oolite, a limestone composed of spherical, sand-sized grains called ooids. Upon deformation, the spherical ooids become ellipsoidal and can be measured directly.

Do problem 14-8 on page 191.

Further Reading

Compton, R. R. 1985. *Geology in the Field*. New York: Wiley. One chapter is devoted to field analysis of deformed rocks, including analysis of strain.

Ramsay, J. G., 1967. *Folding and Fracturing of Rocks*. New York: McGraw-Hill. A standard reference on strain analysis.

Ramsay, J. G., and Huber, M. I. 1983. *The Techniques of Modern Structural Geology, Volume 1: Strain Analysis*. London: Academic Press. A comprehensive coverage of strain analysis, suitable for the advanced structure student. Contains problems with answers and discussions.

Problem 14-6

Figure 14-30 is a photograph of deformed pygidia (posterior sections) from several trilobites.

1. Determine the strain ellipse for this rock, and draw it in the upper right corner of the page (include your tracing paper with your solution).
2. Determine the $1 + e_1{:}1 + e_2$ ratio, and the orientation of the maximum principal strain with respect to north.

Figure 14-30
Deformed trilobite pygidia for use in Problem 14-6.

Problem 14-7

On the four drawings of deformed rocks in Figure 14-31 indicate the orientations of the principal strain axes. Below each drawing indicate the relative lengths of the axes (as was done in Figure 14-29).

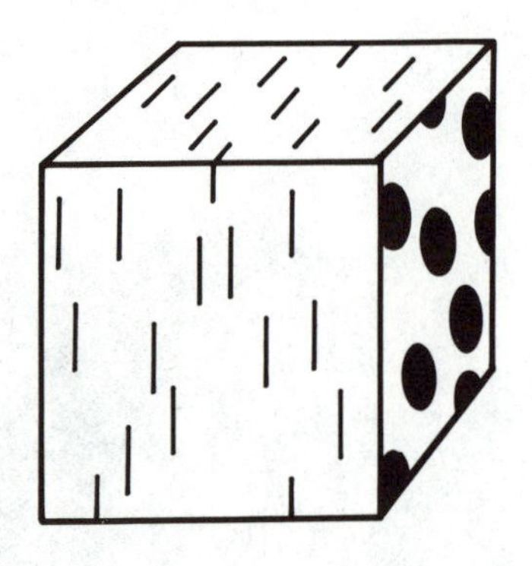

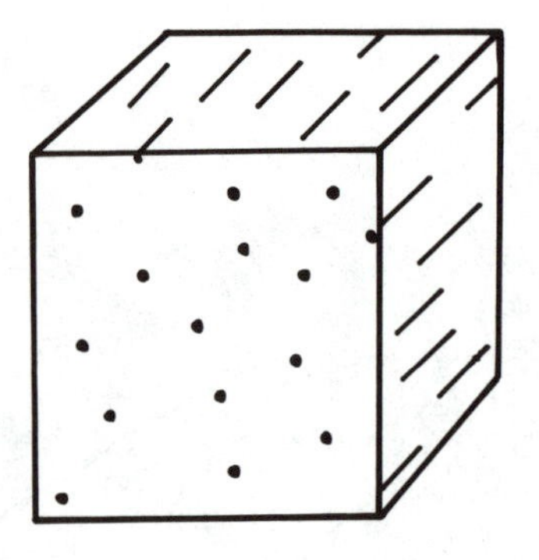

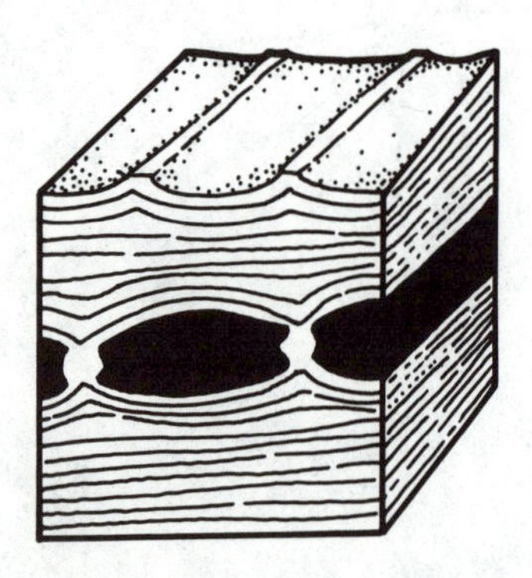

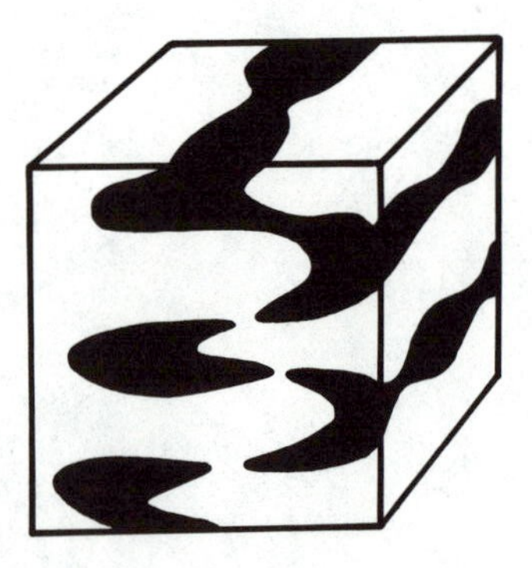

Figure 14-31
Four block diagrams of deformed rocks for use in Problem 14-7.

Problem 14-8

In Figure 14-32a is a sketch of a hand specimen of oolite. The orientations of the principal strain axes have been determined in the field on the basis of lineations, cleavages, and the shapes of the ooids. Two thin sections have been cut perpendicular to two of the principal strain axes, and in Figure 14-32b are microscopic fields of view of each of the two thin sections.

1. Measure the dimensions of several ooids and determine the arithmetic mean ($\bar{x}$) of the major and the minor axis in each field of view. Indicate the major axis-minor axis ratio for each field of view, and combine these ratios to find the $1 + e_1$:$1 + e_2$:$1 + e_3$ ratio for the strain ellipsoid.
2. Plot the strain ellipsoid on the Flinn diagram in Figure 14-32c.

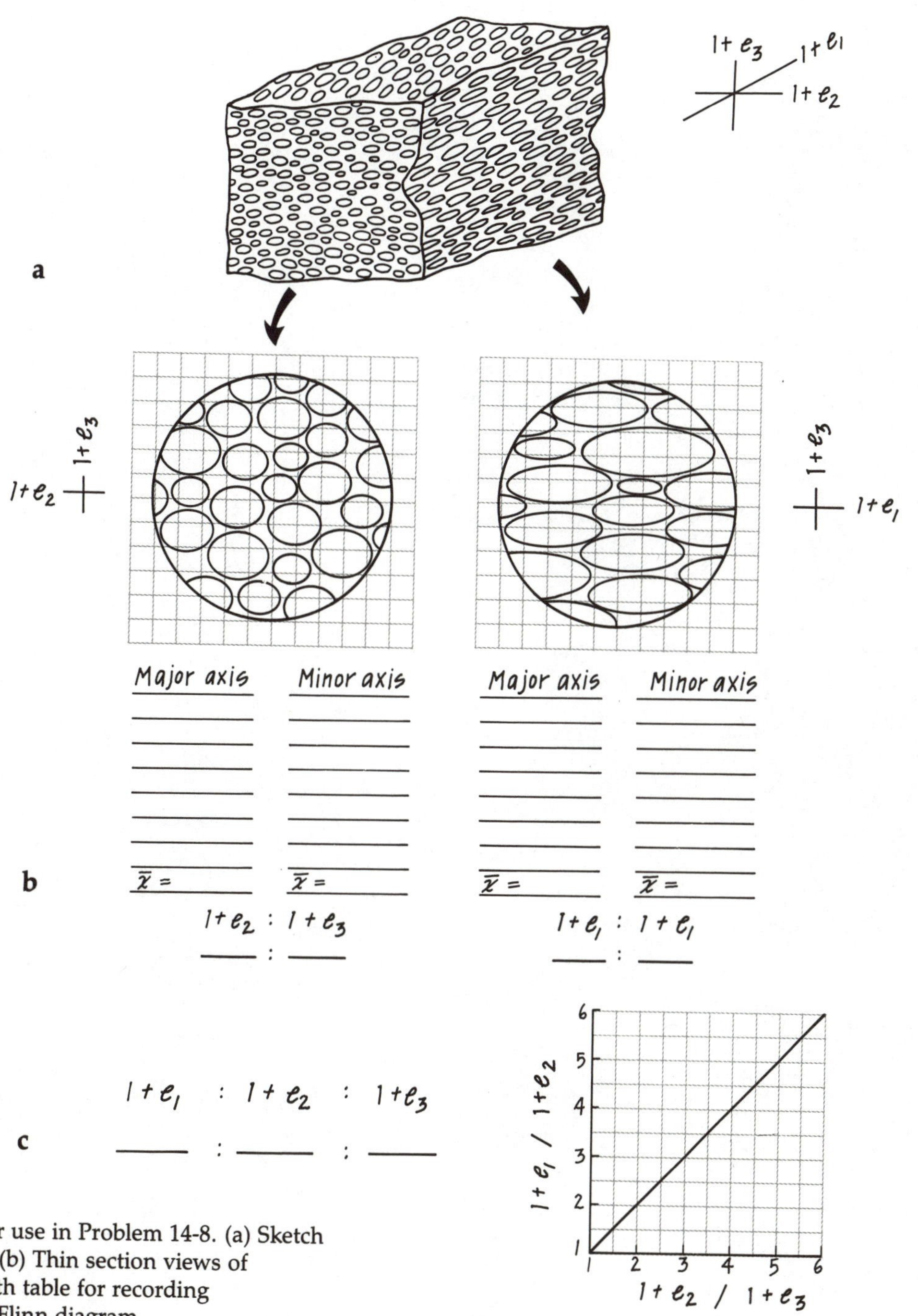

Figure 14-32
Deformed oolite for use in Problem 14-8. (a) Sketch of hand specimen. (b) Thin section views of deformed ooids with table for recording measurements. (c) Flinn diagram.

Notes

Table 14-1
Results of Experiment 14-1.

Time	Length of semi-major axis	e_1	$1 + e_1$	Length of semi-minor axis	e_2	$1 + e_2$	$1 + e_1 : 1 + e_2$
t_0	$l_0 =$	0	1.0	$l_0 =$	0	1.0	1.0:1.0
t_1	______	___	______	______	___	______	______
t_2	______	___	______	______	___	______	______
t_3	______	___	______	______	___	______	______
t_4	______	___	______	______	___	______	______
t_5	______	___	______	______	___	______	______
t_6	______	___	______	______	___	______	______

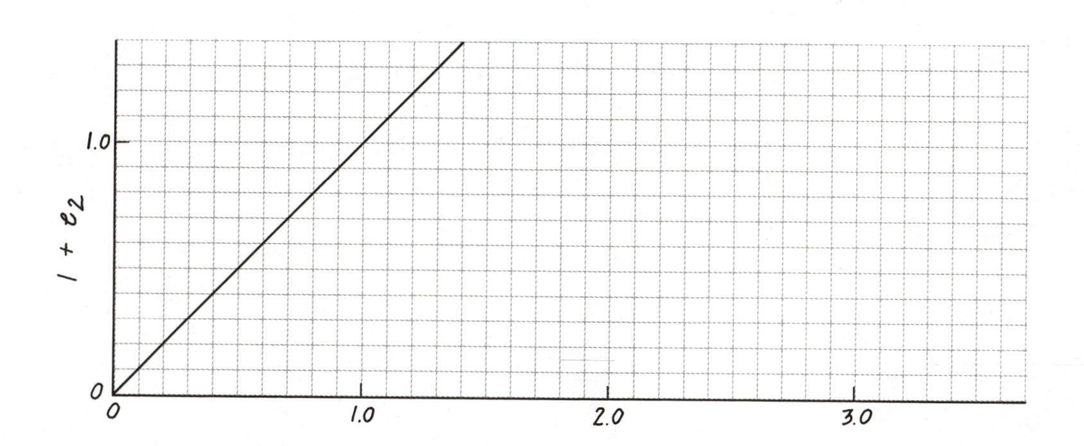

Figure 14-33
Graph for plotting coaxial deformation path in Experiment 14-1.

Notes

Table 14-2
Results of Experiment 14-4.

Time	Length of semi-major axis	e_1	$1 + e_1$	Length of semi-minor axis	e_2	$1 + e_2$	$1 + e_1{:}1 + e_2$	Angle of rotation
t_0	$l_0 =$	0	1.0	$l_0 =$	0	1.0	1.0:1.0	0°
t_1	______	___	______	______	___	______	______	______
t_2	______	___	______	______	___	______	______	______
t_3	______	___	______	______	___	______	______	______
t_4	______	___	______	______	___	______	______	______
t_5	______	___	______	______	___	______	______	______
t_6	______	___	______	______	___	______	______	______

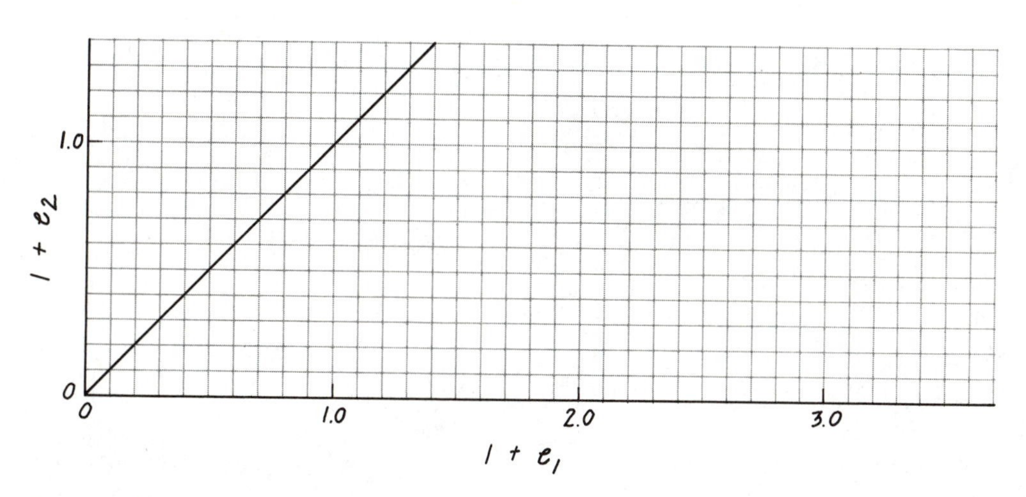

Figure 14-34
Graph for plotting noncoaxial deformation path in Experiment 14-4.

Notes

Appendices

Appendix A:

Measuring Attitudes with a Brunton Compass

It is often in their structural geology course that students get their first experience with geologic mapping. This appendix is included to facilitate the inclusion of field work in the structure course. I recommend that students practice in the lab before going to the field. This can be done with "portable outcrops" (Rowland, 1978).

In North America the traditional instrument for measuring attitudes is the Brunton compass, which requires the strike and dip to be measured separately. There is an instrument, called the Clar compass, that measures the dip direction (90° from strike) and angle of dip in a single operation (see Suppe, 1985, Fig. 2-2). The following explanation assumes the use of the Brunton compass. Only the relatively simple cases of measuring strike and dip of an exposed plane and trend and plunge of lineations on an exposed plane will be considered here. For other aspects of compasscraft, refer to a book on field geology (e.g., Compton, 1985).

Strike is measured by placing an *edge* (not a side) of the compass along the plane and leveling the bull's eye level (Fig. A-1). Either end of the needle may be used to read the strike. It is often useful to place a map board or field book against the plane to flatten out irregularities.

Dip is measured by placing the flat *side* of the Brunton against the plane and rotating the arm on the back of the compass until the tube level is level. The face of the compass *must* be vertical. A common error is failure to make the face vertical. The angle of dip is read on the inner scale of the compass face.

Orientations of lineations in a plane are most easily determined by first measuring the attitude of the plane. Then draw a horizontal line on the plane, and, with a protractor, measure the pitch of the lineation within the plane (Fig. A-2). The trend and plunge of each lineation is then determined with a stereonet (see Chapter 5, Fig. 5-11). Alternatively, the trend and plunge of a lineation may be measured directly by placing your map board vertically and coplanar with the lineation.

When doing field work the attitude should be recorded in the field notes and *immediately* plotted on the map. Orient the map and confirm that it is correct. It is exceedingly easy to plot the strike northwest instead of northeast or to misdirect the dip line. If you are measuring more than one feature at the same outcrop (e.g., orientations of three joint sets), make a neat sketch in your notebook showing the relationships of the various features and their attitudes.

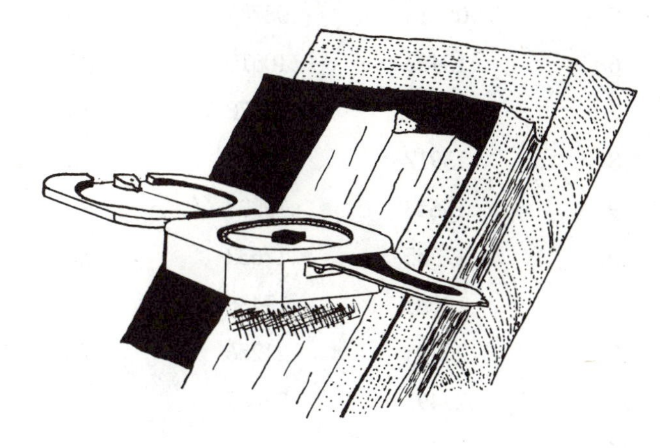

Figure A-1
Measuring strike with a Brunton compass.

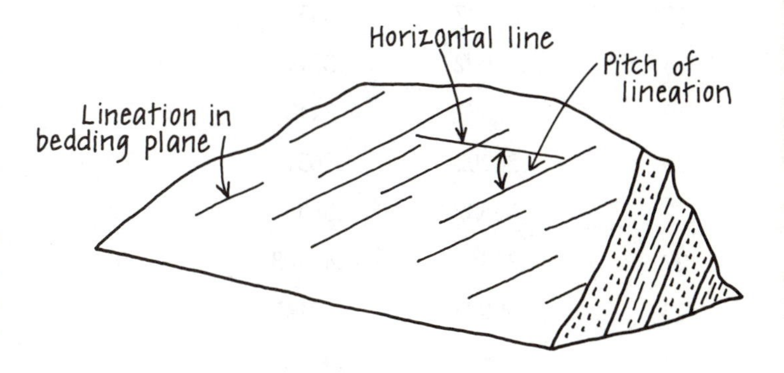

Figure A-2
Measuring pitch of lineations within a dipping plane.

Appendix B:

Trigonometric Functions

Degree	Sin	Tan	
0	.00000	.00000	90
1	.01745	.01746	89
2	.03490	.03492	88
3	.05234	.05241	87
4	.06976	.06993	86
5	.08716	.08749	85
6	.10453	.10510	84
7	.12187	.12278	83
8	.13917	.14054	82
9	.15643	.15838	81
10	.17365	.17633	80
11	.19081	.19438	79
12	.20791	.21256	78
13	.22495	.23087	77
14	.24192	.24933	76
15	.25882	.26795	75
16	.27564	.28675	74
17	.29237	.30573	73
18	.30902	.32492	72
19	.32557	.34433	71
20	.34202	.36397	70
21	.35837	.38386	69
22	.37461	.40403	68
23	.39073	.42447	67
24	.40674	.44523	66
25	.42262	.46631	65
	Cos	Cot	Degree

Degree	Sin	Tan	
26	.43837	.48773	64
27	.45399	.50953	63
28	.46947	.53171	62
29	.48481	.55431	61
30	.50000	.57735	60
31	.51504	.60086	59
32	.52992	.62487	58
33	.54464	.64941	57
34	.55919	.67451	56
35	.57358	.70021	55
36	.58779	.72654	54
37	.60182	.75355	53
38	.61566	.78129	52
39	.62932	.80978	51
40	.64279	.83910	50
41	.65606	.86929	49
42	.66913	.90040	48
43	.68200	.93252	47
44	.69466	.96569	46
45	.70711	1.0000	45
46	.71934	1.0355	44
47	.73135	1.0724	43
48	.74314	1.1106	42
49	.75471	1.1504	41
50	.76604	1.1918	40
	Cos	Cot	Degree

Degree	Sin	Tan	
51	.77715	1.2349	39
52	.78801	1.2799	38
53	.79864	1.3270	37
54	.80902	1.3764	36
55	.81915	1.4281	35
56	.82904	1.4826	34
57	.83867	1.5399	33
58	.84805	1.6003	32
59	.85117	1.6643	31
60	.86603	1.7321	30
61	.87462	1.8040	29
62	.88295	1.8807	28
63	.89101	1.9626	27
64	.89879	2.0503	26
65	.90631	2.1445	25
66	.91355	2.2460	24
67	.92050	2.3559	23
68	.92718	2.4751	22
69	.93358	2.6051	21
70	.93969	2.7475	20
71	.94552	2.9042	19
72	.95106	3.0777	18
73	.95630	3.2709	17
74	.96126	3.4874	16
75	.96593	3.7321	15
	Cos	**Cot**	**Degree**

Degree	Sin	Tan	
76	.97030	4.0108	14
77	.97437	4.3315	13
78	.97815	4.7046	12
79	.98163	5.1446	11
80	.98481	5.6713	10
81	.98769	6.3138	9
82	.99027	7.1154	8
83	.99255	8.1443	7
84	.99452	9.5144	6
85	.99619	11.430	5
86	.99756	14.301	4
87	.99863	19.081	3
88	.99939	28.636	2
89	.99985	57.290	1
90	1.0000		0
	Cos	**Cot**	**Degree**

Appendix C:

Greek Letters and Their Use in This Book

Letter	Use
α (alpha)	Apparent dip (Chapters 1, 2)
β (beta)	Angle between the strike of a plane and the trend of an apparent dip (Chapter 1)
γ (gamma)	Shear strain (Chapter 14)
δ (delta)	Plunge of true dip (Chapters 1, 2, 3)
ϵ (epsilon)	Strain (Chapter 12)
$\dot{\epsilon}$	Strain rate (Chapter 12)
η (eta)	Coefficient of viscosity (Chapter 12)
θ (theta)	Trend of apparent dip (Chapters 1, 2) Angle between a plane and direction of σ_3 (Chapter 13)
μ (mu)	Coulomb coefficient (Chapter 13)
π (pi)	Pole of foliation attitude (Chapter 7)
σ (sigma)	Stress (Chapters 10, 12)
σ_a	Axial load (Chapter 13)
σ_c	Confining pressure (Chapter 13)
σ_n	Normal stress (Chapter 13)
σ_s	Shear stress (Chapter 13)
σ_y	Yield stress (Chapter 12)
τ (tau)	Same as σ_s
ϕ (phi)	Angle of internal friction (Chapter 13)
ψ (psi)	Angular shear (Chapter 14)

Appendix D:

Graph for Determining Exaggerated Dips on Structure Sections with Vertical Exaggeration

Nominal dips are indicated on curved lines. To determine exaggerated dip, carry vertical exaggeration horizontally across graph until it intersects desired dip line. Exaggerated dip lies on horizontal axis directly below. For example, on a structure section with 4.0 vertical exaggeration a 20° dip must be drawn at 55°. After Dennison (1968). Alternatively, the exaggerated dip can be calculated trigonometrically using the following relationship: tan exaggerated dip = (tan δ) (vertical exaggeration).

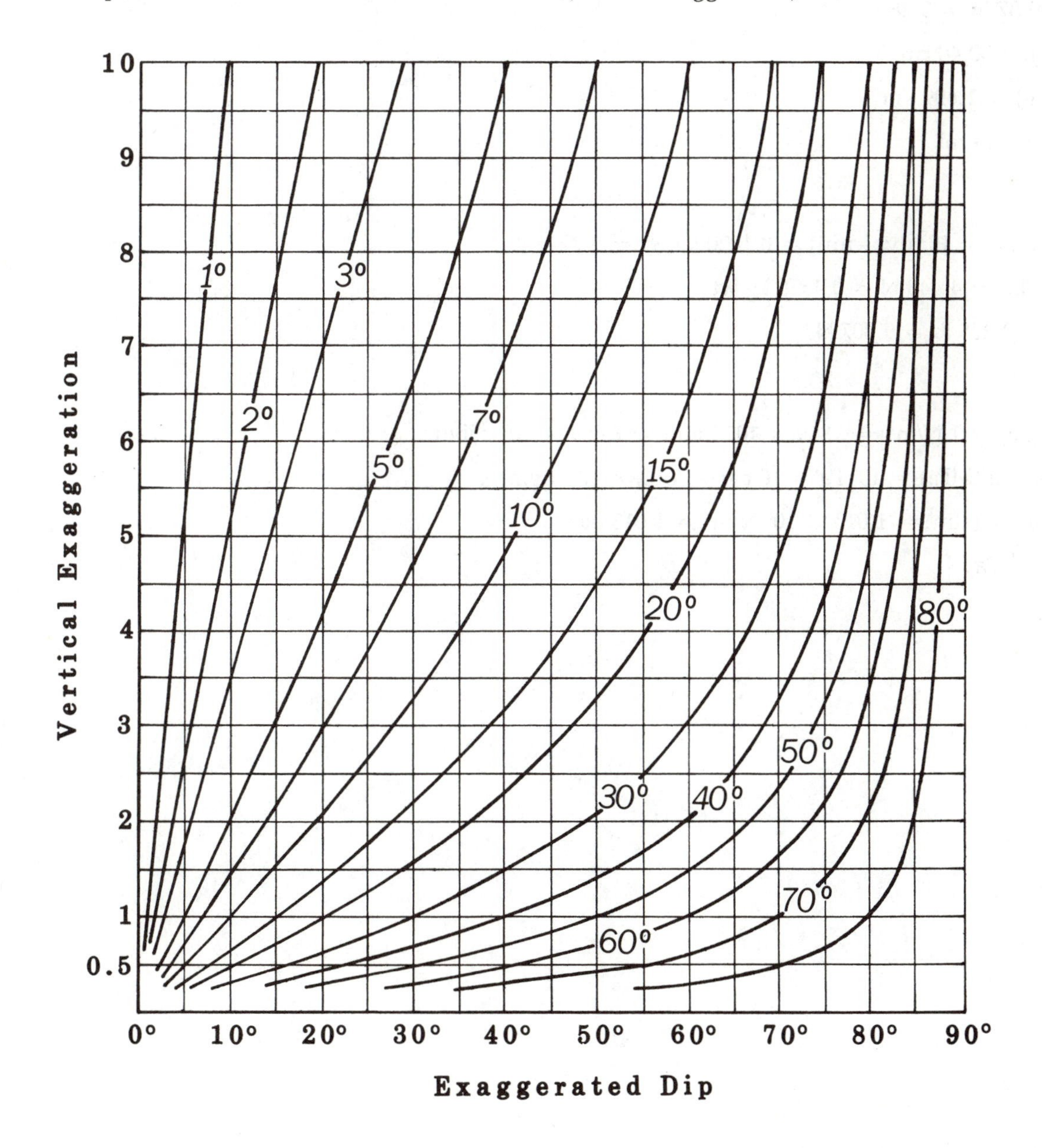

Appendix E:

Conversion Tables

Length

1 inch (in) = 2.540 cm

1 cm = 0.3937 in

1 m = 39.37 in = 3.28 ft

1 foot (ft) = 30.48 cm

1 mile (mi) = 1.609 km

1 km = 0.6214 mi

Force

1 newton (N) = 10^5 dyne (dn) = 0.1020 kg wt = 0.2248 lb

1 pound (lb) = 4.448 N = 0.4536 kg wt

1 kg wt = 2.205 lb = 9.807 N

Pressure

1 pascal (Pa) = 1 N/m^2 = 9.869×10^{-6} atm = 2.089×10^{-2} lb/ft^2

1 megapascal (MPa) = 10^6 Pa = 10 bar = 1.02×10^5 kg/m^2

1 atmosphere (atm) = 1.013×10^5 N/m^2 = 1.013 bar

1 bar = 10^5 Pa

References

Amenta, R. V. 1975. "Multiple deformation and metamorphism from structural analysis in the eastern Pennsylvania piedmont." *Geological Society of America Bulletin,* v. 85, 1647–1660.

Anderson, E. M. 1942 (2nd edition, 1952). *The dynamics of faulting.* Edinburgh, Scotland: Oliver and Boyd, 191 p.

Angelier, J. 1979. "Determination of the mean principal directions of stresses for a given fault population." *Tectonophysics,* v. 56, T17–T26.

Angelier, J., Colletta, B., and Anderson, R. W. 1985. "Neogene paleostress changes in the Basin and Range." *Geological Society of America Bulletin,* v. 96, 347–361.

Aydin, A., and Reches, Z. 1982. "Number and orientation of fault sets in the field and in experiments." *Geology,* v. 10, 107–112.

Bartley, J. M., and Glazner, A. F. 1985. "Hydrothermal systems and Tertiary low-angle normal faulting in the southwestern United States." *Geology,* v. 13, 562–564.

Bengtson, C. A. 1980. "Structural uses of tangent diagrams." *Geology,* v. 8, 599–602.

Bishop, M. S. 1960. *Subsurface Mapping.* New York: Wiley. p. 198.

Blyth, F. G. 1965. *Geological Maps and Their Interpretation.* London: Edward Arnold. p. 48.

Burger, R. H., and Hamill, M. N. 1976. "Petrofabric stress analysis of the Dry Creek Ridge anticline, Montana." *Geological Society of America Bulletin,* v. 87, 555–566.

Cluff, J. L. 1980. "Time and time again." *Journal of Sedimentary Petrology,* v. 50, 1021–1022.

Cochran, W., Fenner, P., and Hill, M., eds. 1984. *Geowriting—A guide to writing, editing, and printing in earth science* (4th edition). Falls Church, VA: American Geological Institute, p. 80.

Compton, R. R. 1985. *Geology in the Field.* New York: Wiley, p. 398.

Dahlstrom, C. D. A. 1954. "Statistical analysis of cylindrical folds." *Bulletin of the Canadian Institute of Mining and Metallurgy,* v. 57, 140–145.

Dalziel, I. W. D., and Stirewalt, G. L. 1975. "Stress history of folding and cleavage development, Baraboo syncline, Wisconsin." *Geological Society of America Bulletin,* v. 86, 1671–1690.

Davis, G. H. 1978. "Experiencing structural geology." *Journal of Geological Education,* v. 26, 52–59.

De Jong, K. A. 1975. "Electronic calculators facilitate solution of problems in structural geology." *Journal of Geological Education,* v. 23, 125–128.

Dennison, J. M. 1968. *Analysis of Geologic Structures.* New York: Norton. p. 209.

Dibblee, T. W., Jr. 1966. *Geologic Map and Sections of the Palo Alto 15' Quadrangle, California.* California Division of Mines and Geology, Map sheet 8.

Donath, F. A. 1970. "Rock deformation apparatus and experiments for dynamic structure geology." *Journal of Geological Education,* v. 18, 3–13.

Hobbs, B. E., Means, W. D., and Williams, P. F. 1976. *An Outline of Structural Geology.* New York: Wiley. 571 p.

Huber, N. K., and Rinehart, C. D. 1965. *Geologic Map of the Devils Postpile Quadrangle, Sierra Nevada, California.* U.S. Geological Survey Map GQ-437.

Mansfield, C. F. 1985. "Modeling Newtonian fluids and Bingham plastics." *Journal of Geological Education,* v. 33, 97–100.

Marjoribanks, R. W. 1974. "An instrument for measuring dip isogons and fold layer shape parameters." *Journal of Geological Education,* v. 22, 62–64.

McLaughlin, R. J. 1974. "The Sargent-Berrocal fault zone and its relation to the San Andreas fault system in the southern San Francisco Bay region and Santa Clara Valley, California." *Journal of Research of the U.S. Geological Survey,* v. 2, 593–598.

Molnar, P., and Tapponier, P. 1975. "Cenozoic tectonics of Asia: Effects of a continental collision." *Science*, v. 189, 419–426.

Moody, G. B. (ed.). 1961. *Petroleum Exploration Handbook*. New York: McGraw-Hill. p. 829.

Murray, M. W. 1968. "Written communication—A substitute for good dialog." *American Association of Petroleum Geologists Bulletin*, v. 52, 2092–2097.

Nelson, C. A. 1971. *Geologic Map of the Waucoba Spring Quadrangle, Inyo County, California*. U.S. Geological Survey Map GQ-921.

Norris, R. M. 1983. "Field geology and the written word." *Journal of Geological Education*, v. 31, 184–189.

Palmer, H. S. 1918. "New graphic method for determining the depth and thickness of strata and the projection of dip." *U.S. Geological Survey Professional Paper* 120-G, 123–128.

Phillips, F. C. 1960. *The Use of Stereographic Projection in Structural Geology*. London: Edward Arnold. p. 86.

Ragan, D. M. 1985. *Structural Geology: An Introduction to Geometrical Techniques* (3rd edition). New York: Wiley. p. 393.

Raleigh, C. B., Healy, J. H., and Bredehoeft, J. D. 1972. "Faulting and crustal stress at Rangely, Colorado." *In:* Heard, H. C., and others, eds., *Flow and Fracture of Rocks*, American Geophysical Union, Geophysical Monograph 16, 275–284.

Ramsay, J. G. 1967. *Folding and Fracturing of Rocks*. New York: McGraw-Hill. p. 568.

Ramsay, J. G., and Huber, M. I. 1983. *The Techniques of Modern Structural Geology, Volume I: Strain Analysis*. London: Academic Press, p. 307.

Roberts, J. L. 1982. *Introduction to Geologic Maps and Structures*. Oxford, England: Pergamon, p. 332.

Rowland, S. M. 1978. "Portable outcrops." *Journal of Geological Education*, v. 26, 109–110.

Secor, D. T. 1965. "Role of fluid pressure in jointing." *American Journal of Science*, v. 263, 633–646.

Suppe, J. 1985. *Principles of Structural Geology*. Englewood Cliffs, NJ: Prentice-Hall. p. 537.

Tapponnier, P., Peltzer, G., Le Dain, A. Y., Armijo, R., and Cobbold, P. 1982. "Propagating extrusion tectonics in Asia: New insights from simple experiments with plasticene." *Geology*, v. 10, 611–616.

Walker, J. 1978. "Serious fun with Polyox, Silly Putty, Slime and other non-Newtonian fluids." *Scientific American*, v. 329, no. 5 (November), 186–196.

Wellman, H. G. 1962. "A graphic method for analyzing fossil distortion caused by tectonic deformation." *Geological Magazine*, v. 99, 348–352.

Weiss, L. E. 1954. "A study of tectonic style: Structural investigation of a marble-quartzite complex in southern California." *University of California Publications in Geological Sciences*, v. 30, no. 1, 1–102.

Whitten, E. H. T. 1966. *Structural Geology of Folded Rocks*. Chicago: Rand McNally and Co. p. 663.

Wright, L. 1976. "Late Cenozoic fault patterns and stress fields in the Great Basin and westward displacement of the Sierra Nevada Block." *Geology*, v. 4, 489–494.

Index

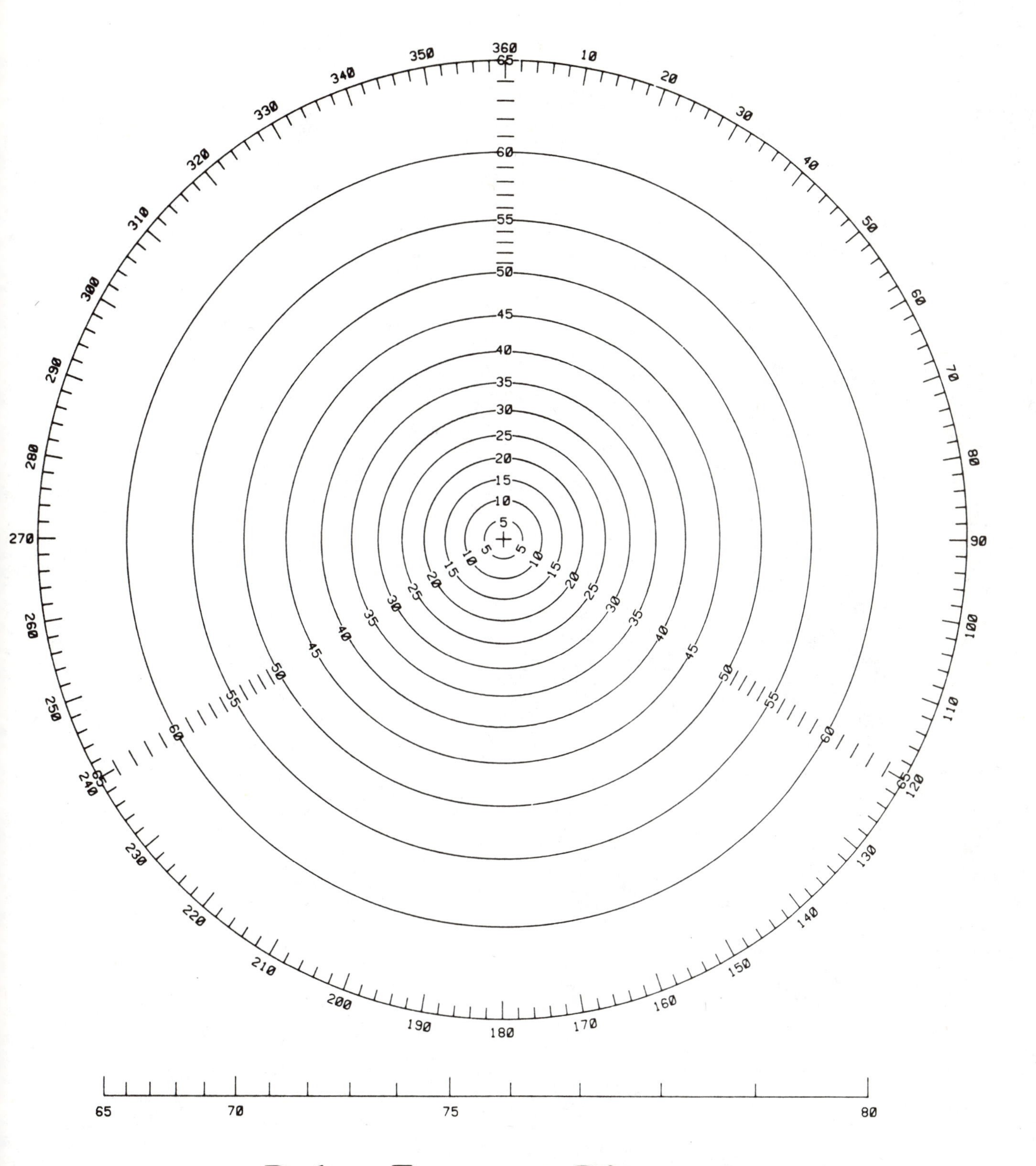

Polar Tangent Diagram

Auxiliary scale extends range to 80°. From Bengtson (1980).

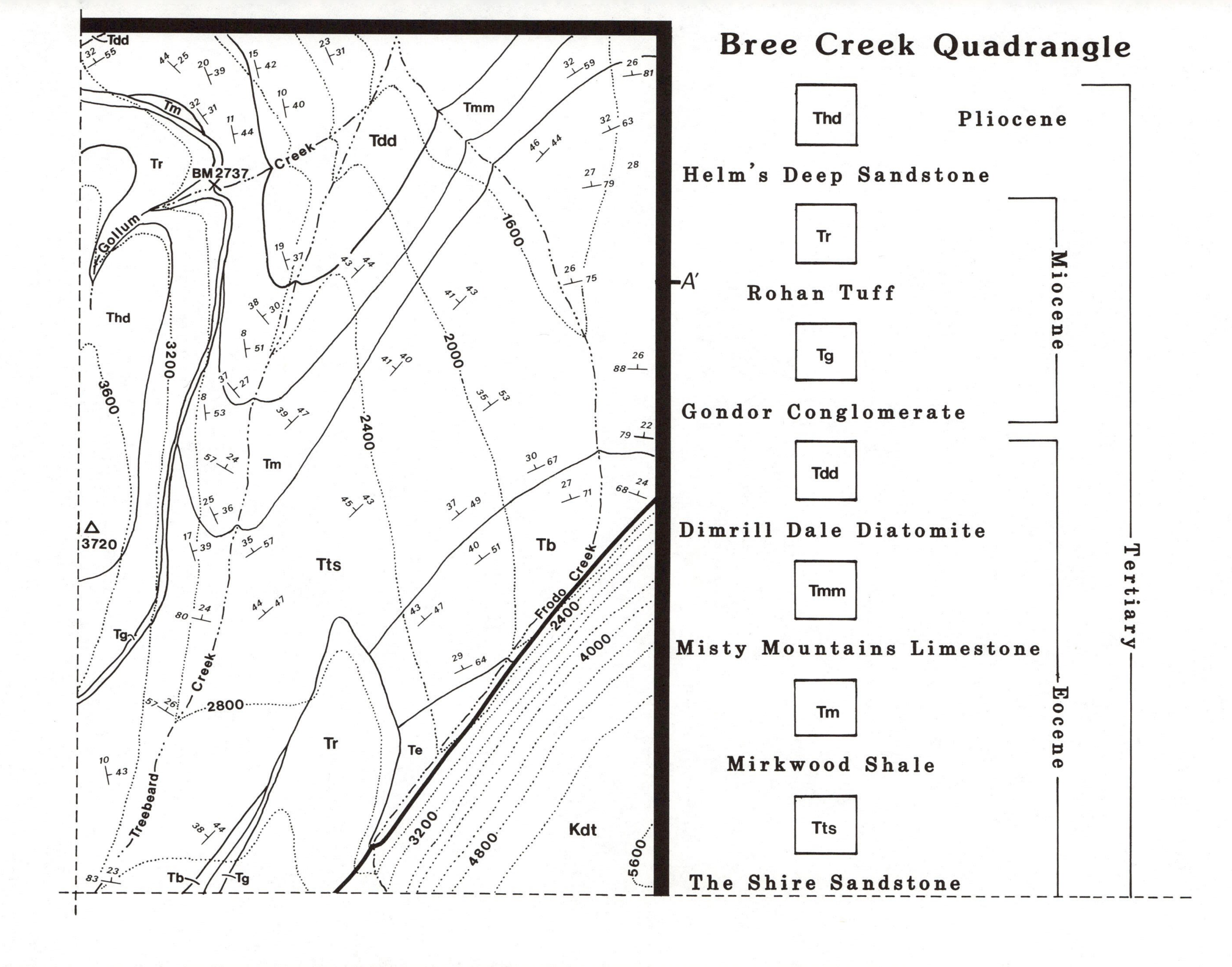
Bree Creek Quadrangle
Thd
Pliocene
Helm's Deep Sandstone
Tr
Miocene
Rohan Tuff
Tg
Gondor Conglomerate
Tdd
Dimrill Dale Diatomite
Tmm
Misty Mountains Limestone
Tm
Eocene
Mirkwood Shale
Tts
The Shire Sandstone
Tertiary
A'
Tdd
Tm
Tr
BM2737
Creek
Tdd
Tmm
Gollum
Thd
3200
3600
1600
2000
2400
Tm
Tts
Tb
Frodo Creek
2400
4000
3720
Tg
Creek
2800
Tr
Te
Treebeard
3200
4800
Kdt
5600
Tb
Tg

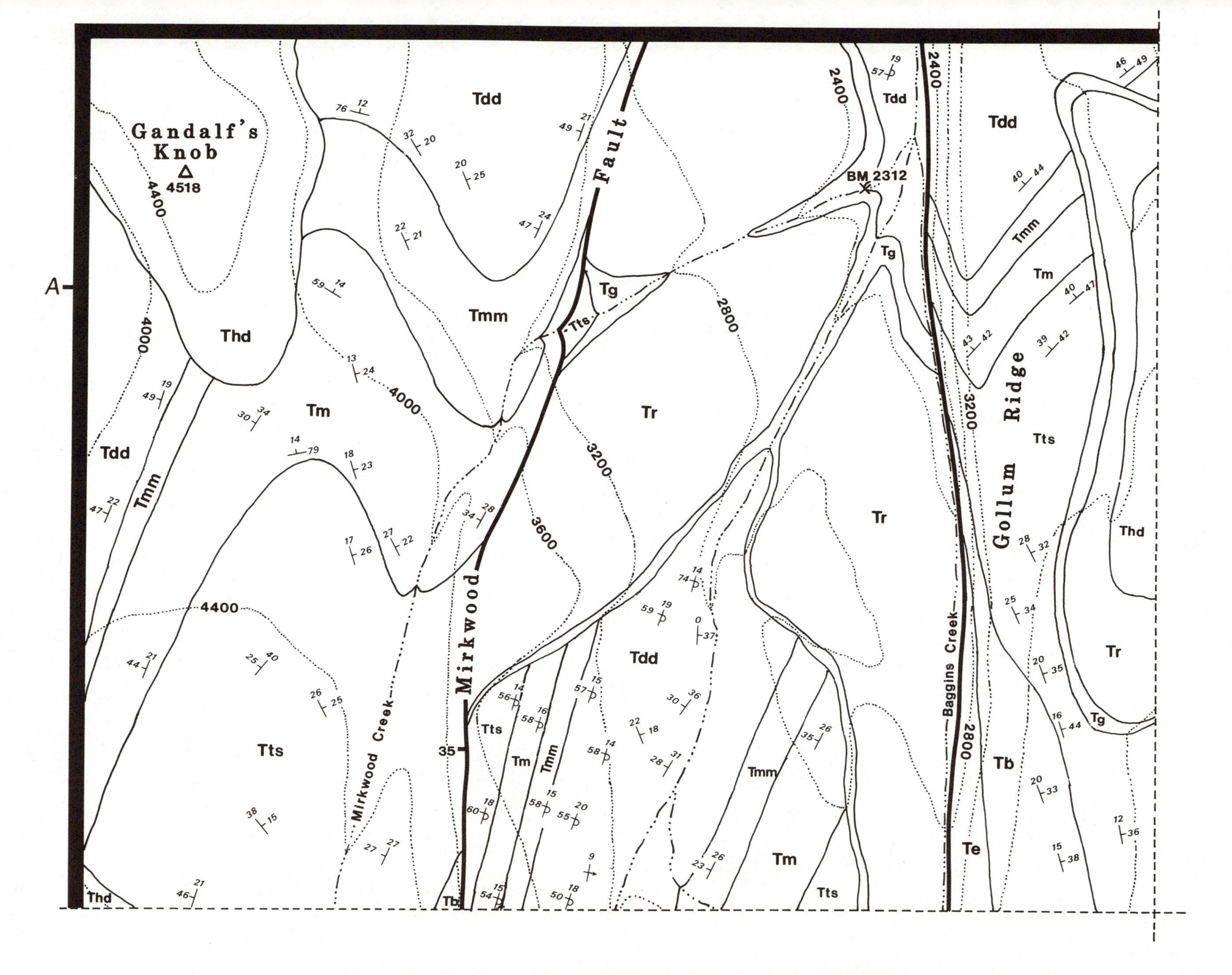

Gandalf's Knob
4518
Mirkwood Fault
Mirkwood Creek
Gollum Ridge
Baggins Creek
BM 2312
2400
2800
3200
3600
4000
4400
35
A
Tdd
Tmm
Tm
Tts
Tr
Tg
Thd
Tb
Te

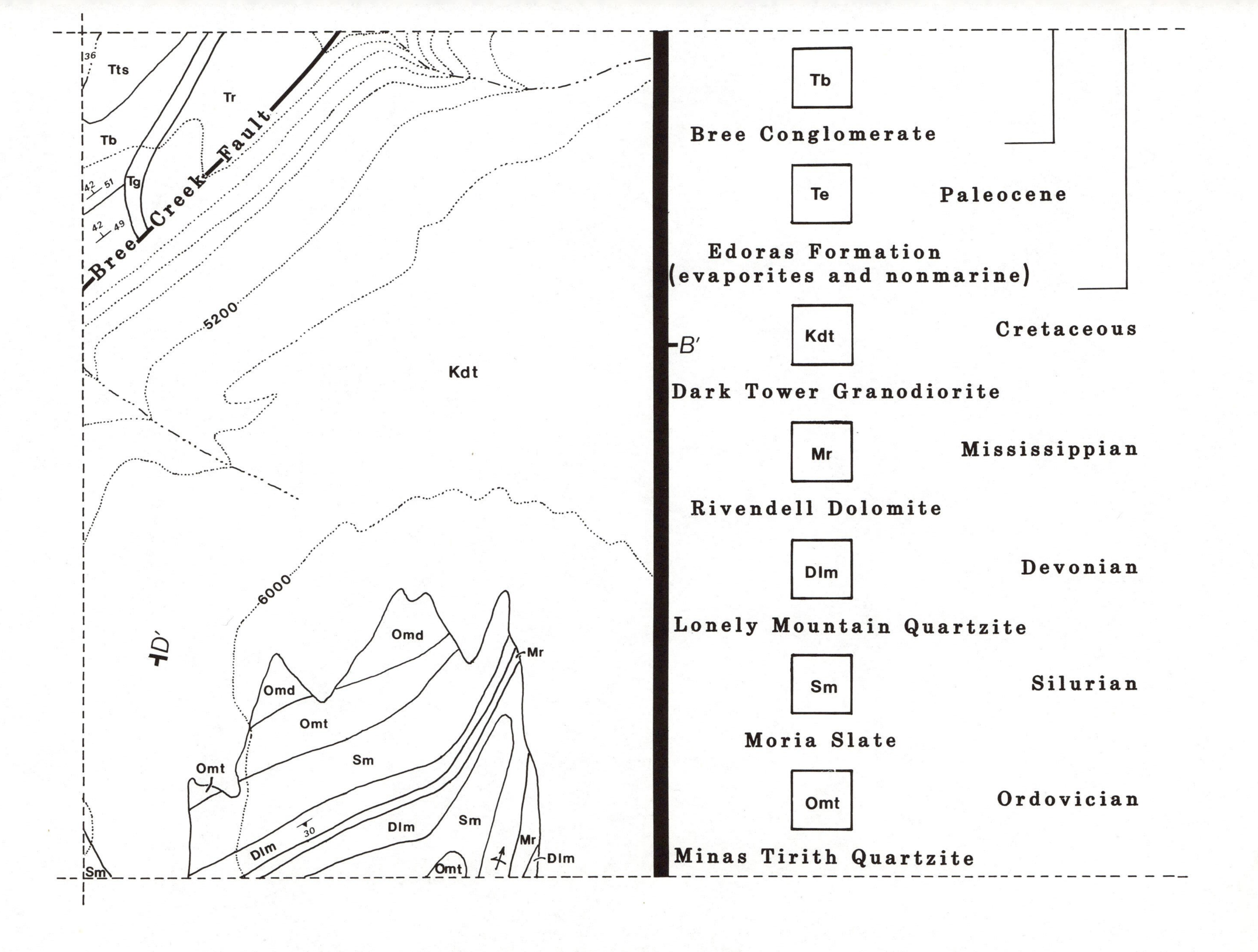
Tts
Tr
Tb
Tg
Bree Creek Fault
5200
6000
Kdt
D′
B′
Omd
Omt
Sm
Dlm
Mr
Sm
Tb
Bree Conglomerate
Te
Paleocene
Edoras Formation
(evaporites and nonmarine)
Kdt
Cretaceous
Dark Tower Granodiorite
Mr
Mississippian
Rivendell Dolomite
Dlm
Devonian
Lonely Mountain Quartzite
Sm
Silurian
Moria Slate
Omt
Ordovician
Minas Tirith Quartzite

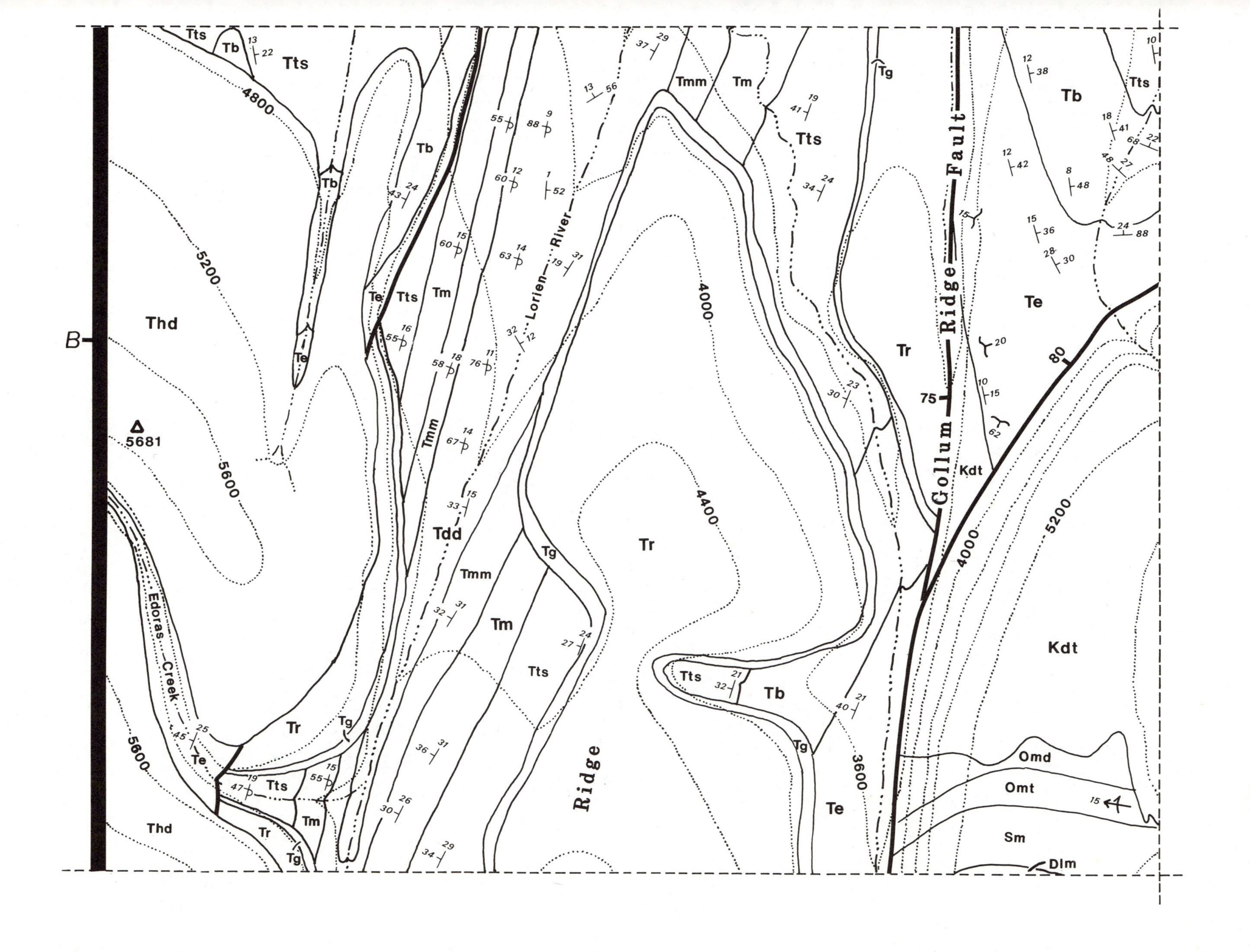

Gollum Ridge Fault
Lorien River
Edoras Creek
Ridge
B
5681
4800
5200
5600
4000
4400
3600

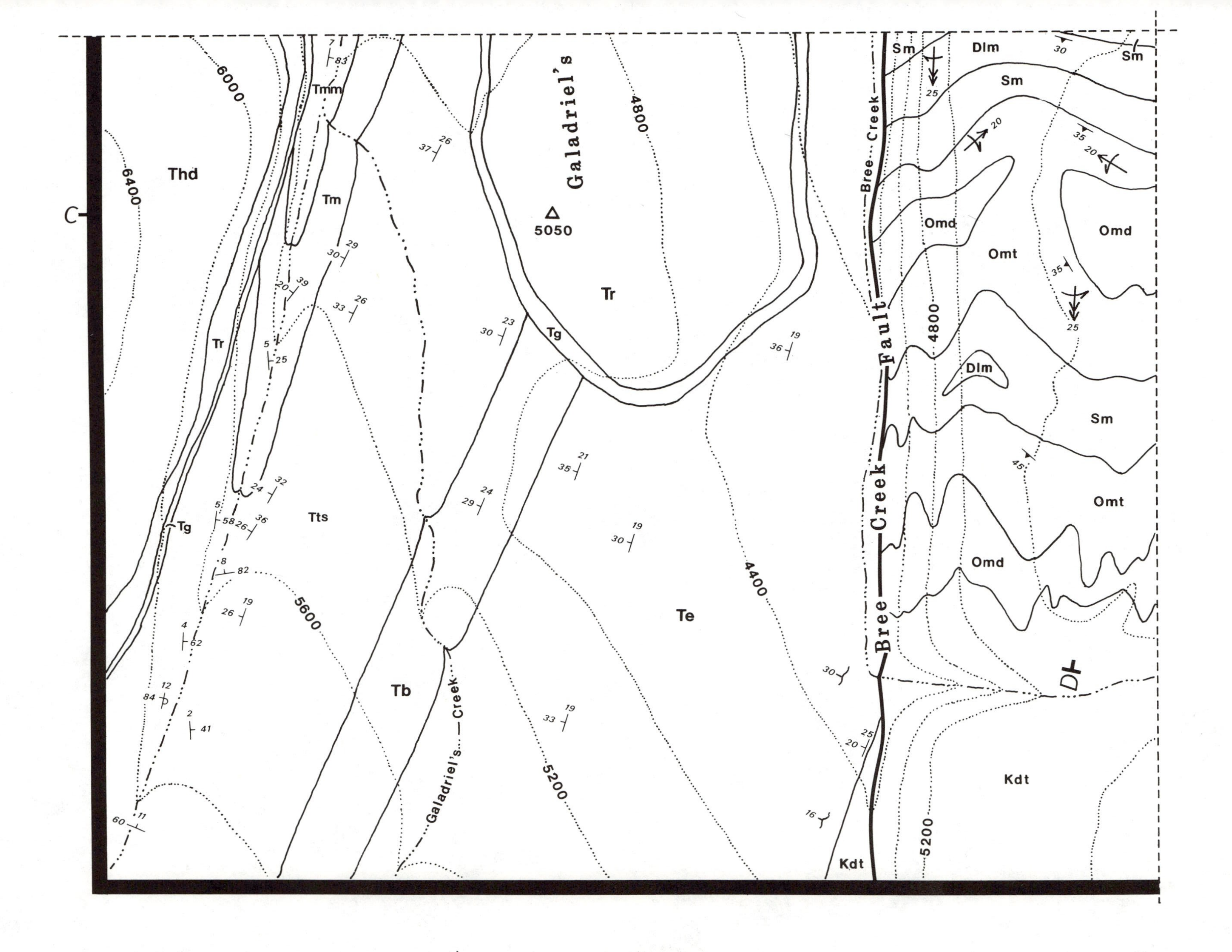

Thd
Tr
Tg
Tmm
Tm
Tts
Tb
Te
Tr
Tg
Galadriel's
5050
Galadriel's Creek
Bree Creek Fault
Bree Creek
Sm
Dlm
Omd
Omt
Kdt
6000
6400
5600
5200
4800
4400
C
D

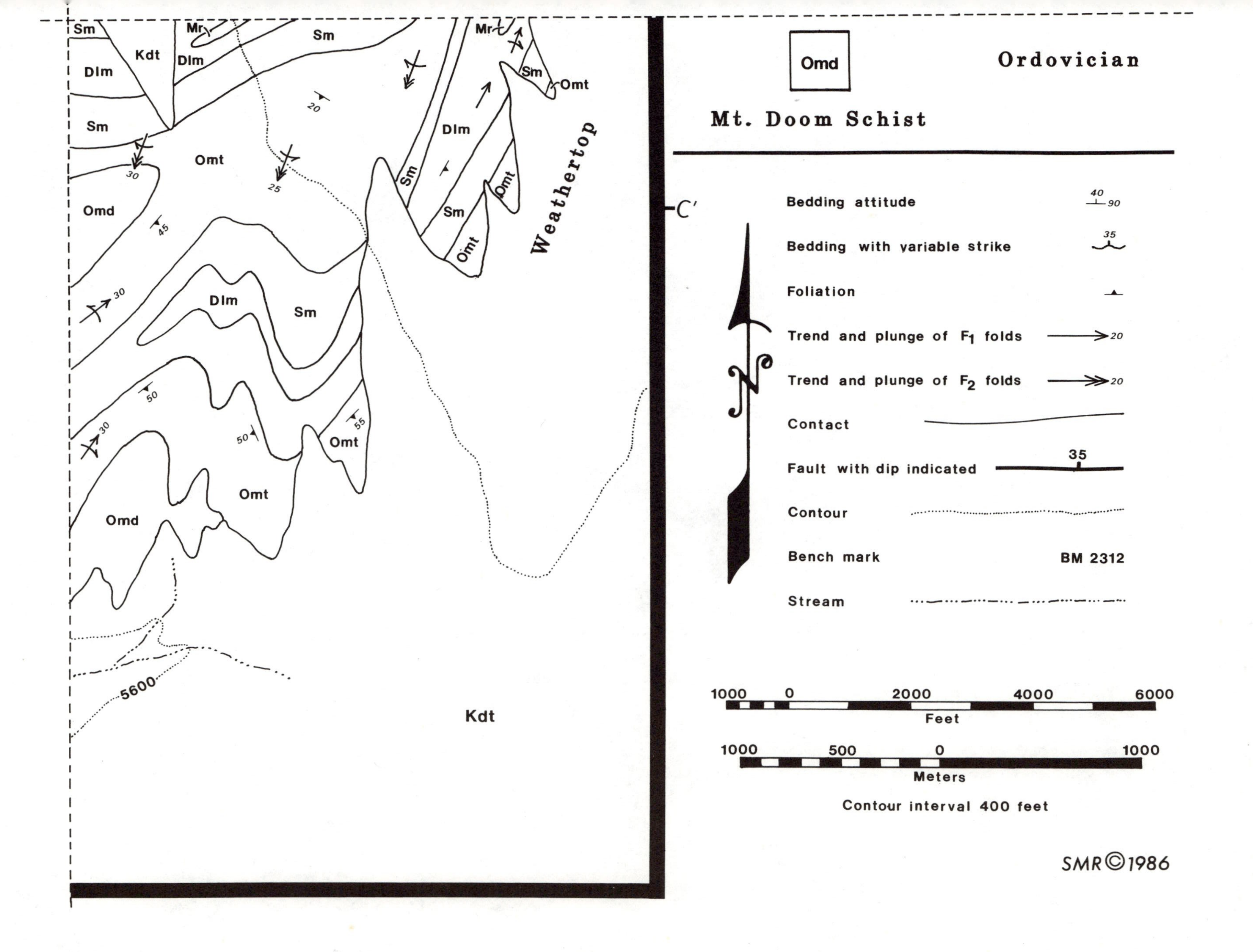
Sm
Kdt
Dlm
Mr
Sm
Omt
Omd
Weathertop
Kdt
5600
C′
Omd
Ordovician
Mt. Doom Schist
Bedding attitude
Bedding with variable strike
Foliation
Trend and plunge of F_1 folds
Trend and plunge of F_2 folds
Contact
Fault with dip indicated
Contour
Bench mark
BM 2312
Stream
1000 0 2000 4000 6000
Feet
1000 500 0 1000
Meters
Contour interval 400 feet
SMR©1986